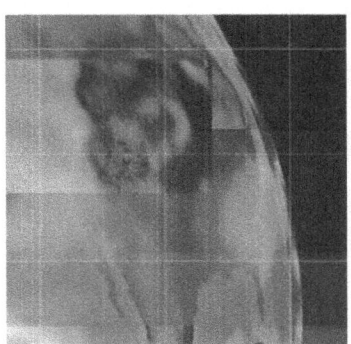

NASA's Implementation Plan for Space Shuttle Return to Flight and Beyond

A document that summarizes NASA's implementation of the Columbia *Accident Investigation Board recommendations and additional, internally generated activities generated to ensure a safe Return to Flight through the second Return to Flight test mission, STS-121.*

May 15, 2007
Volume 1, Final Edition

Final Edition Summary
May 15, 2007

The safe landing of STS-121 on July 17, 2006, marked the successful completion of NASA's second Return to Flight (RTF) Mission. This Final Edition of *NASA's Implementation Plan for Space Shuttle Return to Flight and Beyond* provides an update to NASA's responses to the *Columbia* Accident Investigation Board recommendations and observations, as well as updates to the Space Shuttle Program (SSP) "raising the bar" initiatives through the two RTF missions (STS-114 and STS-121). Also included is an update to RTF-related costs through the end of Fiscal Year 2006.

This Final Edition provides technical updates that reflect the completion and post-flight analysis of STS-121. These include NASA's improvements to ground and on-board ascent and on-orbit imagery capabilities, changes to the External Tank designed to reduce the ascent debris risk, and the growing maturity of NASA's on-orbit thermal protection system (TPS) repair capabilities. These updates describe how swapping out the first STS-114 External Tank, ET-120, led to an opportunity to analyze External Tank foam from a tank that had gone through a complete cryogenic loading cycle. This analysis led to a better understanding of the root cause mechanisms behind foam loss, and specifically resulted in the removal of the protuberance air load (PAL) ramps from the External Tank for STS-121 and all subsequent flights. STS-114 also provided valuable experience and lessons learned for preventing the loose tile gap fillers seen on that flight. Following implementation of improved procedures to validate or replace gap fillers, there were no mission impacts from protruding gap fillers on STS-121.

Several new capabilities were also first demonstrated on STS-121. New cameras with improved viewing angles were installed on the Solid Rocket Boosters, providing engineers with better coverage of the vehicle during ascent. The Orbiter Boom Sensor System's ability to serve as a stable work platform for potential TPS repairs was validated. A TPS repair material, non-oxide experimental adhesive (NOAX) was successfully tested by astronauts during a space walk in the cargo bay, demonstrating that NOAX was effective in repairing small gouges and cracks in reinforced carbon-carbon material. Finally, a late inspection of the TPS two days prior to re-entry was added to the mission timeline to inspect the Orbiter for any significant damage due to micrometeorite and orbital debris impacts. Together with the results from STS-114, STS-121 succeeded in demonstrating the effectiveness of many of the safety improvements made to the Space Shuttle system during RTF.

NASA will continue to pursue safety improvements to the Space Shuttle system until the planned retirement of the Space Shuttle in 2010. Improvements to the Space Shuttle system from STS-115 (the first International Space Station assembly flight after RTF) onwards will be diligently reviewed and accounted for as part of the normal SSP and Agency requirements and resources review process.

As we continue to pursue rigorous engineering analyses across NASA, we are encouraging healthy debate to identify, understand, and eliminate unacceptable risk that remains in the Space Shuttle Program. RTF is not simply a set of objectives or goals, but an evolution and improvement in the way we do business. While we work to safely fly the Space Shuttle, we are also working to efficiently transition those elements that are applicable to our future in exploration and discontinue those elements that will no longer be needed.

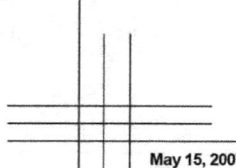 NASA's Implementation Plan for Space Shuttle Return to Flight and Beyond

May 15, 2007

Message from Michael D. Griffin

This final revision of *NASA's Implementation Plan for Space Shuttle Return to Flight and Beyond* summarizes what NASA has done and learned since the tragic loss of the *Columbia* and its crew. The performance of the NASA team on the missions we've conducted since STS-114 in July 2005 has demonstrated the technical excellence and dedication to mission for which we strive. With the two most recent shuttle flights, STS-115 and STS-116, the NASA Shuttle team has conducted some of the most complex and difficult space flights ever attempted. But we must remain vigilant, because what we do is at the edge of what is humanly possible.

No activity is more important to NASA and the future of the space program than safely operating the remaining space shuttle flights on the manifest. This document will serve as a valuable resource to everyone involved in the challenging but rewarding work of expanding the human presence in space.

By working diligently to fly the Shuttle safely until 2010, we will be able to complete the International Space Station, extend the service life of the Hubble Space Telescope, and use this knowledge to plan for an orderly transition to America's next generation of spacecraft.

To the members of the Shuttle team and all those who have assisted with the shuttle's return to flight, you have my deepest respect.

Michael D. Griffin

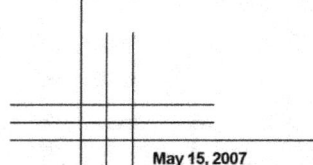

NASA's Implementation Plan for Space Shuttle Return to Flight and Beyond

Return to Flight Update from the Associate Administrator for Space Operations

As *NASA's Implementation Plan for Return to Flight and Beyond* was being developed in late 2003 and early 2004, NASA realized that the breadth and scope of the changes being made to the Space Shuttle system required two dedicated test flights to validate their efficacy. The success of STS-114 in 2005 and STS-121 in 2006 were important milestones on the way toward restoring confidence in the Space Shuttle. More than 116 individual hardware modifications and dozens of procedural changes were tested under actual flight conditions. Most of those changes operated flawlessly. In other areas (such as the performance of the External Tank's protuberance airload and ice frost ramp foam), the crucible of space flight made clear that there were opportunities for further improvement.

With this, the last update to the *Implementation Plan*, we are providing a summary of the actions taken to meet the recommendations of the *Columbia* Accident Investigation Board (CAIB) and our own, internal "raise the bar" initiatives up through the flight of STS-121 and the end of fiscal year 2006. The reader will note that each of the 29 CAIB recommendations, the 15 Space Shuttle Program "raise the bar" initiatives, the 27 CAIB observations, and the 14 supplementary recommendations in Appendix D.a of the CAIB Final Report have been addressed and dispositioned through a detailed program and agency process.

The results of this comprehensive effort were dramatically demonstrated in 2005 and 2006 with STS-114 and STS-121. These return to flight test missions paved the way for resumption of International Space Station assembly flights on STS-115 and STS-116, two of the most complex missions ever attempted in the history of space flight. While these and future flights will continue to stretch the capabilities of this extraordinary vehicle and the men and women who fly it, our emphasis will always be on the safety of those who have dedicated their lives to extending humanity's reach beyond low-Earth orbit and throughout the solar system. When the Space Shuttle's job is successfully completed in 2010 and a new generation of exploration capabilities is brought on line 5 years afterwards, I believe this document will continue to serve as a vital reference for technical experts, NASA managers, and all those who are interested in leading ambitious programs through an unyielding focus on technical excellence.

William H. Gerstenmaier
Associate Administrator for Space Operations

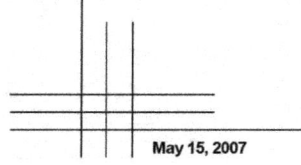

NASA's Implementation Plan for Space Shuttle Return to Flight and Beyond

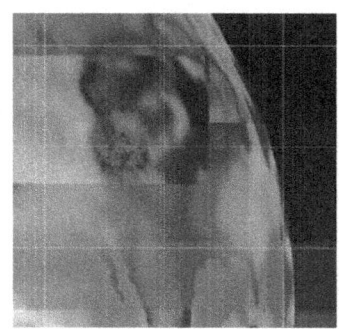

Contents

Introduction
Return to Flight Cost Summary
Part 1 – NASA's Response to the *Columbia* Accident Investigation Board's Recommendations

3.2-1	External Tank Thermal Protection System Modifications [RTF]	1-1
3.3-2	Orbiter Hardening and Thermal Protection System Impact Tolerance [RTF]	1-21
3.3-1	Reinforced Carbon-Carbon Nondestructive Inspection [RTF]	1-27
6.4-1	Thermal Protection System On-Orbit Inspect and Repair [RTF]	1-31
3.3-3	Entry with Minor Damage	1-43
3.3-4	Reinforced Carbon-Carbon Database	1-45
3.3-5	Minimizing Zinc Primer Leaching	1-47
3.8-1	Reinforced Carbon-Carbon Spares	1-49
3.8-2	Thermal Protection System Impact Damage Computer Modeling	1-51
3.4-1	Ground-Based Imagery [RTF]	1-53
3.4-2	External Tank Separation Imagery [RTF]	1-59
3.4-3	On-Vehicle Ascent Imagery [RTF]	1-61
6.3-2	National Imagery and Mapping Agency Memorandum of Agreement [RTF]	1-65
3.6-1	Update Modular Auxiliary Data Systems	1-67
3.6-2	Modular Auxiliary Data System Redesign	1-69
4.2-2	Enhance Wiring Inspection Capability	1-71
4.2-1	Solid Rocket Booster Bolt Catcher [RTF]	1-75
4.2-3	Closeout Inspection [RTF]	1-79
4.2-4	Micrometeoroid and Orbital Debris Risk	1-81

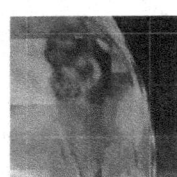

Contents

4.2-5	Foreign Object Debris Processes [RTF]	1-85
6.2-1	Scheduling [RTF]	1-87
6.3-1	Mission Management Team Improvements [RTF]	1-93
9.1-1	Detailed Plan for Organizational Changes [RTF]	1-97
7.5-1	Independent Technical Engineering Authority	1-97
7.5-2	Safety and Mission Assurance Organization	1-97
7.5-3	Reorganize Space Shuttle Integration Office	1-97
9.2-1	Mid-Life Recertification	1-105
10.3-1	Digitize Closeout Photographs [RTF]	1-107
10.3-2	Engineering Drawing Update	1-111

Part 2 – Raising the Bar – Other Corrective Actions

2.1 – Space Shuttle Program Actions

SSP-1	Quality Planning and Requirements Document/Government Mandated Inspection Points	2-1
SSP-2	Public Risk of Overflight	2-3
SSP-3	Contingency Shuttle Crew Support	2-5
SSP-4	Acceptable Risk Hazards	2-9
SSP-5	Critical Debris Sources	2-11
SSP-6	Waivers, Deviations, and Exceptions	2-15
SSP-7	NASA Accident Investigation Team Working Group Findings	2-17
SSP-8	Certification of Flight Readiness Improvements	2-19
SSP-9	Failure Mode and Effects Analyses/ Critical Items Lists	2-21
SSP-10	Contingency Action Plans	2-23
SSP-11	Rudder Speed Brake Actuators	2-25
SSP-12	Radar Coverage Capabilities and Requirements	2-27

Contents

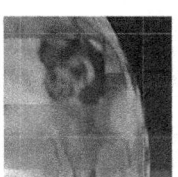

SSP-13	Hardware Processing and Operations	2-29
SSP-14	Critical Debris Size	2-31
SSP-15	Problem Tracking, In-Flight Anomaly Disposition, and Anomaly Resolution	2-33

2.2 – CAIB Observations

O10.1-1	Public Risk Policy	2-37
O10.1-2	Public Overflight Risk Mitigation	2-43
O10.1-3	Public Risk During Entry	2-43
O10.2-1	Crew Survivability	2-45
O10.4-1	KSC Quality Planning Requirements Document	2-47
O10.4-2	KSC Mission Assurance Office	2-49
O10.4-3	KSC Quality Assurance Personnel Training Programs	2-53
O10.4-4	ISO 9000/9001 and the Shuttle	2-55
O10.5-1	Review of Work Documents for STS-114	2-57
O10.5-2	Orbiter Processing Improvements	2-59
O10.5-3	NASA Oversight Process	2-61
O10.6-1	Orbiter Major Maintenance Planning	2-63
O10.6-2	Workforce and Infrastructure Requirements	2-65
O10.6-3	NASA's Work with the U.S. Air Force	2-67
O10.6-4	Orbiter Major Maintenance Intervals	2-69
O10.7-1	Orbiter Corrosion	2-71
O10.7-2	Long-Term Corrosion Detection	2-73
O10.7-3	Nondestructive Evaluation Inspections	2-75
O10.7-4	Corrosion Due to Environmental Exposure	2-77
O10.8-1	A-286 Bolts	2-79
O10.8-2	Galvanic Coupling	2-81
O10.8-3	Room Temperature Vulcanizing 560 and Koropon	2-83
O10.8-4	Acceptance and Qualification Procedures	2-85
O10.9-1	Hold-Down Post Cable System Redesign	2-87
O10.10-1	External Tank Attach Ring	2-91
O10.11-1	Shuttle Maintenance Through 2020	2-95

Contents

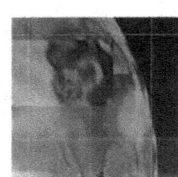

O10.12-1	Agencywide Leadership and Management Training	2-97

2.3 CAIB Report, Volume II, Appendix D.a

D.a-1	Review Quality Planning Requirements Document Process	2-101
D.a-2	Responsive System to Update Government Mandatory Inspection Points	2-102
D.a-3	Statistically Driven Sampling of Contractor Operations	2-103
D.a-4	Forecasting and Filling Personnel Vacancies	2-104
D.a-5	Quality Assurance Specialist Job Qualifications	2-106
D.a-6	Review Mandatory Inspection Document Process	2-107
D.a-7	Responsive System to Update Government Mandatory Inspection Points at the Michoud Assembly Facility	2-108
D.a-8	Use of ISO 9000/9001	2-109
D.a-9	Orbiter Corrosion	2-110
D.a-10	Hold-Down Post Cable Anomaly	2-111
D.a-11	Solid Rocket Booster External Tank Attach Ring	2-112
D.a-12	Crew Survivability	2-113
D.a-13	RSRM Segment Shipping Security	2-114
D.a-14	Michoud Assembly Facility Security	2-115

Appendix A – NASA's Return to Flight Process

Appendix B – Return to Flight Task Group

Appendix C – Return to Flight Summary Overview

Appendix D – The Integrated Accepted Risk Approach for Return to Flight

Appendix E – Return to Flight Suggestions

Appendix F – CAIB Recommendations Implementation Schedule

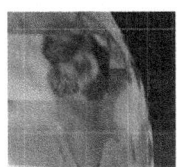

Introduction

THE FIRST RETURN TO FLIGHT MISSIONS: STS-114 and STS-121

On July 26, 2005, NASA successfully launched STS-114, the first of two return to flight (RTF) missions. STS-114 tested new processes, procedures, and capabilities implemented in response to the Columbia Accident Investigation Board (CAIB) Recommendations and Observations and the Space Shuttle Program (SSP) Raising the Bar Actions. The mission was extremely successful and all mission objectives were completed. New ascent and orbit imagery combined with the inspection data obtained from the Orbiter Boom Sensor System (OBSS) provided comprehensive insight into the condition of the Space Shuttle Thermal Protection System (TPS) and identified areas that required further analysis, such as the protruding gap fillers and several small damage sites. These areas were assessed by the SSP Mission Management Team (MMT). During the nine days the Space Shuttle was docked with the International Space Station (ISS), over 12,000 lbs of equipment and supplies were delivered to the ISS; and over 7,000 lbs of experiments and equipment were returned to Earth with Discovery. STS-114 also accomplished several key objectives for continued ISS operations, including delivering the Human Research Facility 2 experiment rack and returning to Earth for analysis a Russian Elektron oxygen generating unit that had malfunctioned. The STS-114 crew also completed three extravehicular activities (EVAs) during the mission. The first EVA accomplished the Tile Emittance Wash and Reinforced Carbon-Carbon (RCC) NOAX Crack Repair Test objectives, and restored power to Control Moment Gyro (CMG) -2. The second EVA removed and replaced CMG-1. The third EVA successfully accomplished the unplanned activity of removing two protruding gap fillers from the Space Shuttle's belly tile.

During STS-114, the new tools and capabilities added to improve our ability to image the Space Shuttle during launch, ascent, and on orbit provided an unprecedented amount of data on the Space Shuttle's performance and the health of the Orbiter's TPS. The Program was able to efficiently organize and analyze the new data, and provide them to the MMT in a way that supported effective, real-time mission decision making. The new structure and training of the MMT also proved effective, and the reorganized MMT was able to draw on numerous engineering resources from across the Agency to make several critical in-flight decisions, such as whether to remove protruding gap fillers on the belly of the Orbiter and whether to attempt to repair a piece of thermal blanket near the Orbiter's window.

Despite the significant work done to modify the External Tank (ET) and reduce the possibility of critical foam loss, STS-114 experienced several unexpectedly large ET foam loss events. Fortunately, the large pieces of foam were released at a time during ascent where they posed little risk to the Orbiter. However, the Program reassessed the remaining risk of foam loss from the ET. After extensive analysis, the ET Project determined the most probable cause of the foam losses and worked to redesign, test, and eliminate those causes. The foam loss from the LH_2 protuberance air load (PAL) ramp was most likely caused by a previously unrecognized failure mechanism related to the pressure and temperature changes associated with tanking.

On July 4, 2006, NASA successfully launched the second return to flight mission STS-121. The STS-121 mission provided a second test mission for the new processes, procedures, capabilities implemented on STS-114 in response to the CAIB Recommendations and Observations and the SSP Raising the Bar Actions. Additionally, the second RTF mission tested the changes made as a result of what was learned from the first test mission. The STS-121 mission was extremely successful and all mission objectives were completed.

The STS-121 mission tested the removal of the PAL ramp from the ET based on extensive analysis and wind tunnel testing done post STS-114. The successful launch of STS-121 demonstrated that the removal of the PAL ramp, along with NASA's continuing improvements to the ET foam, was successful in reducing the debris risk during ascent. As demonstrated on the two RTF missions, the SSP continues to improve its understanding of foam loss mechanisms. Based on that understanding, NASA is moving forward with additional modifications to the ET to reduce foam loss, the most significant being a redesign of the ice/frost ramps to reduce or eliminate foam that might liberate and pose a debris risk to the vehicle during ascent.

As a result of the protruding gap fillers observed during the STS-114 mission, prior to the launch of STS-121, NASA also instituted new procedures to validate or replace gap fillers in critical areas on the underside of the Orbiter. This was done to minimize debris risk and to mitigate the potential for a protruding filler to cause premature boundary layer tripping and subsequent heating on reentry. STS-121 also served to continue the process of refining the capabilities that had been developed and demonstrated during STS-114, including additional tests of RCC and tile repair techniques, use of the Shuttle robotic arm and OBSS as a TPS inspection and repair platform, and collection and analysis of imagery and radar data to understand the performance of the vehicle during ascent and the health of the Orbiter prior to re-entry. Like STS-114, STS-121 also carried essential supplies and hardware to the ISS, while the crew conducted two EVAs (spacewalks) to validate the use of the OBSS for inspection and repair and to prepare the ISS for the resumption of assembly flights starting on the next mission, STS-115.

ADDRESSING THE RETURN TO FLIGHT TASK GROUP ASSESSMENT

Much of NASA's work in preparation for STS-114 was assessed by the Return to Flight Task Group. The NASA Administrator chartered the Task Group in July 2003 to provide an independent evaluation of NASA's work to fulfill the intent of the 15 CAIB RTF Recommendations. The Task Group's assessment of NASA's readiness for RTF was one of several inputs that informed NASA's leadership during the RTF planning and decision-making process. The Return to Flight Task Group completed their assessment of NASA's work on June 27, 2005, and released their final report in early July 2005, prior to the launch of STS-114. The Task Group determined that NASA met the CAIB intent for 12 of the 15 RTF CAIB Recommendations; it found, however, that the intent was not fully met for CAIB Recommendations 3.2-1, External Tank Debris Shedding; 3.3-2, Orbiter Hardening; and 6.4-1, Thermal Protection System Inspection and Repair.

In its final report, the Task Group included a series of observations by individual members. These observations addressed issues that lay outside the scope of the specific CAIB RTF Recommendations, but that the Task Group felt needed additional attention from NASA's leadership. NASA is actively implementing these core observations from several of the Task Group members:

"In order to properly prepare the Agency for the future, including the return to the Moon and journey to Mars, we offer the following suggested actions, all of which must start at the top and flow down to the programs, projects, and workforce:

1) Clearly set achievable expectations and holding people accountable, in addition to the positive consequences, this includes negative consequences for not performing to expectations;

2) Return to classic program management and systems engineering principles and practices (including integrated risk management), and execute these with rigor;

3) Ensure managers at all levels have a solid foundation in these attributes <u>before</u> appointing them to such responsibilities; this requires not only training, but successful demonstration of these skills at a lower level;

4) Eliminate the prejudices and barriers that prevent the Agency, and especially the human spaceflight programs, from learning from their own and others' mistakes." (Return to Flight Task Group, Final Report, July 2005, p. 197)

The Vision for Space Exploration clearly defined the mission of the Space Shuttle Program as the completion of the ISS. It also specified that this mission would be complete by the end of fiscal year 2010. NASA has developed a clear, unambiguous plan to execute this direction that enables our exploration activities. NASA and Space Shuttle Program leadership are being held accountable to these goals.

Similarly, the Space Operations Mission Directorate leadership has reinforced the systems engineering rigor established prior to RTF by ensuring that testing will be used to verify the results of analysis and modeling used in space flight programs The Space Shuttle Program established a plan for verifying the integrity of the ETs that were flown on the second RTF mission, STS-121, and subsequent missions. These tanks were flown without PAL ramps, which have been a standard element of the tanks in the past. This plan involved both analysis and testing to verify analytical results.

Based on the results of the ongoing ET tests and analyses, our integrated hazard analyses are being continually reevaluated and improved to reflect our increased understanding of the root causes and risks of foam loss. The systems engineering rigor that is being used as we evaluate the ET will result in a more accurate integrated risk environment for the Space Shuttle. We also understand the importance of balance among risk management, schedule, and budget. A good risk assessment requires the application of the appropriate amount of rigor in the decision-making process to maintain that balance.

NASA is working to ensure that not only our engineering but our management practices are rigorous and appropriate. In keeping with this commitment to management excellence, the Space Shuttle Program has a management team with a strong mix of skills and experience. The program manager has designated three deputy program managers to assist in overall program management and a lead engineer to direct the technical aspects of this large, complex program, (1) a deputy program manager who will focus on the day-to-day oversight of the Space Shuttle propulsion systems at the Marshall Space Flight Center; (2) a deputy program manager who will focus on the day-to-day control of Space Shuttle resources; and, (3) a deputy program manager who will focus on the day-to-day oversight of Space Shuttle operations. In addition, the program management team includes a lead engineer responsible for assessing technical issues and overseeing the engineering rigor within the SSP. This management team is designed to manage the particular challenges of concurrently operating and shutting down the aging Shuttle system while supporting transition of those elements necessary for exploration.

Finally, we continue to learn from our experiences and are finding ways to share those experiences both within and outside of the SSP. For example, as a result of the discovery of cracks in the PAL ramp foam on ET-120 (which had been cryo-loaded but not flown), we realized that NASA had never examined an ET after loading it with liquid hydrogen and liquid oxygen and running it through pressurization cycles. By dissecting the foam cracks on ET-120, we discovered a new foam failure mode and have a better understanding of foam loss mechanisms. These results have been shared with the Exploration Systems programs responsible for development of the Ares I and Ares V launch vehicles, which themselves are leveraging much of the experience with ET production from the SSP.

While learning from our daily experiences, we will also reflect periodically on our accomplishments and decisions. During future management off-site gatherings, we will take a look back at the decisions made and the information available at the time those decisions were made to assess our effectiveness. This approach was successfully applied in the International Space Station Program and, as a lesson learned from that activity, will be employed in the SSP. The Space Operations Mission Directorate is working closely with the Exploration Systems Mission Directorate, as the latter develops its new vehicles and systems for exploration, to ensure that the knowledge base of the current operational programs is applied. In preparation for the transition to exploration, the SSP has engaged in extensive benchmarking with other industries that have shut down major programs, including the Department of Defense and the aerospace industry. We are actively applying these lessons learned in our own transition planning.

At the Agency level, a number of activities are under way to comprehensively address the four broad areas mentioned by the Task Group. NASA is revising the Independent Technical Authority to separate the responsibility and management of programs from institutional capabilities such as engineering. NASA policies, procedures, requirements, and guidelines have either recently been revised or are currently under revision to improve the governance of conducting NASA business. NASA has recently revised one of its key governing documents, the Strategic Management Handbook (NPD 1000.0). NASA Procedural Requirements (NPR) 1000.3, which defines the roles and responsibilities of the NASA organization and its people, is continually being revised to address changes in organizational relationships. NASA programs and projects operate by the March 6, 2007 release of NPR 7120.5D, and a new Systems Engineering NPR is also in its final stages of development.

Current NASA, Department of Defense, and industry specifications and standards are being evaluated to ensure continued engineering rigor in the development and execution of highly complex space missions. NASA has also developed a competency-based model to provide a unified framework for the professional development of systems engineers and project managers; this professional development model assures that NASA program/project management proficiency meets the needs of its missions. Moreover, NASA certification of program and project managers and systems engineers is being continually evaluated. NASA professional development activities will incorporate lessons learned and case studies to teach participants about past mistakes made by NASA as well as external organizations, to ensure that lessons learned are actively used by management. Finally, NASA is incorporating lessons learned into specifications and standards applicable to all NASA projects and programs.

For NASA and the Space Shuttle Program, return to flight has been a journey more than a milestone, a process of continual learning that will continue through the last flight of the Space Shuttle and beyond. We continue to incorporate the lessons learned from the Columbia accident to ensure that we apply a high level of engineering rigor and that we actively encourage dissent in our program discussions to understand and drive out unacceptable risk in the system. As we work to keep the Space Shuttle safe through its remaining flights, we are simultaneously working to establish efficient processes for transitioning Shuttle assets to exploration programs and to shut down those assets that will no longer be needed.

Return to Flight Cost Summary

NASA estimates of Return to Flight (RTF) costs for fiscal year (FY) 2003 – FY 2005 have changed little since the July 2005 estimate of $1.141B. At the end of FY 2005, the reported costs for RTF are $1.105B; $36M below our estimate. Although the total cost of RTF has changed little, the cost and phasing of several RTF activities were updated for FY 2003 – FY 2005. The majority of these changes are not material changes to work performed. They are adjustments to reflect a consistent reporting of actuals when costed and reconciled, rather than estimates when authorized. Some of the changes reflect a more rigorous categorization of which activities support RTF. Only a few changes since the last update are of a material nature (e.g., External Tank, On-Orbit Thermal Protection System Inspection and Repair, and Orbiter Certification), and their increases are generally offset by decreases in other RTF activities or included in Operations.

NASA entered FY 2006 with a cost estimate of $160M. NASA's current total estimate of RTF costs for FY 2003 – FY 2006 is now $1,184M.

The second RTF mission has been completed and the post-flight analyses have positioned the Shuttle Program to re-assess the RTF work remaining and to adjust the work content. The following RTF activities were completed in FY 2006.

- Orbiter Tile/RCC related modeling
- Emittance Wash tile repair development
- NOAX RCC crack repair development
- Plug repair development for RCC
- GFE overlay tile repair development
- OBSS spares and generic structural flight certification
- ground camera and debris radar procurement
- Wing Leading Edge sensor deliveries for all orbiters with improved batteries
- Front SPAR/ Carrier Panel redesign (hardening)
- EVA IR Camera DTO project
- Complete procurement of high priority STS-107 flight crew equipment

Any minimal, residual RTF project activities will be managed under the Shuttle Operations budget line beginning in FY 2007.

Return to Flight Budget / Implementation Plan Map for FY 2006

(Cost in Millions)

As of September 30, 2006

	FY 03	FY 04	FY 05	FY 06	CAIB #3.2-1	CAIB #3.3-1	CAIB #3.3.2	CAIB #3.3.3	CAIB #3.3.4	CAIB #3.3.5	CAIB #3.4-1	CAIB #3.4-2	CAIB #3.4-3	CAIB #3.6-1	CAIB #3.6-2	CAIB #3.8-1	CAIB #3.8-2	CAIB #4.2-1	CAIB #4.2-3	CAIB #4.2-4	CAIB #4.2-5	CAIB #6.2-1	CAIB #6.3-1	CAIB #6.3-2	CAIB #6.4-1	CAIB #7.5-1	CAIB #7.5-2	CAIB #7.5-3	CAIB #9.1-1	CAIB #9.2-1	CAIB #10.3-1	CAIB #10.3-2	SSP Recommendations
Total Initiated SSP RTF Activities	**28**	**518**	**559**	**80**																													
Orbiter RCC Inspections & Orbiter RCC-2 Shipsets Spares	0	11	11	7		X	X															X	X										X	X
On-orbit TPS Inspection & EVA Tile Repair	0	155	235	11																														
Orbiter Workforce	0	0	0	0								X																					
Orbiter Hardening	0	17	14	1			X																X							X				
Orbiter/GFE	0	2	4	0																														
Orbiter Contingency	0	0	0	0																														
Orbiter Certification / Verification	15	89	56	0		X													X	X			X							X				
External Tank Items (Camera, Bipod Ramp, etc.)	11	111	107	6	X			X										X	X															
SRB Items (Bolt Catcher, Camera, ETA Ring Invest., Camera)	0	29	18	1					X								X																	
Ground Camera Ascent Imagery Upgrade	0	21	32	6						X												X		X	X								X	X
KSC Ground Operations Workforce	0	0	0	0																														
Other (System Intgr, JBOSC Sys, Full Cost, Additional FTEs, etc.)	0	80	80	40	X	X	X	X	X																	X	X	X	X					X
Stafford - Covey Team	1	2	2	0																															

(1) This update includes actual cost for FY 03 through FY 06.

(2) The FY 2006 RTF cost includes activities that have been approved for implementation.

(3) Orbiter Workforce, Orbiter Contingency, and KSC Ground Operations Workforce was absorbed in the baseline budget.

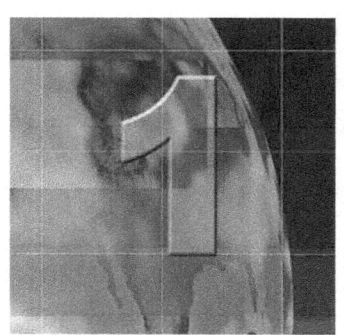

NASA's Response to the *Columbia* Accident Investigation Board's Recommendations

The following section details NASA's response to the CAIB recommendations in the order in which they appear in the CAIB Report. NASA complied with those actions marked "RTF" prior to returning to flight with STS-114 in 2005. This implementation plan documents NASA's implementation of both the CAIB recommendations and NASA's own, internally generated Program milestones through the second Return to Flight test mission, STS-121.

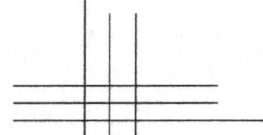 NASA's Implementation Plan for Space Shuttle Return to Flight and Beyond

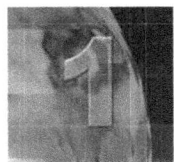

Columbia Accident Investigation Board
Recommendation 3.2-1

Initiate an aggressive program to eliminate all External Tank Thermal Protection System debris-shedding at the source with particular emphasis on the region where the bipod struts attach to the External Tank. [RTF]

Note: NASA has closed this recommendation through the formal Program Requirements Control Board process. The following summary details NASA's response to the recommendations and any additional work NASA performed beyond the Columbia Accident Investigation Board recommendations.

BACKGROUND

The External Tank (ET) requires a Thermal Protection System (TPS) to maintain the cryogenic temperature of the propellants, to protect the tank from atmospheric heating, and to prevent ice formation on the exterior of the tank. The majority of the ET TPS is spray-on foam insulation (SOFI). Foam is the only material that can meet the TPS requirement for a very lightweight, yet highly insulating material. However, foam poses some manufacturing challenges. For example, it is subject to small voids during application, especially around uneven areas such as joints or protrusions. This problem is exacerbated by the fact that foam for complicated areas must be applied manually, rather than with the more consistent automated process that is used for the smooth areas. Quality assurance in the application process is another major challenge because foam encapsulates air, causing voids in the foam. Using nondestructive evaluation (NDE) to find inconsistencies or defects in the foam is an engineering challenge that has eluded a reliable technical solution. NASA has conducted comprehensive searches for NDE techniques in industry and research institutions and has made repeated attempts to develop a method of inspecting the foam for correct application. We continue these efforts. As an alternative to inspection, NASA is reinvigorating strict process controls of both automated and manual foam applications to reduce the likelihood of voids.

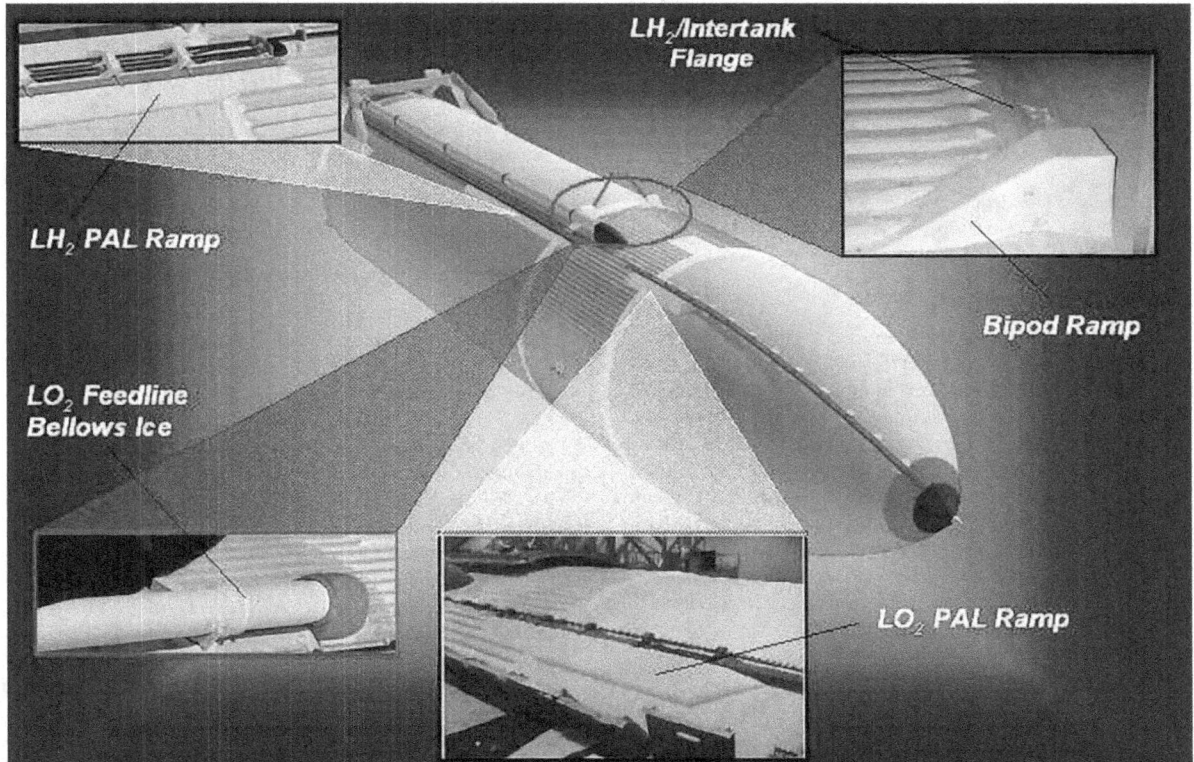

Figure 3.2-1-1. Primary potential ET debris sources being evaluated.

NASA concurred with the *Columbia* Accident Investigation Board (CAIB) findings and recommendations and charged several internal and independent review teams with reviewing the design, manufacturing, process control, and acceptance of the ET. Findings reported by the ET Working Group (ETWG) in June 2003 noted design verification and process validation shortfalls, quality control verification gaps, and a lack of acceptance testing and inspection techniques that are capable of detecting deficient foam applications. Based on the findings of the ETWG, the ET Project concluded that all TPS applications in those areas of the ET where liberated foam poses a risk of significantly damaging the Orbiter, otherwise known as critical debris zones, should be identified for mandatory evaluation. NASA made changes to several areas of the ET to reduce the possibility that critical debris would be shed during ascent. These changes are detailed in the next section.

Figure 3.2-1-1 illustrates the primary areas on the ET that were evaluated as potential debris sources for Return to Flight (RTF).

Despite NASA's efforts to eliminate the possibility of critical debris, on the first RTF test flight, STS-114, the ET shed some unexpectedly large pieces of foam; the largest of which came from one of the protuberance air load (PAL) ramps. The total amount of foam shed on STS-114 was far less than on past missions, and the majority of the debris was shed late in the ascent period when it poses less risk to the Orbiter; however, the debris, particularly from the PAL ramp, was larger than predicted by preflight analysis, and there was more foam loss than expected. As a result, NASA undertook a significant reassessment of the ET and has implemented additional steps to mitigate the risk of foam loss on future flights. The details of these actions are captured in the Post STS-114 and Final Update sections at the end of this write-up.

ET Forward Bipod Background

Before STS-107, several cases of foam loss from the left bipod ramp were documented through photographic evidence. The most significant foam loss events in the early 1990s were attributed to debonds or voids in the "two-tone foam" bond layer configuration on the intertank area forward of the bipod ramp. The intertank foam was thought to have peeled off portions of the bipod ramp when liberated. Corrective action taken after STS-50 included implementation of a two-gun spray technique in the ET bipod ramp area (figure 3.2-1-2) to eliminate the two-tone foam configuration. After the STS-112 foam loss event, the ET Project began developing redesign concepts for the bipod ramp; this activity was still under way at the time of the STS-107 accident. Dissection of bipod ramps conducted for the STS-107 investigation has indicated that defects resulting from a manual foam spray operation over an extremely complex geometry could produce foam loss. Liquid nitrogen (LN_2)

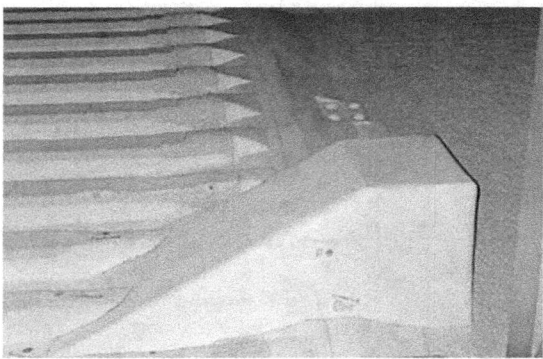

Figure 3.2-1-2. ET forward bipod ramp (foam).

ingestion was also identified as an additional potential load environment that, when combined with a defect in the TPS, could contribute to large-scale foam loss.

Liquid Oxygen (LO_2) Feedline Bellows Background

The LO_2 feedline is the large external pipe that runs the length of the ET. Bellows are located at three joints along this pipe to accommodate thermal expansion and contraction. The bellows shields (figure 3.2-1-3) are covered with TPS foam, but the ends are exposed. Ice and frost form when moisture in the air contacts the cold surface of the exposed bellows. Although Space Shuttle Program (SSP) requirements include provisions for ice on the feedline supports and adjacent lines, ice in this area presents a potential source of debris in the critical debris zone.

Protuberance Airload Ramps Background

The ET PAL ramps were designed to reduce adverse aerodynamic loading on the ET cable trays and pressurization lines (figure 3.2-1-4). The PAL ramps are manually sprayed foam applications (using a less complex manual spray process than that used on the bipod) located adjacent to the cable trays and pressurization lines. Foam from the PAL ramp, if liberated, could become the source of critical debris. Prior to STS-114, PAL ramp foam loss had been observed on STS-4/ET-4 and STS-7/ET-6. The most likely causes of these losses were believed to be repairs and cryo-pumping (air ingestion) into the SuperLight Ablator (SLA) panels under and adjacent to the PAL ramps. Configuration changes and repair criteria were revised early in the Program, in an attempt to preclude recurrence of these failures. Because of their potential for debris, NASA placed the PAL ramps at the top of the priority list for TPS verification reassessment and NDE prior to RTF. Inspections prior to STS-114

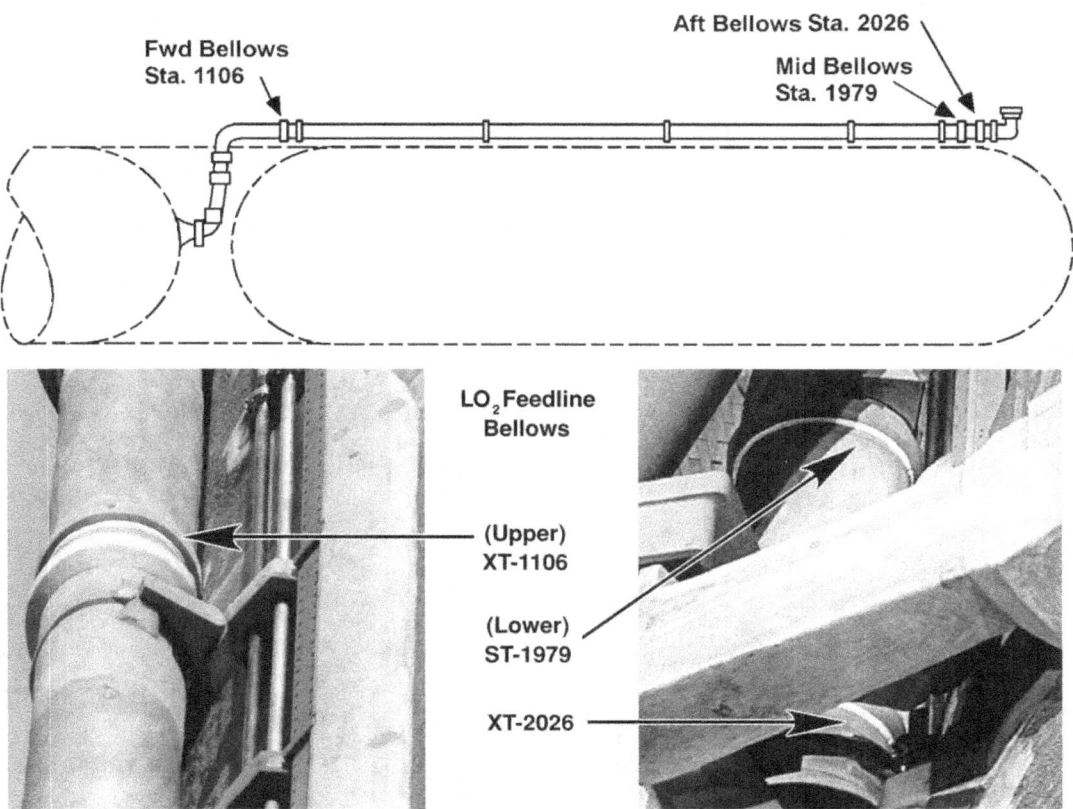

Figure 3.2-1-3. LO₂ feedline bellows.

Figure 3.2-1-4. PAL ramp locations.

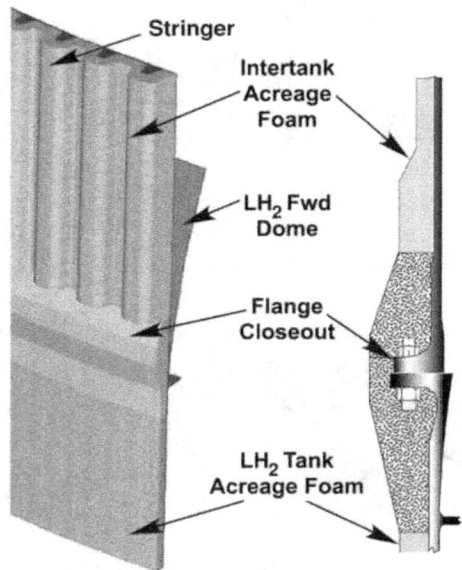

Figure 3.2-1-5. ET LH₂ flange area.

revealed no unacceptable defects. After the PAL ramp foam loss on STS-114, NASA shipped two ETs back to the Michoud Assembly Facility (MAF) for destructive evaluation and NDE. These inspections revealed that ET-119, which had not been through tanking and thermal cycles, did not have cracks in the PAL ramps foam. In contrast, ET-120, which had been through two tanking and thermal cycles, had large, deep cracks in the liquid hydrogen (LH$_2$) PAL ramp foam. These cracks most likely occurred during thermal cycling, and similar cracks were the most likely cause of the foam loss on STS-114/ET-121.

ET Liquid Hydrogen Intertank Flange Background

The intertank separates the LO$_2$ tank from the LH$_2$ tank. The area where the intertank connects to the pressurized hydrogen tank is called the LH$_2$/intertank flange (figure 3.2-1-5). ET separation imagery has shown repeated losses of the foam overlying this flange. Foam divots from the LH$_2$/intertank flange emanate from within the critical debris zone, which is the area of the ET where debris loss could adversely impact the Orbiter or other Shuttle elements.

NASA IMPLEMENTATION

In preparation for RTF, NASA initiated a three-phase approach to eliminate the potential for debris loss from the ET. Phase 1 included those activities implemented prior to RTF to control critical debris on tanks already constructed.

Phase 2 activities were not required for the first RTF mission, but rather focused on continuous improvement. Phase 2 included debris elimination enhancements that could be incorporated into the ET production line as they become available, but were not considered mandatory for RTF. Implementation of Phase 2 activities is incorporating lessons learned from STS-114. Phase 3 is comprised of long-term development activities that would eliminate TPS foam on the vehicle. However, this phase of the plan will not be implemented because the Shuttle will be retired at the end of the decade after completion of the International Space Station.

Phase 1

A NASA/Lockheed Martin ET Project team conducted the Phase 1 improvements. This team included membership from multiple centers and technical disciplines, including the Marshall Space Flight Center (MSFC) (structures, thermal, materials, and processes), Langley Research Center (LaRC) (structures), Stennis Space Center (safety and mission assurance), and Glenn Research Center (GRC) (structures). The team identified potential sources of ascent debris, assessed those that could cause critical damage, determined the failure mode and root cause(s) for each potential debris site, proposed corrective design or manufacturing process changes for each site, and selected and implemented the best alternative.

The first step for the ET Project Team was to identify potential sources of ascent debris. It did this by reviewing historical flight data and through engineering analysis. By thoroughly reviewing all ascent and ET separation imagery available since the beginning of the SSP, the team identified the areas that historically liberated the most debris. By applying engineering analysis, knowledge of ET design, and engineering judgment, the team identified other areas for additional investigation. These activities were accelerated by the results of inquiries and engineering studies into ascent debris that had been conducted over the life of the Program.

The next step was to assess which sources of debris, if liberated, could cause critical damage to the Space Shuttle. The Space Shuttle Systems Engineering and Integration Office (SEIO) took the lead on this activity (details may be found in SSP-5 and R3.3-2). SEIO performed aerodynamic debris transport modeling for each debris piece at a number of representative ascent flow fields across the vehicle. For those pieces of debris impacting flight hardware, SEIO provided to the hardware projects the impact locations, masses, and velocities. The hardware projects then applied modeling validated by actual impact testing data to determine the extent of damage. If this damage was deemed critical (defined as potentially catastrophic), SEIO added the debris to a database that eventually provided a matrix of debris allowables for the ET. The debris allowables matrix for ET segmented the ET into zones, and each zone was assigned an appropriate maximum allowable mass for debris liberation. The first ET debris allowable matrix for foam was approved by the Program Requirements Control Board (PRCB) in November 2004. A similar ET debris allowable matrix for ice liberation has been baselined in SSP documentation (NSTS-60559, Expected Debris Generation and Impact Tolerance Requirements, Groundrules, and Assumptions).

Using the TPS debris allowables matrix, the ET Project identified TPS debris sources that could potentially cause critical damage and therefore required mitigation. For Phase 1, the areas identified were the bipod ramp foam, the hydrogen tank/intertank flange foam, and the PAL foam ramps. The ET Project also recognized the ice formation at the LO_2 feedline bellows as a potentially critical debris source to be addressed during Phase 1. Each of the critical foam debris areas was assigned to a focus team that dissected other ET foam applications and then conducted tests and analysis to understand the root causes of the debris generation. Analyses included inquiries into the basic molecular and cellular structure of foam and the physical failure mechanisms to which ET foam is vulnerable.

NASA developed alternative solutions for correcting the failure causes, including design changes and manufacturing process improvements. The selected improvements were implemented using the same verification processes applied to all Space Shuttle flight hardware, assuring rigor in configuration management, satisfaction of technical standards and requirements, understanding of any risk to be accepted, and avoidance of unintended consequences.

In addition to the four focus teams discussed above, the ET Project assigned another team to survey all foam application processes, both robotic and manual, on the remaining areas of the ET. This was necessary because the STS-114 ET TPS was applied prior to the *Columbia* accident. This team reviewed the specific procedures, manufacturing data, and available acceptance testing or inspection techniques for each application. The results of this survey have verified the acceptability of the robotically applied foam that makes up a large percentage of the overall foam acreage. The survey of more than 300 manually applied foam closeouts was reported at the Phase 2 ET Design Certification Review (DCR) in March 2005. This survey identified the rear attachment longeron manual closeouts as particularly vulnerable to defects. Although there is no significant history of liberation of foam from the longeron area, the ET Project took the conservative path of removing and reapplying this area of foam with an improved process.

As part of the Phase 1 effort, NASA enhanced or redesigned the areas of known critical debris sources (figure 3.2-1-1). This includes redesigning the forward bipod fitting and associated TPS closeout, redesigning the LH_2/intertank flange TPS closeout, and reducing ice from the LO_2 feedline bellows. In addition to these known areas of debris, NASA has reassessed all TPS areas to verify the TPS configuration, including both automated and manual spray applications. Special consideration was given to the LO_2 and LH_2 PAL ramps due to their size and location. This task included assessing the existing verification data, establishing requirements for additional verification data, conducting tests to demonstrate performance against the divoting (cohesive strength failure) failure mode, and evaluating methods to improve process control of the TPS application for re-sprayed hardware.

NASA also pursued a comprehensive testing program to identify and understand the root causes of foam shedding and developed alternative design solutions to reduce the debris loss potential. Additionally, NASA further developed two NDE techniques, terahertz imaging and backscatter radiography, to conduct ET TPS inspection without damaging the fragile insulating foam. During Phase 1, NDE was used on the LO_2 and LH_2 PAL ramps as engineering information only; certification of the foam was achieved primarily through verifying the foam design and application.

NASA also invited a number of different assessments and reviews of our ET modification efforts. These included a Lockheed Martin independent review team, an independent team of retired contractor and NASA experts, a NASA Engineering and Safety Center review team, and a Quality and Safety Assurance team chartered by the Director of MSFC. These teams provided insight and suggestions that have been incorporated into ET Project corrective measures.

Phase 2

Phase 2 efforts included redesigning or eliminating the LO_2 and LH_2 PAL ramps and enhancing NDE technology with the goal of using NDE as an acceptance tool. TPS application processes were enhanced as appropriate and more stringent process controls were incorporated. Another Phase 2 effort was to enhance the TPS thermal analysis tools to potentially reduce the amount of TPS on the vehicle.

Phase 3

Because the Space Shuttle will be retired at the end of the decade, NASA will not implement Phase 3. Phase 3 activities would have included examining additional means of further reducing ET debris potential. Proposed concepts included:

- Rotating the LO_2 tank 180 deg to relocate manually applied TPS closeouts outside of the critical debris zone.
- Developing a "smooth" LO_2 tank without external cable trays or pressurization lines.
- Developing a smooth intertank using an internal orthogrid instead of external stringers.
- Developing a protuberance tunnel in the LH_2 tank that would have provided a tank with a smooth outer mold line and eliminated complex TPS closeouts and manual sprayers.

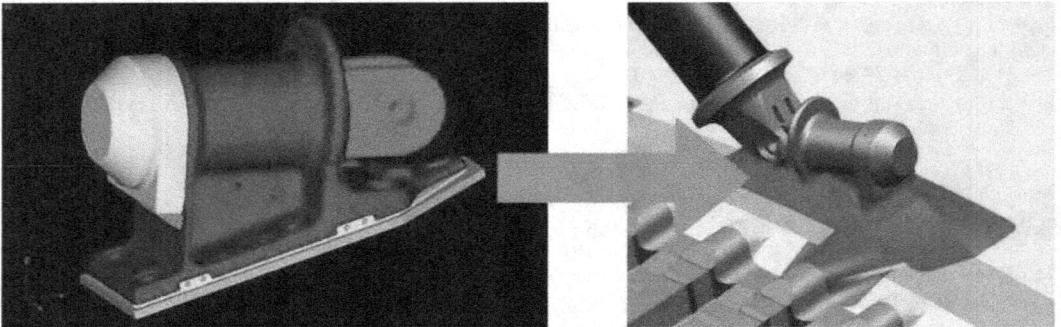

Figure 3.2-1-6. ET forward bipod redesign.

ET Forward Bipod Implementation Approach

NASA redesigned the ET forward bipod fitting (figure 3.2-1-6) to incorporate redundant heaters in the base of the bipod fittings, which preclude ice formation formerly prevented by the foam ramps. This redesign addresses the specific proximate cause of the *Columbia* accident. The results of wind tunnel tests conducted prior to the STS-114 mission confirmed the acceptability of the increased aero-heating loads that result from this new configuration.

LO$_2$ Feedline Bellows Implementation Approach

NASA evaluated several concepts to reduce ice formation on the forward bellows (figure 3.2-1-7). An initial trade study included a heated gaseous nitrogen (GN$_2$) purge, a flexible boot over the bellows, heaters at the bellows opening, and other concepts. Analysis and testing eliminated the flexible bellows boot as a potential solution because it could not eliminate ice formation within the available volume. Heaters at the bellows opening were analyzed and determined to have a potential impact on adjacent hardware and propellant quality. The heated GN$_2$ or gaseous helium purge options were eliminated due to implementation issues and debris potential for purge hardware.

It was during development testing that NASA identified the condensate drain "drip lip" as a solution that could reduce the formation of ice. Since the drip lip alone was not sufficient to completely eliminate the ice, NASA continued to develop alternate mitigations and determined that the most effective solution to eliminate ice at the forward bellows location was an alternate heater system installed inside the bellows cavity.

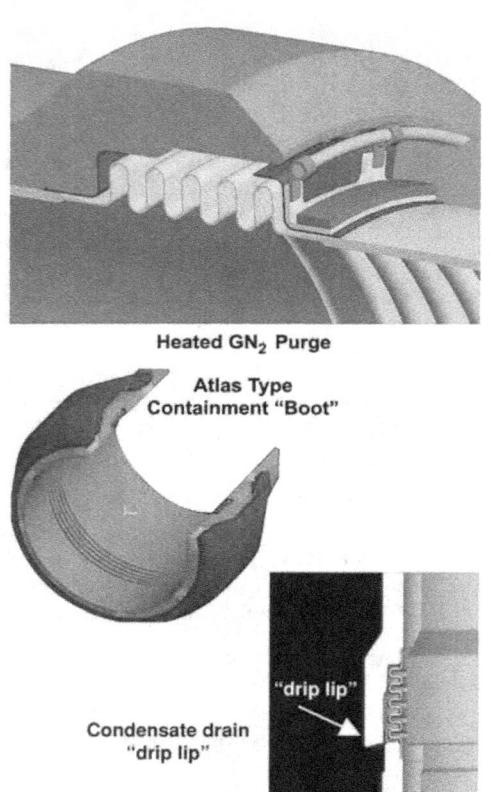

Figure 3.2-1-7. LO$_2$ feedline bellows design concepts.

LH₂/Intertank Flange Closeout Implementation Approach

NASA has conducted tests to determine the cause of foam liberation from the LH$_2$/intertank flange area. These tests revealed that foam loss from this area was the outcome of a multistep causal chain: GN$_2$, which was used to purge any leaking hydrogen from the intertank, condensed into a liquid upon contact with the cold LH$_2$ tank dome. The LN$_2$ then pooled around the perimeter of the dome and seeped through the flange joint, fasteners, vent paths, and other penetrations to the underside of the foam layer. If foam voids were in the seep paths, the LN$_2$ collected in the voids. During ascent the rapidly decreasing atmospheric pressure and increasing environmental temperature caused the LN$_2$ to boil into a gas, expand, and blow off a foam divot.

NASA's corrective measures interrupt this failure chain. First, seepage of LN$_2$ was significantly reduced by sealing the flange bolts and injecting foam into the intertank stringers. Second, NASA removed all existing foam from the flange area and replaced it by using an enhanced three-step foam application process to reduce both the volume and quantity of voids.

An update to the original Level 2 debris transport analyses expanded the critical debris zone that must be addressed and significantly reduced the allowable debris mass in this region. The critical debris zone was expanded from ±67.5 deg from the top of the ET (the top of the tank directly faces the underside of the Orbiter) to greater than ±100 deg from the top of the tank. As a result, a new closeout process for the thrust panel of the intertank flange region has been developed and applied to the entire thrust panel, expanding the enhanced closeout region to ±112 deg from the top of the tank (figure 3.2-1-8).

PAL Ramps Implementation Approach

Prior to STS-114, NASA assessed the verification data for the existing PAL ramps and determined that the existing verification was valid. To increase our confidence in the verification data, NASA dissected similar hardware to determine TPS application process performance. However, NASA determined after STS-114 that one contributor to the PAL ramp foam loss was likely cracks resulting from

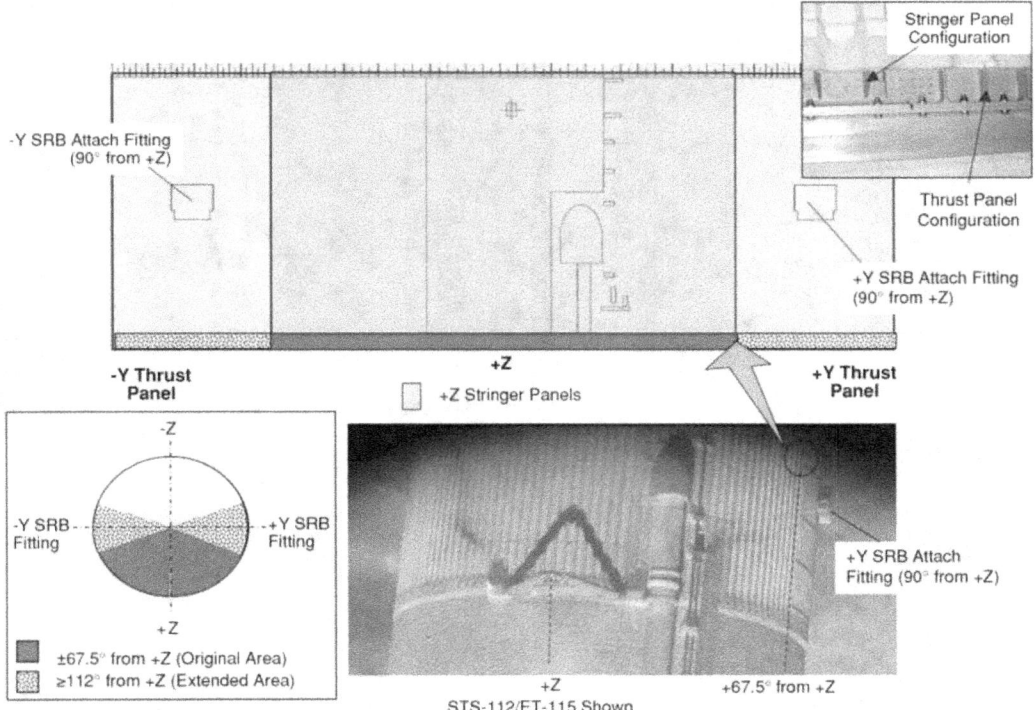

Figure 3.2-1-8. LH₂ intertank flange expanded debris zone.

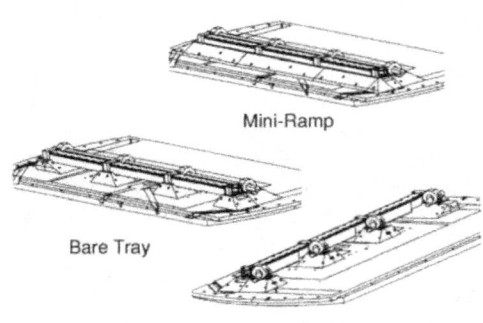

Figure 3.2-1-9. Phase 2 minimal debris ET – PAL ramp redesign solutions.

the thermal cycles associated with tanking. As a result, the inspections conducted prior to tanking would not have revealed the potential for foam loss.

Following STS-114, NASA developed plans for the redesign or removal of the PAL ramps. Three redesign solutions were proposed (figure 3.2-1-9): eliminating the ramps, reducing the size of the ramps, and redesigning the cable tray with a trailing edge fence. A wind tunnel test was used to evaluate the potential for aerodynamic instabilities of the basic cable trays and associated hardware due to the proposed redesigns. The test articles were instrumented with pressure transducers, strain gauges, and accelerometers to measure the aero-elastic effect on the test articles. After review of flight and wind tunnel test data and related analyses, SSP determined that eliminating the PAL ramps provided the best means of reducing the risk of foam debris from the area.

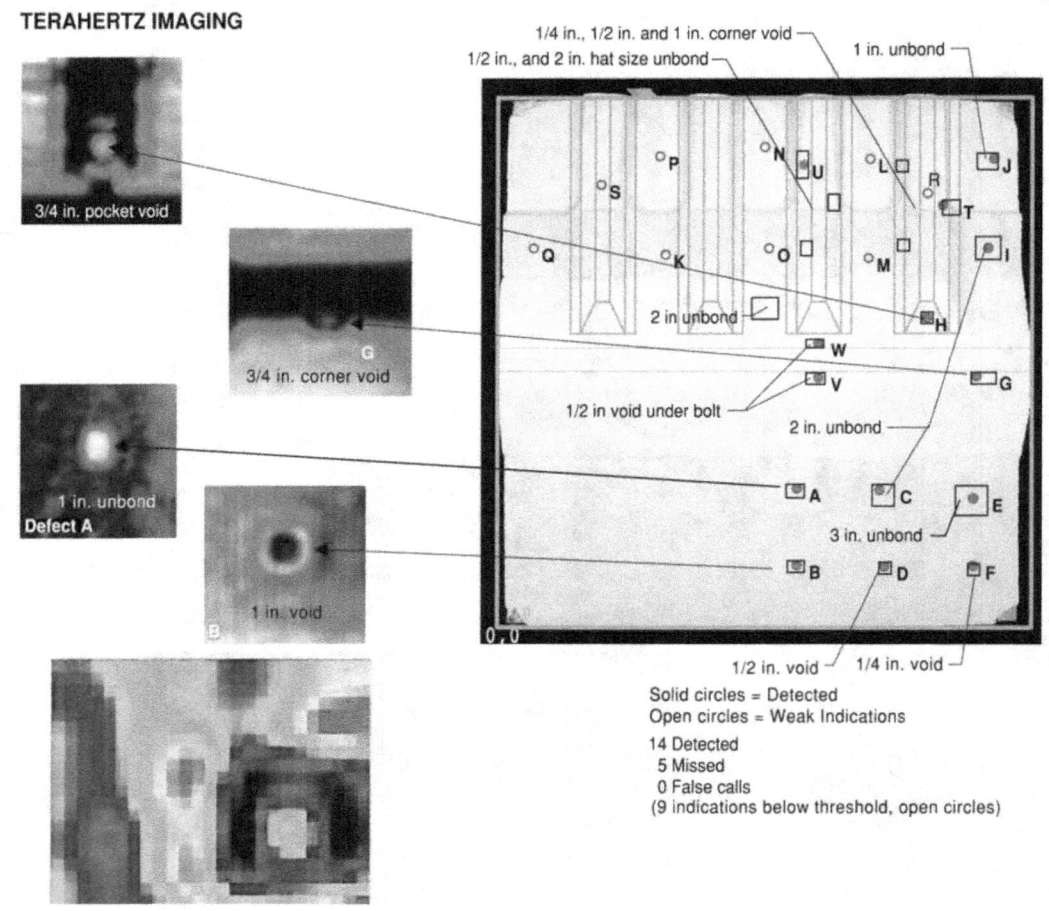

Figure 3.2-1-10. Terahertz images.

TPS (Foam) Verification Reassessment Implementation Approach

NASA has performed an assessment for both manual and automated TPS applications in the critical debris zones. This assessment was performed using the same approach applied to the PAL ramps: evaluating existing verification data, performing additional tests and analyses to demonstrate performance against critical failure modes, and reviewing and updating of the process controls applied to re-sprayed TPS applications.

NASA enhanced the TPS application processes and implemented more stringent process control measures for reapplied TPS hardware and future TPS applications as appropriate. These enhancements include requiring at least two certified production operations personnel attend all final closeouts and critical hand-spraying procedures to ensure that proper processing and updates to the process controls are applied to the foam applications (ref. CAIB Recommendation 4.2-3).

NDE of Foam Implementation Approach

During Phase 1, NASA surveyed state-of-the-art technologies, evaluated their capabilities, down-selected options, and began developing a system to detect critical flaws in ET insulation systems. At an initial screening, test articles with known defects, such as voids and delaminations (figure 3.2-1-10), were provided to determine detection limits of the various NDE methods.

NASA pursued development of TPS NDE techniques to verify proper application of foam without defects, such as unacceptable volume of voids or poor subsurface adhesion. After the initial screening, NASA selected the tetrahertz and backscatter radiation technologies and conducted more comprehensive probability of detection (POD) tests for those applicable NDE methods. Although NASA originally planned during Phase 2 to optimize and fully certify the selected technologies for use on the ET, these new NDE techniques were only used for engineering information on the RTF ET PAL ramps. The NDE techniques were used on the LO_2 and LH_2 PAL ramps and detected no anomalous conditions above the certification acceptance criteria.

STATUS

ET Forward Bipod Status

NASA successfully completed a Systems Design Review and a Preliminary Design Review. The Critical Design Review (CDR) was held in November 2003, with a Delta CDR in June 2004. The Delta CDR Board approved the bipod redesign. A Production Readiness Review (PRR) was held in June 2004. The PRR Board gave approval for manufacturing operations to proceed with the bipod wedge foam spray on ET-120, which is now complete. The wedge spray is a foam closeout that serves as a transition area for routing of the heater harnesses from the fitting base into the intertank. The wedge is applied prior to fitting installation; after the fitting installation is complete, the final bipod closeout is performed. The final closeout application process has been verified and validated (figure 3.2-1-11).

The bipod fitting design, fitting closeout, and heater system were reviewed during the ET DCR. This hardware, as well as the other ET redesigns, was verified to meet the current SSP performance requirements.

The bipod fitting redesign verification is complete. The verification included thermal tests to determine the capability of the design to preclude prelaunch ice, with an automated heater control baselined and validated based on bipod web temperature measurements. Structural verification tests have confirmed the performance of the modified fitting in flight environments. Wind tunnel testing has verified the TPS closeout performance when exposed to ascent aerodynamic and thermal environments. The system verification included a full-scale integrated bipod test using hydrogen, the tank fluid, a prototype ground control system to demonstrate system performance, and a thermal-vacuum test with combined prelaunch and flight environments to demonstrate TPS performance. Post STS-114 performance reviews indicated foam loss due to cryo-pumping occurred in the bipod region. Plans and updates in response to this foam loss are included in the post STS-114 update section of this write-up.

LO_2 Feedline Bellows Status

NASA implemented a TPS "drip lip" to reduce ice formation on the three LO_2 feedline bellows. The drip lip diverts condensate from the bellows and significantly reduces ice formation. Because the drip lip alone was insufficient to completely eliminate ice on the bellows, NASA continued to pursue solutions to complement the TPS drip lip. Analysis of the residual ice formation, estimates of the liberated ice, and transport analyses identified the residual ice at the forward LO_2 feedline bellows location as an unacceptable debris source.

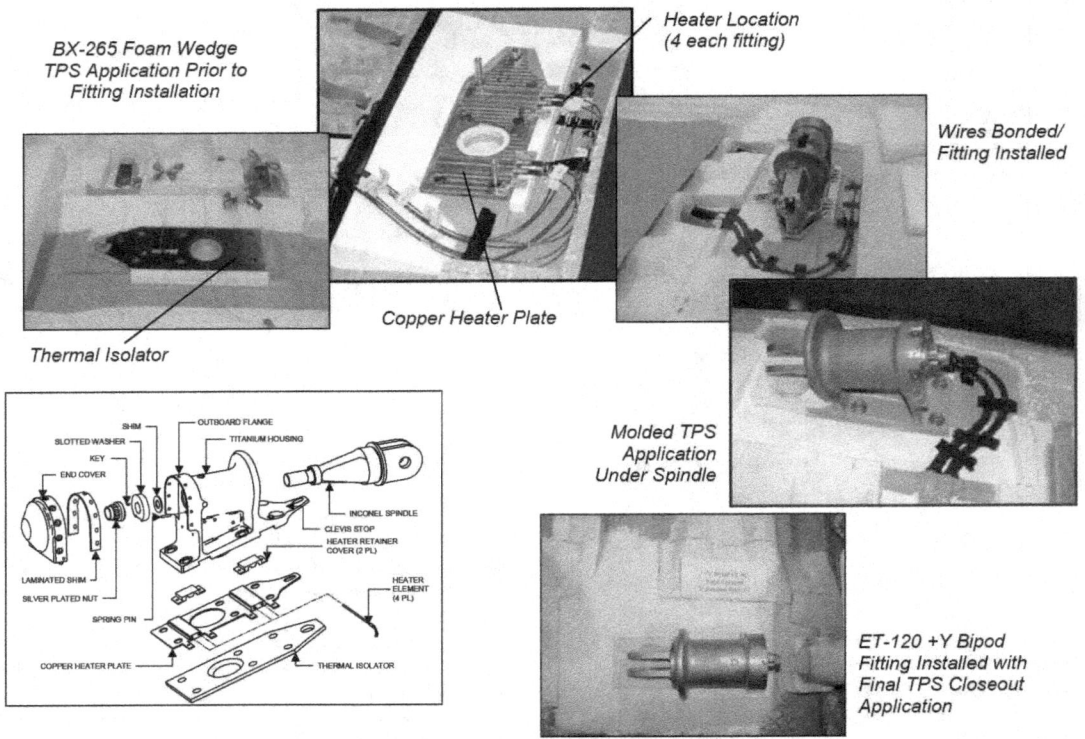

Figure 3.2-1-11. Bipod fitting redesign and TPS closeout.

NASA determined that a heater installed in the bellows cavity would reduce ice formation to an acceptable level. The design change included a redundant heater system with a silicon rubber gasket to join heater elements. The heater is bonded to the bellows rain shield and convolute shield; heater wires are bonded to the external LO_2 feedline substrate. Bonding of the heaters requires removal and replacement of a three-inch width of TPS along the existing drip lip and LO_2 feedline surface. Verification of the heater system design included testing and analysis, which qualified the system and ensured that the system did not result in a debris source.

NASA evaluated other ice mitigation techniques, such as the infrared projector, that could be implemented at the launch pad if necessary. However, the bellows heater performed successfully and other ice mitigation techniques were not required.

NASA also determined that, if liberated, the ice at the two aft bellows locations would not impact the Orbiter Reinforced Carbon-Carbon or critical belly tile areas; therefore, no additional ice mitigation was required for those locations.

LH₂/Intertank Flange Closeout Status

NASA successfully determined the primary root cause of foam loss in the intertank/LH_2 tank flange area. When the GN_2 used as a safety purge in the intertank came into contact with the extremely cold hydrogen tank dome, the GN_2 condensed into LN_2. The LN_2 migrated through intertank joints, fasteners, vent paths, and other penetrations into the foam and then filled voids in the foam caused by unacceptable variability in the manual foam application. During ascent, the LN_2 returned to a gaseous state, pressurizing the voids and causing the foam to detach.

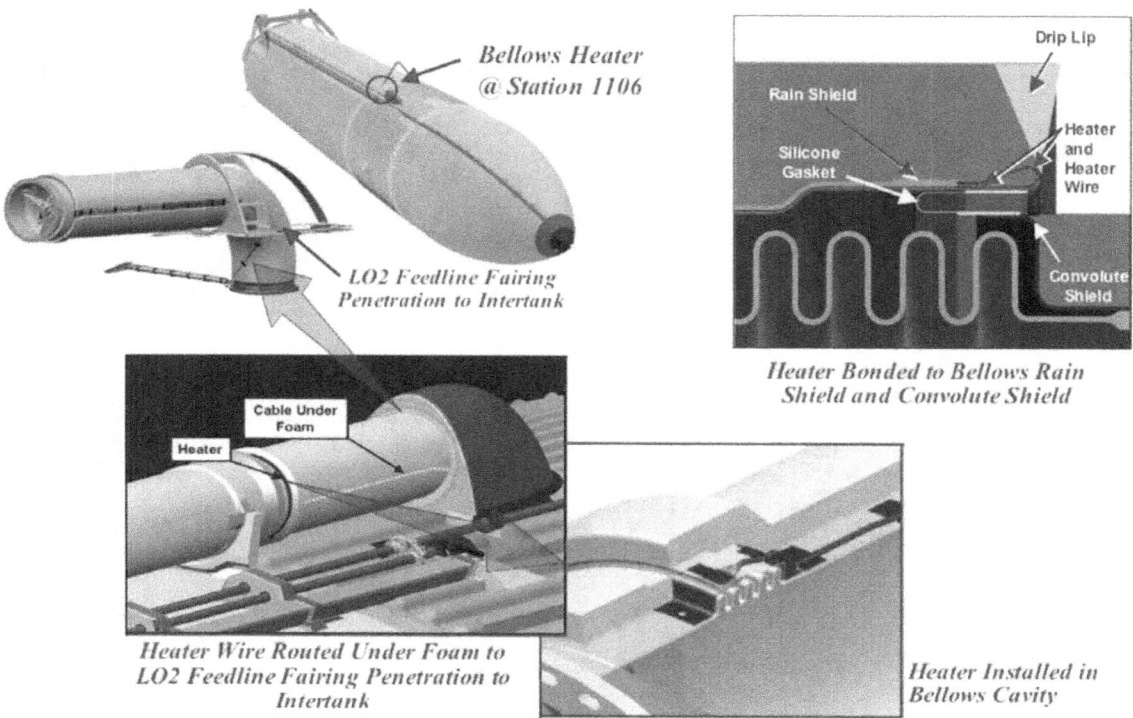

Figure 3.2-1-12. LO$_2$ feedline bellows condensate drain "drip lip" with heater.

NASA evaluated the foam loss in this region through rigorous testing and analysis. First, a series of one-foot-square aluminum substrate panels with induced voids of varying diameters and depths below the foam surface was subjected to the vacuum, heat profiles, and backface cryogenic temperatures experienced during launch. These tests were successful at producing divots in a predictable manner.

Follow-on testing was conducted on panels that simulated the LH$_2$ intertank flange geometry and TPS closeout configuration to replicate divot formation in a flight-like configuration. Two panel configurations were simulated: (1) a three-stringer configuration and (2) a five-stringer configuration. The panels were subjected to flight-like conditions, including vacuum front face heating, backface cryogenics (consisting of a 1.5-hour chill-down, a 5-hour hold, and an 8-minute heating), ascent pressure profile, and flange deflection. These tests were successful at demonstrating the root cause failure mode for foam loss from the LH$_2$ tank/intertank flange region.

With this knowledge, NASA evaluated the LH$_2$/intertank closeout design to minimize foam voids and LN$_2$ leakage from the intertank into the foam (figure 3.2-1-5). Several design concepts were initially considered to eliminate the causes of debris from the area, including incorporating an active helium purge of the intertank crevice to eliminate the formation of LN$_2$ and developing enhanced foam application procedures.

NASA rejected helium purges when testing indicated that a helium purge would not completely eliminate the formation of foam divots since helium could produce enough pressure in the foam voids to cause divot formation. NASA also tried applying a volume fill or barrier material in the intertank crevice to reduce or eliminate nitrogen condensation migration into the voids. However, analyses and development tests showed that the internal flange seal and volume fill solution might not be totally effective on tanks that had existing foam applications. As a result, NASA also rejected this solution.

NASA focused on the enhanced TPS closeout in the LH$_2$ intertank area to reduce the presence of defects within the foam by using a three-step closeout procedure. This approach greatly reduces or eliminates void formations in the area of the flange joining the LH$_2$ tank to the intertank. The flange bolts in this area are reversed to put the lower bolt head profile at the lower flange. The LH$_2$ tank side of flange (figure 3.2-1-13) provides the foam application technician a much less complex configuration for the foam spray application and subsequently reduces the potential for void formation behind the bolt head. The higher profile (nut end) is encapsulated in the stringer or rib pocket closeout prior to final closeout application. The application process for the intertank stringer panels is shown in figure 3.2-1-14. The stringer panels are the intertank panels ±67.5 deg from the centerline of the tank directly below the Orbiter.

The areas beyond ±67.5 deg that remain in the critical debris zone are the intertank thrust panels. The geometry of these panels is simplified by hand-spraying the thrust panel pockets prior to applying the final closeout shown in Steps 2 and 3 of figure 3.2-1-14.

Testing performed on eight panels using the enhanced closeout configuration demonstrated the effectiveness of the closeout; there were no divots formed in any of the tests.

NASA now has a much better understanding of the foam failure mechanism in the intertank area and has implemented the appropriate solutions. The baseline flange closeout enhancement (±112 deg from the +Z, excluding area under LO$_2$ feedline and cable tray) uses a multi-pronged approach. The baseline includes the external three-step closeout, point fill of the structure, reversal of the flange bolts, and sealant on the threads of the bolts. The external three-step enhanced procedure reduces foam loss to a level within acceptable limits by removing critical voids in the foam. The newly enhanced ET-120 closeout was applied using a verified and validated TPS application process. During production of the ET-120 flange closeout (and all subsequent flange closeouts), a series of high-fidelity production test articles was used to demonstrate the application on the flight hardware. The acceptability of the closeout is demonstrated through a series of mechanical property tests and dissection of the foam to determine process performance. Defect tolerance of the flange closeout design was demonstrated in a combined environment test.

In addition to hardware redesign, the Integration Cell access platforms at Kennedy Space Center (KSC) were modified to provide a more accessible work environment in which to reduce the potential for TPS damage during ground processing of hardware near the intertank/LH$_2$ tank flange and bipod area.

PAL Ramp Status

Because the PAL ramps had an excellent flight history without observed foam loss since the last configuration change after STS-7, NASA's baseline approach for RTF was to develop sufficient certification data to accept the minimal debris risk of the existing design. Evaluating the available verification data and augmenting them with additional tests, analyses, and inspections was believed to have accomplished this. NASA also dissected several existing PAL ramps to understand the void sizes produced by the existing PAL ramp TPS process. Prior to STS-114, NASA believed it had obtained sufficient data to proceed to launch with the existing LO$_2$ and LH$_2$ PAL ramps.

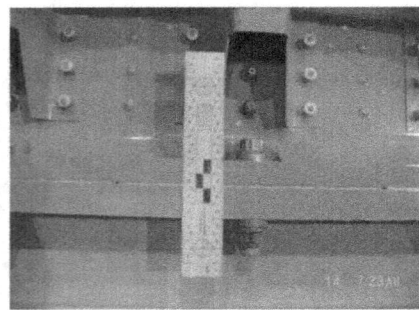   

Previous orientation – bolt head forward (top)   *New orientation – bolt head aft (bottom)*

Figure 3.2-1-13. Flange bolt reversal.

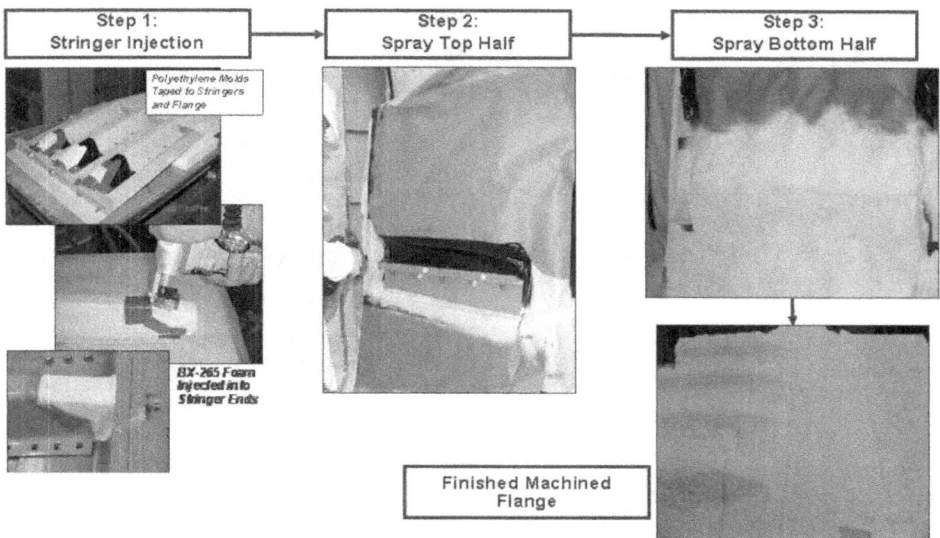

Figure 3.2-1-14. Three-step closeout for LH₂ tank/intertank.

The LH₂ PAL ramp is approximately 38 ft in length; the forward 10 feet of the ramp spans the high-risk LH₂ flange closeout. For STS-114, this portion of the LH₂ PAL ramp was removed to access the underlying intertank/LH₂ tank flange closeout. Removing the 10-ft section allowed an enhanced LH₂/intertank flange closeout to be performed. The removed portion of the LH₂ PAL ramp was replaced with an improved process manual spray application. Inspection of both the LO₂ and LH₂ PAL ramps using NDE technologies detected no anomalous conditions above the certification acceptance criteria prior to STS-114. Despite this, on STS-114 the LH₂ PAL ramp lost a large piece of foam late in ascent.

As a part of the Phase 2 activities prior to STS-114, NASA had developed concept designs to eliminate the large PAL ramps. Redesign options included eliminating the PAL ramps altogether, implementing smaller mini-ramps, or incorporating a cable tray aero-block fence on either the leading or trailing edge of the tray. NASA performed analysis of the aerodynamic loading on the adjacent cable trays and conducted subscale and full-scale wind tunnel testing of the cable trays to determine the aerodynamic and aero-elastic characteristics of the trays. The tests provided sufficient confidence in the analysis to continue pursuit of ramp elimination. This analysis formed the basis for removal of the PAL ramps. Concurrent with this work, SSP Systems Engineering and Integration (SEIO) analyzed the aerodynamic and structural loads effects of removing the PAL ramps.

TPS (Foam) Verification Reassessment Status

The SSP established a TPS Certification Plan for the ET RTF efforts. This plan has been applied to each TPS application within the critical debris zone by evaluating the available verification data and augmenting them with additional tests, analyses, and/or inspections. The plan also included dissection of TPS applications within the critical debris zone to understand the void sizes produced by TPS processes that existed when the TPS was applied to the RTF tanks.

The TPS applications underwent visual inspection, verification of the TPS application to specific acceptance criteria, and validation of the acceptance criteria. A series of materials properties tests was performed to provide data for analysis. Acceptance testing, including raw and cured materials at both the supplier and the Michoud Assembly Facility, were used to demonstrate that as-built hardware integrity is consistent with design requirements and test databases. Mechanical property tests, including plug pull, coring, and density, are being performed on the as-built hardware.

NASA also conducted stress analysis of foam performance under flight-like structural loads and environmental conditions, with component strength and fracture tests grounding the assessments. Dissection of equivalent or flight hardware was performed to determine process performance. TPS defect testing was conducted to determine the critical defect sizes for each application. In addition, various bond adhesion, cryoflex, storage life verification, cryo/load/thermal tests, and acceptance tests were performed to fully certify the TPS application against all failure modes. Finally, a Manual Spray Enhancement Team has been established to provide recommendations for improving the TPS closeout of manual spray applications.

Production-like sprays on mockups were performed to verify and validate the acceptability of the production parameters of redesigned or re-sprayed TPS applications.

NDE of Foam Status

Activities to develop NDE techniques for use on ET TPS included the following prototype systems then under development by industry and academia:

- Backscatter Radiography: University of Florida
- Microwave/Radar: MSFC, Pacific Northwest National Labs, University of Missouri, Ohio State
- Shearography: KSC, Laser Technology, Inc.
- Terahertz Imaging: LaRC, Picometrix, Inc., Rensselaer
- Laser Doppler Vibrometry: MSFC, Honeywell

The terahertz imaging and backscatter radiography systems were selected for further POD testing based on the results of the initial proof-of-concept tests. The microwave system was evaluated during the Phase 2 development activity and the results were analyzed. The preliminary results indicated that these technologies were not yet reliable enough to certify TPS applications over complex geometries such as the bipod or intertank flange regions.

Prior to STS-114, NASA employed a lead tank/trail tank approach for RTF to mitigate risk in the event that NASA identified any changes required on the lead tank due to the evolving debris analysis. The "trail" or second tank (intended for STS-121 or a launch-on-need rescue mission) was not planned for shipment to the KSC for mating to the Orbiter until after the final ET DCR. The SSP decided to ship the second RTF tank prior to the final ET DCR, which was rescheduled to a date after the required ship date. This decision was based on the ET DCR Pre-board identifying no significant issues, as well as the capability to do certain types of work on the ET at KSC should it be required. Additionally, NASA redefined the trail tank as the third tank (ET-119), once the second RTF tank (ET-121) was shipped. ET-119 was retrofitted with the required design changes to eliminate critical TPS debris and eliminate critical ice debris by installing a heater system at the forward LO_2 feedline bellows. Final certification of the heater system and shipment of ET-119 was completed in June 2005.

POST STS-114 UPDATE
Overview

Following the launch of the STS-114 mission, photo and video analyses revealed multiple foam losses from the ET; of those losses, 11 were classified as in-flight anomalies (IFAs). Although the majority of these foam losses took place late in the ascent and therefore posed less risk to the Orbiter, the size of some of the foam losses caused concern because they were much larger than analysis had predicted was likely. In October 2005, NASA returned to MAF the two ETs (ET-119 and ET-120) that had been previously shipped to KSC. At MAF, the ETs underwent destructive evaluation and NDE. NASA created two teams to investigate the 11 foam loss events: the ET Independent IFA Investigation Team (including a number of sub-teams) and the ET Tiger Team. The 11 foam losses investigated occurred in six areas on ET-121 (figure 3.2-1-15):

1. LH_2 PAL ramp (one loss)
2. Left bipod fitting closeout (two losses)
3. Ice/frost ramps (three losses)
4. LH_2 tank to intertank flange (two losses)
5. LH_2 tank acreage foam (two losses)
6. +Y thrust strut flange (one loss)

The IFA Independent Investigation Teams worked through a fault-tree analysis for the foam losses to determine possible root causes, to examine whether these root causes were unique to STS-114/ET-121, and to make recommendations to minimize the likelihood of recurrence. Those recommendations resulted in testing programs to investigate the root causes of foam loss on STS-114 and to validate and certify new processes and hardware configurations. The IFA Independent Team also made 35 recommendations for the ET Project related to the foam loss events on STS-114.

Investigating teams found no off-nominal ascent flight environment effects. The ET Tiger Team noted and investigated a potential correlation between areas of the ET with high processing traffic (adjacent to work platforms, mats, etc.) and foam loss locations on the STS-114 ET (ET-121) and ETs flown on previous missions. The ET Tiger Team also noted that ET-121 was extensively reworked to include RTF modifications, and that STS-114 was the first flight in which BX-265 foam was used for LO_2 and LH_2 PAL ramps.

To support the IFA teams' recommendations, investigate root causes, and evaluate solutions, the ET Project established an IFA testing program that completed its work in February 2006. The IFA teams' final reports are complete. Concurrent with the IFA testing, final RTF and certification tests were conducted. All testing was completed prior to the scheduled STS-121 FRR on June 16, 2006.

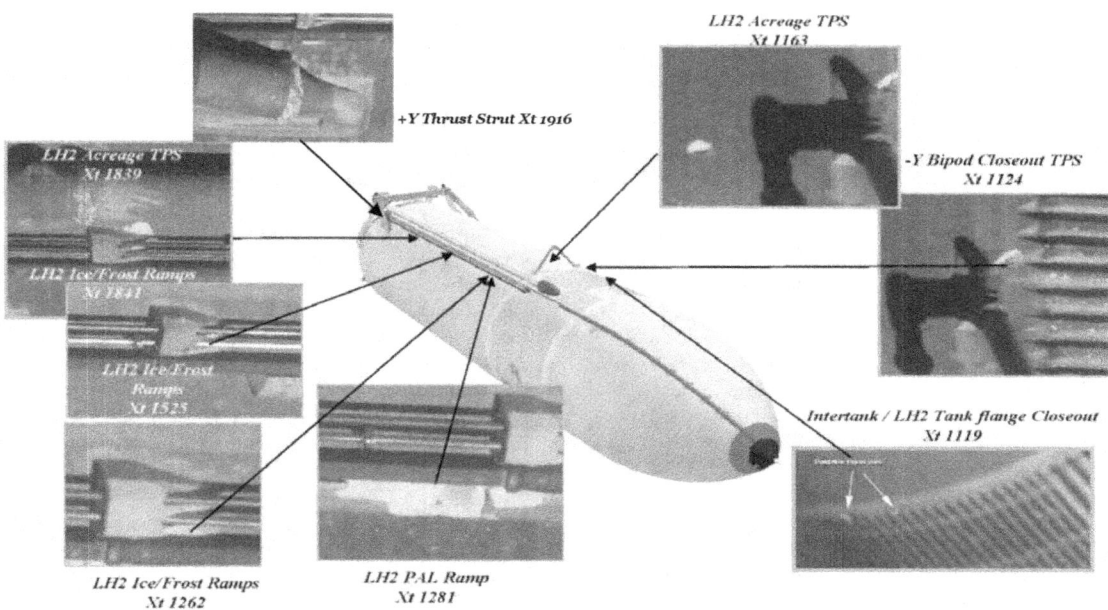

Figure 3.2-1-15. Areas of foam loss on ET-121 investigated by the IFA teams.

Data were also generated to support probabilistic and debris analyses to determine overall flight vehicle performance.

Foam Loss Risk Assessment

The Independent IFA Investigation Team did not conclusively determine the release time of all debris pieces. Foam losses occurring later than 135 seconds after launch were not considered a critical threat due to reduced aerodynamic pressure, which results in an impact velocity too low to cause critical TPS damage. Three of the STS-114 failures occurred after 135 seconds. The PAL ramp loss occurred at 127.1 seconds. Initial indications were that as many as six of the losses occurred late in ascent and one after ET separation; however, the time accuracy of these time estimates was limited by camera angles, resolution, and frame speed (1/30 second).

The preflight predictions of foam loss probability addressed only the foam divoting failure mode. A number of other ET TPS foam failure modes were observed experimentally and in historical post-separation imagery. These include substrate debond, popcorning (small divots), delamination, transverse cracking, fragmentation/crushing, strength failure, aero-shear failure, and fatigue. The STS-114 probabilistic risk assessments were updated to include additional foam dissection data and potential contributions from crushed foam and popcorning. As additional quantitative information from flights or ground testing became available regarding other failure modes, the risk assessment was updated.

The post STS-114 Independent IFA Investigation Team efforts focused on two key failure modes: divoting and fragmentation or crushing. Divoting may be caused by voids created during foam spraying or formed by debonding at the substrate or knitlines. Ascent aerodynamic heating, the drop in propellant level, the influx of hot ullage gas, and a reduction in external pressure can cause an expansion of trapped gas or cryogenic liquid in voids. This build-up of pressure may overstress any shear foam, resulting in divots. Foam crushing or fragmentation can be caused either during processing or by debris impacts during ascent. This failure mode depends on the incident angle and geometric shape of the impacting debris, the area of incidental processing contact, and the magnitude and duration of the loads. Crushing damages the foam's cellular structure and can create large voids that subsequently divots or weakens the foam.

Post IFA Investigation Activity

PAL Ramp

The IFA Investigation PAL Ramp Team determined that the PAL ramp foam loss had a variety of possible root causes, including: a cohesive failure generating a cavity, a leak path to the atmosphere, external forces, and application process deficiencies. These causes led to both fracture and cryo-pumping. During NDE of ET-119 (STS-121) and ET-120 (originally STS-115, but changed to STS-120) using terahertz imaging and backscatter radiography techniques, subsurface cracks in ET-120's LH_2 PAL ramp were discovered. The ET-120 PAL ramps were dissected and the cracks analyzed further to determine the root cause, which was attributed to thermal cycling of the foam during cryogenic tanking. Analyses demonstrated that these subsurface PAL ramp cracks could result in foam losses similar to the PAL ramp foam loss on STS-114 (ET-121). At a Technical Interchange Meeting in December 2005, the SSP determined that a "No PAL Ramp" design was low risk and most likely would not result in loads or environments that were unacceptable for flight. Late in 2005, a PRCB Directive authorized the "No PAL Ramp" design for the ET, shown in figure 3.2-1-16.

SSP SEIO revised the ET external environments analysis for a "No PAL Ramp" case and released new ascent environments to the ET Project so the Project could determine whether the "No PAL Ramp" design would adversely affect the ET's performance and structure.

The LO_2 and LH_2 PAL ramps were removed down to the NCFI 24-124 base foam. The removal of the LO_2 and LH_2 PAL ramps resulted in a reduction of critical debris mass (LO_2: ~14 lbm / LH_2 ~21 lbm). Elimination of the PAL ramps required assessment of the foam configuration at the underlying and adjacent areas of the PAL ramps. This resulted in changes to the surrounding TPS configuration. Key requirements guiding the final configuration included:

- Structural performance
- Ice/frost prevention
- TPS/ice debris prevention
- Aerothermal performance

Acoustic environments were not affected by removing the PAL ramps. Because cable tray, pressline, and feedline protuberance hardware are considered at risk for increased flight environments, additional loads analyses for all aero-sensitive protuberances and margin impact were assessed in areas where PAL ramps were removed. Critical components with margins of less than 20% (cable trays and presslines) and 10% (LO_2 feedline) were identified, and loads analyses were reassessed for potential margin improvement. Hardware was certified to Level II design loads.

Following the PAL ramp removals, inspections verified that the TPS thickness requirement was met. Where additional acreage TPS was applied to maintain the outer mold line (OML), TPS process application verification and validation was performed. The LO_2 and LH_2 ice/frost ramps were extended outboard with manually applied BX-265 foam to achieve the required OML. The LO_2 Barry Mount and the outboard face of the LO_2 cable tray fairing were closed out using hand-packed SLA. Hot gas and combined environments testing demonstrated that the BX-265 foam over Conathane and NCFI foam configuration is not a debris concern. Previous testing demonstrated the performance of the streamlined ice/frost ramps.

NASA also approved the use of ET flight instrumentation on STS-121 to obtain data to validate the flight environments used in the test and analyses for the "No PAL Ramp" design.

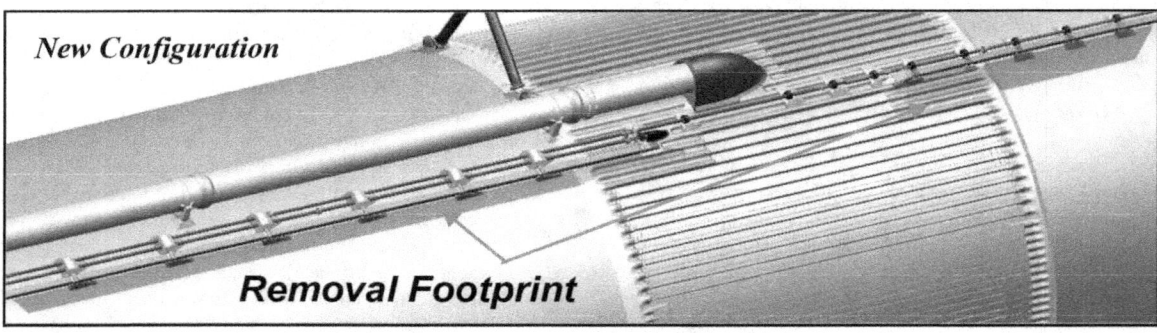

Figure 3.2-1-16. PAL ramp removal configuration.

The "No PAL Ramp" design verification was completed in June 2006 and included:

- TPS: Combined environment tests (cryogenic, axial load, and thermal vacuum), hot gas tests, wind tunnel tests at GRC and Arnold Engineering Development Center (AEDC), and prelaunch cryogenic exposure to assess ice.
- Structural hardware capability testing including bolt tension and bending, simulated service, and capability demonstrations.
- The assessment of system environment updates, which was completed prior to STS-121.

LH_2 Intertank Flange

The IFA Investigation LH_2 Intertank Flange Team determined that voids were the root cause of the two divots released during ascent. These divot releases occurred at approximately liftoff plus 270 seconds, which was safely beyond the 135-second requirement. The voids were possibly caused by a combination of process deficiencies and collateral damage caused by RTF rework.

Testing both verified the release time for the flange foam and examined the effects of substrate geometry on divot mass for small voids. Future enhancements to LH_2 intertank flange foam will be identified, evaluated, and implemented through the standard production channels to improve the quality of the product.

Bipod Closeout

The IFA Bipod Closeout Investigation Team determined that cryo-ingestion (through and into the heater cable) and cryo-pumping (void in intimate contact with the heater cable run) were the root causes of the foam loss in the bipod heater installation closeout. These events caused a cohesive failure.

- The ET Project sealed and filled the wire leak path and eliminated void volume to prevent cryo-pumping through bipod heater wiring and cryo-ingestion beneath the cables for both the +Y and –Y bipod fitting installations.
- The heater wire bonding process was also changed to eliminate the voids associated with the previous process.

For future production activities, the ET Project assessed foam repair processes to ensure the structural strength of TPS repairs and the integrity of the surrounding TPS. Procedures were implemented to reduce processing damage, increase oversight, and heighten awareness of foam fragility during hardware repair.

The TPS Process Control Board implemented procedures on all TPS applications to ensure the integrity of the TPS on the ET and mitigate failures that could lead to catastrophic events. The Control Board continues to assess and propose process improvements for additional manual spray foam applications.

LH_2 Ice/Frost Ramp

While three LH_2 ice/frost ramps experienced foam loss on STS-114, the SSP determined that ice/frost ramp modification was not required for ET-119 based on flight history and bounding risk assessment.

On ET-120, which had experienced two cryo loading cycles, horizontal cracks were observed in dissected NCFI acreage under the PDL ice/frost ramps. The cause of these cracks was determined to be thermally induced loads experienced during cryo loading.

The IFA Investigation Ice/Frost Ramp Team testing determined that, at Station 1262, root causes for foam loss were either over-pressurization of a defect resulting in a divot or impact. At Station 1525, root causes were adhesive debonds, defect over-pressurization and impact. At Station 1841, defect over-pressurization was identified as the root cause.

The ET Project investigated venting and configuration enhancements of the ice/frost ramps to reduce the potential for divoting of subsurface defects and to reduce the debris size. The results of ice/frost ramp testing were provided to SEIO. These data were used by SEIO to update the Level II risk assessment models. Testing established the release times of foam losses.

The ET Project continues to pursue innovative near-term and long-term design solutions to mitigate the possibility of foam loss from the ice/frost ramps. An interim ice/frost ramp modification will fly on STS-120/ET-120 and a final design is planned for STS-124/ET-128. The final design is intended to be flown on all flights throughout the remaining life of the Program.

LH_2 Acreage

The IFA Investigation LH_2 Acreage Team determined that, at Station 1160, a defect over-pressurization was the root cause of the foam loss and, at Station 1851, material property degradation and cryo-pumping were root causes.

To address these root causes, the ET Project and the TPS Process Control Technical Subcommittee continue assessing NDE for the ET TPS (i.e., backscatter X-ray, terahertz imaging and shearography). Testing was conducted to re-verify the effectiveness of red dye to determine the extent of any potential crushed TPS. Testing did show that red

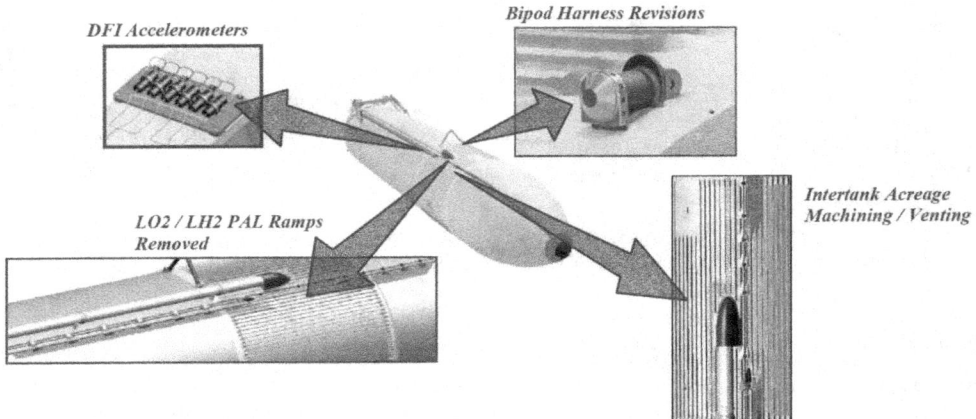

Figure 3.2-1-17. ET-119 changes.

dye effectively locates crushed foam. Hot gas testing characterized the size of foam loss during ascent due to processing damage and demonstrated that crushed foam does not pose a debris risk. These test data were provided to SEIO to update the Level II risk assessment models.

Other Activities

Defect documentation

The ET Project developed an improved in-process data collection and documentation system, including digital photos of damaged TPS and standardized procedures for recording the location of pre- and post-repair TPS damage.

Ice elimination

The ET Project developed techniques to reduce or eliminate ice formation on ET surfaces. Concepts evaluated for the most problematic areas included heaters for aft feedline bellows locations and titanium feedline brackets.

ET TPS elimination

The ET Project performed analyses to refine thermal protection and aerodynamic requirements with the goal of reducing and eliminating where possible the need for ET TPS. Additionally, refined thermal environments have been evaluated to identify reduced aero/thermal requirements, which could result in resizing or elimination of ET TPS.

FINAL UPDATE

Debris Mitigation

STS-121/ET-119 included significant efforts to mitigate the debris risk from the ET (figure 3.2-1-17). This included sanding and venting additional Intertank NCFI 24-124 acreage TPS. STS-121 was the first "No PAL Ramp" mission; the LO_2 and LH_2 PAL ramps were removed and BX-265 foam was manually resprayed over the acreage/ adhesive in the PAL ramp removal footprint to restore the OML as required. Ice/frost ramps, the cable tray fairing, and supports were modified. The development flight instrumentation (DFI) suite was revised and data from accelerometers in the LO_2 cable tray and the LH_2 cable tray were used to verify the aerodynamic effects of removing the PAL ramps on adjacent hardware. The bipod heater wires were sealed and filled to reduce the risk of cryoingestion. Another enhancement was the validated heater wire harness bonding process used to reduce the risk of a void volume under the harnesses. Except for DFI, all of these modifications were implemented into ET production for the remainder of the SSP.

Defect Documentation

Defect Mapping

NASA implemented the first TPS repair map on ET-119 (which flew on STS-121). These maps are generated using nonconformance documents and in-process rework defect data to locate and document repairs to the TPS of the ET. The maps are used by analysts in the postflight TPS performance assessment to determine whether a TPS repair is associated with a preflight or postflight observation. This mapping effort will be continued throughout the remainder of the SSP.

STS-121/ET-119 experienced less than a 0.2% debris liberation rate. The most probable causes of the foam losses were void delta pressure, void/cryopumping, and in-process damage. Since the flight of ET-119 on STS-121, the maps have been used to assess TPS poured foam repairs for acceptability of debris liberation.

NDE Inspections

NASA implemented NDE inspections on ET-119/STS-121. These inspections were not certified and were used for engineering information only. The areas of inspections implemented for ET-119 and all subsequent ETs are shown in figure 3.2-1-18.

NDE inspections are also a critical part of the postflight evaluation process. Any observations originating in the NDE inspection areas are reviewed to determine whether further action is required in the affected hardware/location. NASA plans to continue performing these NDE inspections for the remainder of the Shuttle's operational life.

Postflight Evaluation

ET-119/STS-121 was the first ET for which postflight analysis was performed using the Postflight Engineering Evaluation Plan (PEEP) at MAF. The ET postflight analysis process is structured to provide an engineering assessment of ET performance. The review process begins with cryogenic loading and continues through ET disposal. The evaluation is performed against both the baseline PEEP and special interest items identified in a Special Interest Items Document.

The postflight assessments are made using a variety of imagery assets. These include on-pad and ascent imagery; radar; ET, Solid Rocket Booster and Orbiter mounted cameras; and astronaut handheld imagery. The ET Postflight Assessment Team (EPAT) Board reviews the postflight analysis results and determines whether further assessments are required.

The imagery assessment teams use a conservative TPS divot modeling approach to estimate TPS debris sizes. As-built TPS thicknesses and stereoscopic imaging are used to determine loss depths. Material specification density, a conservative calculation, and a rigorous modeling approach are used to determine loss mass. Assessment dimensions are reviewed prior to acceptance by the EPAT Board.

The postflight assessment process is complete when the Launch Plus 30-Day Postflight Report is completed. This report includes a complete assessment of preflight and postflight activities.

SUMMARY

Based on the flight experience of two RTF missions (STS-114 and STS-121), as well as extensive testing and analysis, NASA now has a much better understanding of the various causes of foam loss and how to mitigate losses from critical areas on the ET. We believe the residual risk of foam loss has been reduced to an acceptable level through the mitigation techniques described above. The Program continues to work on process and design improvements to reduce critical foam loss, as evidenced by the redesign effort on the ice/frost ramps. NASA is committed to this continual improvement effort throughout the remaining life of the SSP.

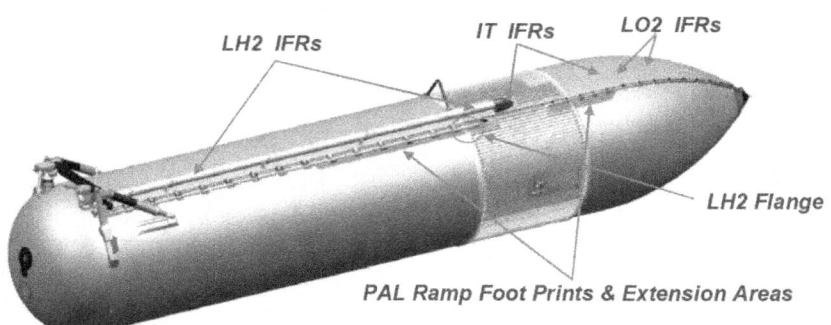

Figure 3.2-1-18. NDE inspection areas on ET-119 and subsequent ETs.

SCHEDULE

Responsibility	Due Date	Activity/Deliverable
SSP	Jun 04 (Completed)	Complete bipod redesign Delta CDR Board
SSP	Apr 04 (Completed)	Perform NDE of PAL ramp on ET-120 (1st RTF tank)
SSP	Jul 04 (Completed)	Complete validation of LH_2/intertank stringer panel closeout

SCHEDULE

Responsibility	Due Date	Activity/Deliverable
SSP	Aug 04 (Completed)	Complete validation of LH_2/intertank thrust panel closeout
SSP	Aug 04 (Completed)	Complete bipod TPS closeout validation
SSP	Nov 04 (Completed)	Complete bellows "drip lip" validation
SSP	Nov 04 (Completed)	Complete bipod retrofit on ET-120
SSP	Nov 04 (Completed)	Complete flange closeout on ET-120
SSP	Dec 04 (Completed)	Critical debris characterization initial phase testing
SSP	Dec 04 (Completed)	Phase 1 ET DCR
SSP	Dec 04 (Completed)	Ready to ship ET-120 to KSC
SSP	Mar 05 (Completed)	Phase 2 ET DCR
SSP	Mar 05 (Completed)	Critical debris characterization final phase testing
SSP	Mar 05 (Completed)	Final External Tank Certification (DCR Board)
SSP	Jun 05 (Completed)	Complete bellows heater verification
SSP	Jun 05 (Completed)	Complete bellows heater DCR
SSP	Jun 05 (Completed)	Complete bellows heater implementation
SSP	Dec 05 (Completed)	PAL Ramp Technical Interchange Meeting
SSP	Dec 05 (Completed)	PRCB Decision on PAL ramp
SSP	Jan 06 (Completed)	Level II Preliminary Design Review Environments
SSP	Feb/Mar 06 (Completed)	GRC wind tunnel test
SSP	Mar 06 (Completed)	STS-114 IFA Testing Complete
SSP	Mar 06 (Completed)	Ship ET-119 to KSC
SSP	Apr 06 (Completed)	Critical Design Review (No PAL Ramp)
SSP	Jun 06 (Completed)	External Tank DCR
SSP	Jun 06 (Completed)	STS-114 IFA Closure Final Report
SSP	Aug 06 (Completed)	AEDC wind tunnel test

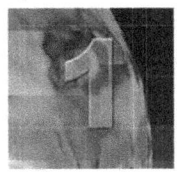

Columbia Accident Investigation Board
Recommendation 3.3-2

Initiate a program designed to increase the Orbiter's ability to sustain minor debris damage by measures such as improved impact-resistant Reinforced Carbon-Carbon and acreage tiles. This program should determine the actual impact resistance of current materials and the effect of likely debris strikes. [RTF]

Note: NASA has closed this recommendation through the formal Program Requirements Control Board process. The following summary details NASA's response to the recommendations and any additional work NASA performed beyond the *Columbia* Accident Investigation Board recommendation.

BACKGROUND

The STS-107 accident clearly demonstrated that the Space Shuttle's Thermal Protection System (TPS) design, including the Reinforced Carbon-Carbon (RCC) panels and acreage tiles, was too vulnerable to impact damage from the existing debris environment. As a result, NASA initiated a broad array of projects to define critical debris (explained in NASA's response to the *Columbia* Accident Investigation Board (CAIB) Return to Flight (RTF) Recommendations 3.3-1 and 6.4-1), to work aggressively to eliminate debris generation (CAIB Recommendation 3.2-1), and to harden the Orbiter against impacts.

NASA chose to address the CAIB requirement by (1) initiating a program of Orbiter hardening and (2) determining the impact resistance of current materials and the effect of likely debris strikes. NASA's Orbiter hardening program is mature and well defined. Four modifications to the Orbiter were implemented for the STS-114 RTF mission. Impact tolerance testing is also a well-defined effort that has identified impact and damage tolerance data for all portions of the TPS for use by all elements of the Space Shuttle Program (SSP). This effort has provided the SSP with a better understanding of the Orbiter's capability to withstand damage relative to the expected debris environment at RTF. This improved understanding, paired with efforts to reduce the generation of critical debris, allowed us to make an informed decision to accept the residual risk posed by ascent debris prior to Return to Flight.

NASA IMPLEMENTATION
Orbiter Hardening

NASA formed an Orbiter Hardening Team to identify options for near-term TPS improvements in critical locations. Initially, the SSP categorized Orbiter hardening into eight candidate design families with 17 design options for further assessment. Each TPS enhancement study was evaluated against the damage history, vulnerability, and criticality potential of the area and the potential safety, operations, and performance benefits of the enhancement. The team focused on those changes that achieve the following goals: increase impact durability for ascent and micrometeoroid orbital debris impacts; increase temperature capability limits; reduce potential leak paths; selectively increase entry redundancy; increase contingency trajectory limits; and reduce contingency operations such as on-orbit TPS repair. These candidates were presented to the SSP Program Requirements Control Board (PRCB), which prioritized them. The result was a refined set of 16 Orbiter hardening options in eight different design families.

The 16 selected Orbiter hardening options were implemented in three phases. Based on maturity of design and schedule for implementation, four projects were identified as Phase I options for implementation before STS-114. These included: front spar "sneak flow" protection for the most vulnerable and critical RCC panels 5 through 13; main landing gear corner void elimination; forward Reaction Control System carrier panel redesign to eliminate bonded studs; and replacing side windows 1 and 6 with thicker outer thermal panes. All four modifications have been implemented on all of the Orbiters. These changes increase the impact resistance of the Orbiter in highly critical areas such as the wing spar, main landing gear door (MLGD), and windows to reduce existing design vulnerabilities.

NASA also selected two Phase II options for implementation after RTF: "sneak flow" front spar protection for the remaining RCC panels 1 through 4 and 14 through 22, and MLGD enhanced thermal barrier redesign. Implementation of the Phase II "sneak flow" modification is in work on all three vehicles. MLGD-enhanced thermal barrier redesign was re-scoped to replace current tiles surrounding the MLGDs with the toughened tiles developed under Phase III. The technical challenges of redesigning the MLGD thermal barrier, the MLGD displacement during high loading, and the excessive maintenance downtime required for implementation drove the SSP to re-scope this modification. Additionally, the SSP approved a study to assess the benefit of minimizing the thermal flow path in the MLGD thermal

Family	Redesign Proposal	Phase
WLESS	"Sneak Flow" Front Spar Protection (RCC #5 – 13)	I
	"Sneak Flow" Front Spar Protection (RCC # 1 – 4, 14 – 22)	II
	Lower Access Panel Redesign/BRI 20 Tile Implementation	III
	Insulator Redesign	III
	Robust RCC	III
Landing Gear and ET Door Thermal Barriers	Main Landing Gear Door Corner Void	I
	Main Landing Gear Door Enhanced Thermal Barrier Redesign	II
	Nose Landing Gear Door Thermal Barrier Material Change	III
	External Tank Door Thermal Barrier Redesign	III
Vehicle Carrier Panels – Bonded Stud Elimination	Forward RCS Carrier Panel Redesign – Bonded Stud Elimination	I
Tougher Lower Surface Tiles	Tougher Periphery (BRI 20) Tiles around MLGD, NLGD, ETD, Window Frames, Elevon Leading Edge and Wing Trailing Edge	III
	Tougher Acreage (BRI 8) Tiles and Ballistics SIP on Lower Surface	III
Instrumentation	TPS Instrumentation	III
Elevon Cove	Elevon Leading Edge Carrier Panel Redesign	III
Tougher Upper Surface Tiles	Tougher Upper Surface Tiles	III
Vertical Tail	Vertical Tail AFSI High Emittance Coating	III

Table 3.3-2-1. Eight Design Families Targeted for Enhancement.

barrier area. This study investigated the feasibility of creating a heat sink or a flow inhibitor by adding material resistant to high temperatures in the area of the MLGD to structure interface, without contacting adjacent MLGD tiles during nominal flight and ground operation.

Finally, NASA designated as Phase III the remaining options that were less mature but held promise for increasing the impact resistance of the Orbiter over the longer term. These options are being implemented as opportunities that become available during processing flows between flights. For instance, NASA has developed new toughened tiles for the Orbiter TPS (BR-18 tile). These tiles are being installed around more critical areas such as the landing gear and External Tank (ET) umbilical doors, and less critical areas, such as wing leading edge (WLE) carrier panels.

Impact and Damage Tolerance

Using both test and analysis, the Orbiter Damage Impact Assessment Team (ODIAT) determined the impact and damage tolerance of TPS tile, RCC, and the Orbiter windows to ET foam, ice, and ablator impacts. Impact tolerance is the ability of the TPS materials to withstand impacts before damage occurs. Damage tolerance is defined as the level of damage from a debris strike that can be tolerated while still safely completing the mission; i.e., safe entry. In general, although tile is not very *impact* tolerant (it is damaged easily with very low levels of kinetic energy), both impact tests and flight history show tile to be very *damage* tolerant (even with significant damage, it resists entry heating well). Conversely, RCC is very impact tolerant but is not damage tolerant, since even minor cracks or coating loss can be critical and prevent safe entry.

Preliminary impact tolerance data generated early in the Program were used by SSP project offices to modify hardware as necessary to reduce the risk that critical debris (debris whose impact would preclude safe entry) would be released. These preliminary data were the basis for the ET Project's work to certify the ET for foam releases. Subsequent test and analysis confirmed that the actual damage tolerance of the tile and RCC was less than the ET certification limit. However, further testing and analysis resulted in a probabilistic risk assessment that showed the risk of exceeding the damage tolerance limit for tile and RCC was acceptable to the Program.

Tile Impact and Damage Tolerance

Tests to determine TPS tile impact tolerance—using foam, ice, and ablator projectiles—are complete. NASA performed impact and damage tolerance testing at several field centers and other test facilities using both acreage and special configuration tiles, and both new and aged tiles. These tests indicated that, although tile is not very resistant to impact, it tolerates entry heating well even with significant damage. Overall, testing shows tile to be tolerant to moderate levels of impact damage; tile damage tolerance depends on tile thickness, which varies by location. As a result, certain areas of reduced thickness, such as those tiles adjacent to the MLGDs, are more susceptible to critical damage. Based on tests and on flight history, NASA developed zone and cavity definitions for 31 areas of tile with similar structural and thermal characteristics to determine the depth of allowable damage penetration into the tile before critical damage occurs and repair is necessary. These zones take into account aeroheating, impact angle, and tile thickness. NASA also completed thermal analyses that encompass all nominal entry trajectories for each of these areas. Besides developing "certification-level" damage thresholds, NASA also developed a set of probabilistic curves that represented a mean and a three-sigma level of acceptable damage.

In addition, analysis of the Space Shuttle's flight history indicated that tile damage fell into three impact classes: (1) numerous, shallow impacts primarily on the forward chine and fuselage; (2) fewer, deeper impacts primarily on the lower surface; and (3) umbilical area impacts. The majority of historical damage observed fell into the first category, and was likely caused by foam popcorning rather than large foam divots. ET intertank venting has reduced popcorning masses in the ET foam. At RTF, the risk of significant damage of this nature was categorized as "remote-catastrophic."[1] The second category of damage, with fewer deeper impacts, was most likely the result of ET foam divots and ice from the ET bellows and brackets. This category of damage is the most likely to require repair. Its likelihood has been reduced through redesign and was categorized as "infrequent-catastrophic" at RTF. Finally, the umbilical area had a mixture of both small and large impacts from a unique subset of sources including ET umbilical ice, baggies, Kapton tape, and ET fire detection paper. Debris transport analysis suggests that most the impacts came from "local" sources rather than from the forward ET. As a result, we expected little change to the damage in this area. At RTF, the risk of damage in this area was classified as "remote-catastrophic."

RCC Impact and Damage Tolerance

Impact and damage tolerance testing on the RCC was performed at several NASA field centers and other test facilities, using both RCC coupons and full-scale RCC panels. Structural and thermal testing of damaged RCC samples established the allowable damage to still maintain a safe return for the crew and vehicle. Test-verified models established impact tolerance thresholds for foam and ice against tile and RCC. These impact tolerance thresholds are the levels at which detectable damage begins to occur, and vary depending on impact location on a panel and the RCC panel location. These thresholds were provided to the Program for risk assessment of the TPS capability against the expected debris environment. As with tile, probabilistic mean and three-sigma damage levels were provided for use in the risk assessment.

Testing shows that the RCC cannot tolerate any significant loss of coating from the front surface in areas that experience full heating on entry. Testing indicates that loss of front-side coating in areas that are hot enough to oxidize and/or promote full heating of the damaged substrate can cause unacceptable erosion damage in delaminated areas. This is of concern because impacts can create subsurface delamination of the RCC that is undetectable through imaging scans. However, the amount of front-side coating loss that would lead to a concern, when coupled with subsurface delamination, can be detected with the Orbiter Boom Sensor System, thus eliminating the concern of "hidden, undetectable" damage. Further testing and modeling have shown that, although the hottest areas on the WLE (the bottom and apex surfaces) cannot tolerate any significant coating loss, other cooler areas (such as the top surface of the WLE) can tolerate some amount of coating loss and subsurface delamination. Testing and model development work has converged to fully map the damage tolerance capabilities of the WLE RCC depending on panel and location (top, apex, or bottom surface).

Overall, the nose cap, chin panel, and WLE panels do not have the capability to withstand predicted worst-case foam impacts in the certified (worst-on-worst) ET debris environment. However, the Program was able to determine that the risks were acceptable when assessed end-to-end in a probabilistic manner that accounted for the probability of a certain size of debris, the probability of its release at a given time during ascent, the probability that it will do damage, and the probability that the damage is critical.

---

[1] The risk categories referenced in this document refer to the SSP Risk Assessment methodology documented in NSTS 22254, "Methodology for Conduct of Space Shuttle Hazard Analyses," and NSTS 07700 Volume I, Chapter 5, "Risk Management."

Window Impact and Damage Tolerance

Testing is complete on window impact from debris, including butcher paper, ablator material, foam, Tyvek, aluminum oxide, and small/fast ogive foam. NASA's debris transport analysis suggests that very small ogive foam has the potential to impact the Orbiter windows, but impact tolerance tests indicated that the windows can withstand these impacts without sustaining critical damage. Testing also indicated that butcher paper causes unacceptable damage to the windows. As a result, NASA replaced the forward Reaction Control System jet butcher paper covers with Tyvek covers that will not cause critical damage.

Expected Damage

Tile

To build a reasonable picture of expected tile damage, NASA undertook a four-stage analysis. This analysis began with a wide-ranging effort to understand the Space Shuttle's flight history and a significant review of the history of tile damage throughout the SSP. The historical damage was then characterized and a database of historical tile damage was created that will be used to support assessments of any damage sustained on future flights. These data were used to demonstrate the capability of our debris models relative to flight history. Finally, NASA captured the improvements expected in the debris environment based on changes made to the Space Shuttle system. The ET Project provided both possible and likely debris sources, masses, and locations that were used to generate a series of possible debris scenarios. Together, these elements provided the expected foam debris environment at RTF.

The damage generated by foam popcorning is acceptable based on the allowable damage map created with certification rigor (worst-on-worst). However, changes to the ET have significantly reduced the size and quantity of foam debris shed during ascent. These changes include: a redesign of the LH_2 flange closeout (the source of the majority of divots over the history of the SSP); the removal of the bipod foam ramp; a significant increase in intertank venting to reduce the size of foam popcorning; and a drip lip for the bellows foam to reduce ice formation.

At RTF, there was still the potential for foam to cause damage to tile that exceeded safe entry limits, as discussed above, but this potential was significantly reduced and the risk was judged acceptable.

RCC

NASA's tests and analysis demonstrate that, in the worst-case environment, potential impacts to the RCC exceed the RCC's ability to withstand damage, in many cases by a significant margin. However, extensive debris transport analysis and probabilistic analysis of expected ascent debris indicate acceptable risks. Overall, NASA's risk assessment for significant foam damage to the RCC is a remote likelihood and catastrophic consequence. The risk of critical damage is greatest in panels 10 through 12. Based on this assessment, the Program has chosen to accept the remaining risk.

The risk assessment for foam impacts to the nose cap is the same as for the RCC overall, with one exception. Our assessments indicate that there is an increased risk for LO_2 intertank flange foam to impact the nose cap, placing it in the infrequent likelihood category.

STATUS

Orbiter Hardening

NASA completed implementation of the four Orbiter hardening Phase I options that were mandatory before RTF. Since RTF, NASA has continued to pursue Phase II and III hardening options and continues to implement all feasible options at the earliest possible opportunity during the Orbiter processing between flights.

The SSP has reviewed and approved the corrective measures taken in response to this recommendation. The SSP Manager has reviewed the suite of activities summarized above and concluded that, taken as an integrated plan, it fully satisfies the CAIB RTF recommendation to initiate a program to increase the Orbiter's ability to sustain minor debris damage. As NASA's analysis has become more defined, we have continued to enhance the steps taken to improve the Orbiter's resistance to potential impact damage beyond RTF.

NASA acknowledges that the elimination of all critical debris is not attainable, and has analyzed and formally accepted the remaining risk as a condition for RTF. Improved nondestructive evaluation (NDE) capabilities will provide greater knowledge of the condition of the ET foam in critical areas and the integrity of the Orbiter RCC prior to launch. Although a dramatic improvement, these capabilities use the best available technology to provide a view of what is beneath the surface, but do not allow us to verify the precise condition of foam and RCC. We accept the risk associated with the limitations of our available NDE capabilities.

Impact and Damage Tolerance

Prior to the first RTF mission, NASA conducted two Design Verification Reviews (DVRs) for Debris in March 2005 and April 2005, and conducted a third DVR in June 2005. The purpose of the DVR for Debris is to present and document the data used to perform the Systems Engineering and Integration assessment of the debris environment with respect to impact tolerance of Orbiter TPS. Participants of the DVR include members of SSP management, NASA Engineering Safety Center, Independent Technical Authority, and Safety and Mission Assurance.

In preparation for the Delta DVR for Debris in June 2005, NASA had two major objectives to support our rationale for RTF. First, NASA developed an end-to-end estimate of the Orbiter's capability to withstand damage relative to the ascent environment. To do this, we used the ET Project's best estimate of the foam and/or ice debris that may be liberated, and worst-case assumptions about the potential of that debris for transport to the Orbiter. Second, for those bounding cases in which the initial assessment indicated that the Orbiter could not withstand the potential impact, NASA performed a probabilistic-based risk assessment to determine the acceptability of a critical debris release potential. NASA assessed four foam transport cases and two ice transport cases. These cases represent the worst potential impacts of a general category and were used to bound the similar, but less severe, transport cases from the same areas. This analysis quantified the risk posed by the debris environment in our formal risk analysis, aided in the determination of the ascent debris risk remaining, and helped formulate flight rationale for risk acceptance by NASA Senior Management.

Impact and Damage Tolerance Verification

The first two Shuttle test flights enabled us to verify the new Space Shuttle ascent debris environment with both imagery and radar and determine the actual size of damage sustained during ascent—rather than the size of the damage after entry heating—through on-orbit inspections. These data allowed us to determine the growth of ascent and on-orbit damage during entry by comparing the on-orbit inspections and postlanding inspections. The result was that the damages were essentially unaffected by entry heating. These data, paired with our test of tile and RCC repair methods, provided us with a more rigorous understanding of the risk of TPS damage and the Orbiter's damage tolerance.

Additional Monte Carlo analyses will support the flight rationale for the debris sources identified with the largest uncertainty in risk, in particular those risks in the "infrequent" category, including foam from LO_2 ice/frost ramps and ice from the LO_2 feedline bellows and brackets.

POST STS-114 UPDATE

On STS-114, as expected, some small foam divots impacted the tile and caused minor, acceptable damage. NDE of the RCC showed the wings to be in overall excellent condition, except for three areas. These areas are located on right-hand RCC panel 8, right-hand rib splice 6, and left-hand rib splice 7 (a complete discussion of these anomalies is found in R3.3-1). The SSP now has a much greater understanding of the causes, consequences, and likelihood of TPS damage, and is treating this as an accepted risk for continued flight.

Prior to the second RTF mission, NASA conducted an additional DVR in May 2006. The purpose of the review was to close out all remaining debris-related action items prior to launch of STS-121, and review all changes and updates to ascent and entry debris risk assessments since the first RTF mission, STS-114. Participants of this review included members of SSP management, NASA Engineering and Safety Center, NASA Chief Engineer, and Safety and Mission Assurance.

The data obtained from STS-114 and from STS-121 allowed us to continually assess the risk of an unacceptable debris impact from foam, ice, or any other source. Follow-on testing and analysis allowed refinement of models and incorporation of flight-derived data. As the debris environment changes, and as probabilistic analysis continues to improve, we will constantly reassess the risk from foam and ice debris.

FINAL UPDATE

Beginning with STS-121, NASA has been aggressively replacing the existing FRCI-12 belly tiles with BRI-18 tiles around the MLGD, nose landing gear door (NLGD), ET umbilical doors, LESS carrier panels, and windows during Orbiter processing between flights. The installed BRI-18 tiles are significantly tougher (impact resistant) than the replaced FRCI-12 tiles. BRI-18 tile replacement priorities are followed, with MLGD perimeter being the highest, followed by ET umbilical door perimeter, NLGD perimeter, LESS carrier panels, and windows.

An MLGD cavity protection modification preliminary design and concept feasibility study is in work as of May 2007. The initial study effort to minimize flow path into MLGD thermal barrier area resulted in a downselect of concepts for the MLGD cavity modification. The selected design incorporates a redundant thermal barrier that is "activated" and expands upon exposure to high heating condition. Final proof-of-concept will be demonstrated by successful arc-jet testing of the mock-up test article, currently planned for end of May 2007. Following successful proof-of-concept testing, engineering and parts will be available by end of contract year 2007 for implementation on all three Orbiters.

SCHEDULE

Responsibility	Due Date	Activity/Deliverable
SSP	Jun 03 (Completed)	Initial plan reported to PRCB
SSP	Aug 03 (Completed)	Initial Test Readiness Review held for Impact Tests
ODIAT	Oct 03 (Completed)	Initial Panel 9 Testing
SSP	Nov 03 (Completed)	Phase I Implementation Plans to PRCB (MLGD corner void, FRCS carrier panel redesign—bonded stud elimination, and WLE impact detection instrumentation)
SSP	Jan 04 (Completed)	Phase II Implementation Plans to PRCB (WLE front spar protection and horse collar redesign, MLGD redundant thermal barrier redesign)
ODIAT	Aug 04 (Completed)	Panel 16R Testing
SSP	Sep 04 (Completed)	Finalize designs for modified wing spar protection between RCC panels 1–4 and 14–22 on OV-103 and OV-104
SSP	Oct 04 (Completed)	Conclude feasibility study of the Robust RCC option
SSP	Jan 05 (Completed)	Complete analysis and preliminary design phase for robust RCC
SSP	Feb 05 (Completed)	Complete modification of wing spar protection behind RCC panels 5–13 on OV-103
ODIAT	Mar 05 (Completed)	Tile Impact Testing Complete
ODIAT	Mar 05 (Completed)	RCC Impact Testing Complete
SSP	Apr 05 (Completed)	Damage Tolerance Test and Analysis Complete (SSP baseline of models and tools)
SSP	Jun 05 (Completed)	Delta DVR for Debris
ODIAT	Jun 05 (Completed)	Final Tile and RCC Model Verification (Project baselining of models and tools)
ODIAT	Jul 05 (Completed)	RCC Materials Testing

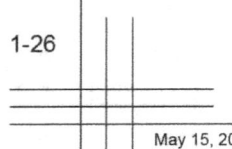

Columbia Accident Investigation Board
Recommendation 3.3-1

Develop and implement a comprehensive inspection plan to determine the structural integrity of all Reinforced Carbon-Carbon system components. This inspection plan should take advantage of advanced non-destructive inspection technology. [RTF]

Note: The Stafford-Covey Return to Flight Task Group held a plenary session on February 17, 2005, in Houston, Texas. NASA's progress toward answering this recommendation was reviewed and the Task Group agreed that the actions taken were sufficient to change this recommendation from conditionally closed to fully closed.

BACKGROUND

At the time of the *Columbia* accident, on-vehicle inspection techniques were inadequate to assess the structural integrity of Reinforced Carbon-Carbon (RCC) components and attachment hardware. There were two aspects to the problem: (1) how we assessed the structural integrity of RCC components and attach hardware throughout their service life, and (2) how we verified that the flight-to-flight RCC mass loss caused by aging did not exceed established criteria. At present, structural integrity is assured by wide design margins; comprehensive nondestructive evaluation (NDE) was conducted only at the time of component manufacture. Mass loss is monitored through a destructive test program that periodically sacrifices flown RCC panels to verify by test that the actual material properties of the panels are within the predictions of the mission life model.

The RCC NDE techniques currently certified include X-ray, ultrasound (wet and dry), eddy current, computer-aided tomography (CAT) scan, and infrared (IR) thermography. Eddy current testing is useful for assessing the health of the RCC outer coating and detecting possible localized subsurface oxidation and mass loss, but it reveals little about a component's internal structure. Since the other certified NDE techniques require hardware removal, each presents its own risk of unintended damage. The vendor and United Space Alliance are fully equipped and certified to perform RCC X-ray and ultrasound. Shuttle Orbiter RCC components are pictured in figure 3.3-1-1.

NASA IMPLEMENTATION

The Space Shuttle Program (SSP) pursued inspection capability improvements using newer technologies to allow comprehensive NDE of the RCC without removing it from the vehicle. A technical interchange meeting held in May 2003 included NDE experts from across the country. This meeting highlighted five techniques with potential for near-term operational deployment: (1) flash thermography, (2) ultrasound (wet and dry), (3) advanced eddy current, (4) shearography, and (5) radiography. The SSP assessed the suitability of commercially available equipment and standards for flight hardware and selected IR thermography as its method for routine post-flight inspections to positively verify the structural integrity of RCC hardware without risking damage by removing the hardware from the vehicle. RCC post-flight inspection requirements now consist of visual, and IR thermography on the installed (i.e., in-situ) RCC components (wing leading edge (WLE) panels, nose cap, chin panel). Contingency inspections (eddy current, ultrasonic, and tactile) will be invoked if there are any suspicions of damage to the RCC by virtue of instrumentation, photographic, thermography, or visual postflight inspection. At the discretion of the Leading Edge Subsystem Problem Resolution Team (LESS PRT), contingency inspections will also be performed to address any damage/impact suspicions identified during ground processing, ascent, on-orbit, entry, atmospheric flight, or landing.

NASA cleared the RCC by certified inspection techniques before Return to Flight. For the long term, a Shuttle Program Requirements Control Board (PRCB) action was assigned to review inspection criteria and NDE techniques for all Orbiter RCC nose cap, chin panel, and WLE system components. Viable NDE candidates were reported to the PRCB in January 2004, and specific options were chosen.

RCC structural integrity and mass loss estimates were validated by off-vehicle NDE of RCC components and destructive testing of flown WLE panels. All WLE panels, seals, nose caps, and chin panels were removed from Orbiter Vehicles (OV)-103, OV-104, and OV-105 and returned to the vendor's Dallas, Texas facility for comprehensive NDE. Inspections included a mix of ultrasonic, X-ray, and eddy current techniques. In addition, NASA introduced off-vehicle flash thermography for all WLE panels and accessible nose cap and chin panel surfaces; any questionable components were subjected to CAT scan for further evaluation. Data collected were used to support development of in-place NDE techniques and established the baseline for future IR thermography inspections.

The health of RCC attach hardware was assessed using visual inspections and NDE techniques appropriate to the critical flaw size inherent in these metallic components. This NDE was performed on select components from OV-103 and OV-104. Destructive evaluation of select attach hardware from both vehicles was also undertaken. No additional requirements were considered necessary based on results of inspections.

STATUS

Advanced On-Vehicle NDE: Near-term advanced NDE technologies were presented to the PRCB in January 2004. Thermography, contact ultrasonics, eddy current, and radiography were selected as the most promising techniques to be used for on-vehicle inspection that could be developed in less than 12 months. The PRCB approved the budget for the development of these techniques. IR thermography was selected as the RCC in-situ NDE method for post-flight inspection. The requirement for IR thermography (and the associated contingency methods consisting of ultrasonics, eddy current, and tactile evaluations) was established in the Operations and Management Requirements Specifications Document on November 20, 2006.

OV-104: The nose cap, chin panel, and all WLE RCC panel assemblies were removed from the vehicle and shipped to the vendor for complete NDE. The data analysis from this suite of inspections was completed in March 2004. Vendor inspection of all WLE panels was completed. Eddy current inspections of the nose cap and chin panel were completed before these components were removed, and the results compared favorably to data collected when the components were manufactured, indicating mass loss and coating degradation were within acceptable limits. Off-vehicle IR thermography inspection at KSC was performed to compare with vendor NDE. All findings were cleared on a case-by-case basis through the Material Review Board (MRB) system.

OV-103: As part of the OV-103 Orbiter maintenance down period (OMDP), WLE panels were removed from the vehicle, inspected by visual and tactile means, and then shipped to the vendor for NDE. The analysis of the inspection results was completed in July 2004. X-ray inspection of the RCC nose cap, which was already at the vendor for coating refurbishment, revealed a previously undocumented 0.025 in. × 6 in. tubular void in the upper left-hand expansion seal area. While this discrepancy did not meet manufacturing criteria, it was located in an area of the panel with substantial design margin (900% at end of panel life) and was deemed acceptable for flight. The suite of inspections performed on the OV-103 nose cap has confirmed the Orbiter's flight worthiness and revealed nothing that might call into question the structural integrity of any other RCC component.

Off-vehicle IR thermography inspection at KSC was performed to compare with vendor NDE. All findings were cleared on a case-by-case basis through the MRB system.

OV-105: All OV-105 RCC components (WLE, nose cap, and chin panel) were removed and inspected during its OMDP, which began in December 2003. Off-vehicle IR thermography inspection at KSC was performed to compare with vendor NDE. All findings were cleared on a case-by-case basis through the MRB system.

RCC Structural Integrity: NASA used destructive mechanical properties testing to evaluate the potential loss of RCC strength due to oxidation as postulated by *Columbia* Accident Investigation Board (CAIB) finding F3.3-4. The latest addition to this data set was obtained from RCC panel 9R, which has a flight history of 30 flights. These results were compared to previously obtained mechanical properties from Panels 12R (15 flights), 10L (19 flights), 8L (26 flights), and 9L (27 flights). Results from the evaluation of these panels did not show any mass loss or aging degradation of the RCC mechanical properties beyond the technical expectations that serve as the basis for mission life determination. In addition, all tested values were substantially greater than the design stress allowables for mechanical strength, which were used to size the WLE panels and are still used today for mission life assessment. These analyses demonstrated that RCC components have not been weakened by mass loss caused by oxidation or any other phenomenon beyond the expected values; they have retained their minimum design strength despite more than 20 years of space operations.

RCC Attach Hardware: The LESS PRT was given approval for a plan to evaluate attach hardware through NDE and destructive testing. Detailed hardware NDE inspection (dye penetrant, eddy current) to address environmental degradation (corrosion and embrittlement) and fatigue damage concerns was performed on selected OV-103/104 WLE panels in the high heat and fatigue areas. No degradation or fatigue damage concerns were found.

The Stafford-Covey Return to Flight Task Group agreed to conditionally close this recommendation in April 2004. In February 2005, they granted full closure to the recommendation, indicating that NASA had fully complied with the CAIB recommendation.

FINAL UPDATE

After STS-114, the Orbiter Project performed in-situ NDE of the WLE support structure components. The NDE showed the system to be in overall excellent condition, except for three areas. These areas were located on right-hand (RH) RCC panel 8, RH rib splice 6, and

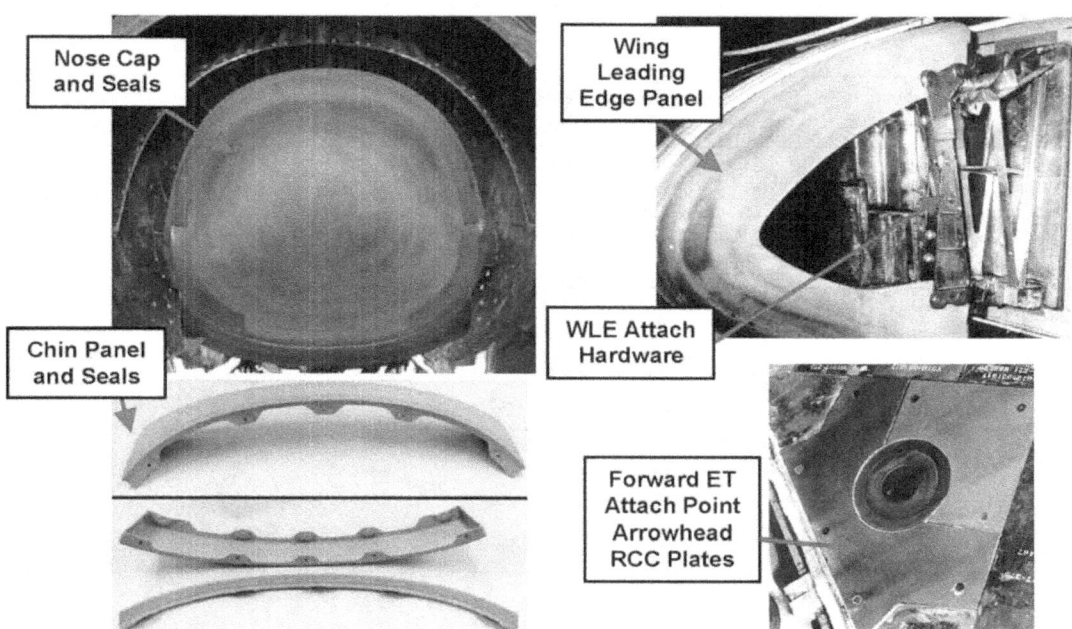

Figure 3.3-1-1. Shuttle Orbiter RCC components.

left-hand (LH) rib splice 7. Preliminary NDE and analysis results indicated that LH rib splice 7 T-seal damage occurred during the mission. Although the WLE sensor system registered an impact indication in the area of LH rib splices 6 and 7, these sensor data were not corroborated with other debris tracking methods. Postflight NDE and analysis results of the damage concluded that LH rib splice 7 T-seal was acceptable on a flight-to-flight basis. This T-seal has been inspected after each flight and no defect growth has been observed. Livermore Laboratory completed NDE evaluation (CT scan) of RH rib splice 6 T-seal. Comparison of preflight and postflight NDE results of the RH rib splice 6 T-seal discrepancy indicated that the area of concern did not grow. Analysis determined that this T-seal was acceptable for flight. Evaluation of RH RCC panel 8 was completed following its removal from service and replacement by a spare. The LESS PRT assessment indicated that larger than normal craze cracks on the panel surface may have contributed to subsurface oxidation and, consequently, higher mass loss.

Panel 8RH is being repaired by the vendor for contingency use, and an additional spare panel 8RH will be delivered to KSC in February 2008.

Several near-term advanced on-vehicle NDE techniques are in development, as are processes and standards for their use. Decisions on long-term NDE techniques (those requiring more than 12 months development) will be made after inspection criteria are better established. Data storage, retrieval, and fusion with CATIA CAD models to enable easy access to NDE data for archiving and disposition purposes are being considered; however, the completion of these tasks has not been funded. The RCC components have been re-baselined according to the Lockheed Martin Mission and Fire Control techniques and the data captured and cataloged for future reference. This catalog of data is unique to each discrete RCC component and serves as the standard for future inspections.

SCHEDULE

Responsibility	Due Date	Activity/Deliverable
SSP	Sep 03 (Completed)	OV-104 WLE RCC NDE analysis complete
SSP	Oct 03 (Completed)	Completion of NDE on OV-104 WLE panel attach hardware
SSP	Dec 03 (Completed)	OV-103 chin panel NDE
SSP	Jan 04 (Completed)	Report viable on-vehicle NDE candidates to the SSP
SSP	Jan 04 (Completed)	Completion of NDE on OV-103 WLE attach hardware
SSP	Feb 04 (Completed)	OV-103 nose cap NDE analysis
SSP	Feb 04 (Completed)	OV-104 chin panel NDE analysis
SSP	Apr 04 (Completed)	OV-104 nose cap NDE analysis
SSP	Jul 04 (Completed)	OV-103 WLE RCC NDE analysis

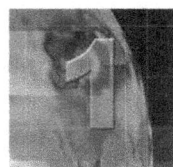

Columbia Accident Investigation Board
Recommendation 6.4-1

For missions to the International Space Station, develop a practicable capability to inspect and effect emergency repairs to the widest possible range of damage to the Thermal Protection System, including both tile and Reinforced Carbon-Carbon, taking advantage of the additional capabilities available when near to or docked at the International Space Station.

For non-Station missions, develop a comprehensive autonomous (independent of Station) inspection and repair capability to cover the widest possible range of damage scenarios.

Accomplish an on-orbit Thermal Protection System inspection, using appropriate assets and capabilities, early in all missions.

The ultimate objective should be a fully autonomous capability for all missions to address the possibility that an International Space Station mission fails to achieve the correct orbit, fails to dock successfully, or is damaged during or after undocking. [RTF]

Note: NASA has closed this recommendation through the formal Program Requirements Control Board process. The following summary details NASA's response to the recommendations and any additional work NASA performed beyond the *Columbia* Accident Investigation Board recommendations.

BACKGROUND

Modification of the Space Shuttle External Tank (ET) to reduce critical debris liberation was a key element of NASA's Return to Flight (RTF). NASA's flight rationale also required a capability to inspect the Orbiter during missions to detect critical damage if any were sustained during ascent. Detailed on-orbit inspection capability allows NASA to make informed, timely decisions about operational measures to preserve the safety of the Orbiter and its crew. NASA also pursued a limited Thermal Protection System (TPS) repair capability for RTF. Finally, in the extremely unlikely event that the Orbiter sustains critical damage that we are unable to repair or compensate for on entry, we have the capability to implement an emergency, limited safe haven on the International Space Station (ISS) with the Contingency Shuttle Crew Support (CSCS) capability (CSCS is addressed in SSP-3). In NASA's formal risk hierarchy, TPS repair and CSCS activities are considered special procedures for additional risk mitigation.

The Space Flight Leadership Council directed the Space Shuttle Program (SSP) to focus its efforts on developing and implementing an inspection and repair capability appropriate for the first RTF missions using ISS resources as required. NASA focused its efforts on mitigating the risk of multiple failures by maximizing the Shuttle's ascent performance margins to achieve ISS orbit, using the docked configuration to maximize inspection and repair capabilities, and flying protective attitudes following undocking from the ISS. In addition, NASA has continued to analyze the relative merit of different approaches to mitigating the risks identified by the *Columbia* Accident Investigation Board (CAIB).

NASA also recognizes that autonomous, non-ISS missions carry a higher risk than missions flown to the ISS. NASA anticipates only one non-ISS mission before the Shuttle is retired: the Hubble Servicing Mission. The additional risks associated with autonomous, non-ISS missions are described below:

1. *Lack of Significant Safe Haven.* The inability to provide a "safe haven" while inspection, repair, and potential rescue from a damaged Orbiter are undertaken creates additional risk in autonomous missions. On missions to the ISS it may be possible to extend time on orbit to mount a well-planned and equipped rescue mission. NASA has continued to study this contingency scenario. For autonomous missions, however, the crew would be limited to an additional on-orbit stay of no more than two to four weeks, depending on how remaining consumables are rationed. The Contingency Shuttle Crew Support concept is discussed in detail in SSP-3.

2. *Double Workload for Ground Launch and Processing Teams.* Because the rescue window for an autonomous mission is only two to four weeks, NASA would need to process two vehicles for launch simultaneously to ensure timely rescue capability. Any processing delays to one vehicle would require a delay in the second vehicle. The launch countdown for the second launch would begin before the launch of the prime mission vehicle. This short time period for assessment is a serious

concern; it also requires two highly complex processes to be carried out simultaneously.

3. *No Changes to Cargo or Vehicle Feasible.* Because of the very short time between the launch of the first vehicle and the requirement for a rescue flight, no significant changes could reasonably be made to the second vehicle. This means that it would not be feasible to change the cargo on the second Space Shuttle to support a repair to the first Shuttle, add additional rescue hardware, or make vehicle modifications necessary to address the cause of the first failure. Not having sufficient time to make the appropriate changes to the rescue vehicle or the cargo could add significant risk to the rescue flight crew or the crew transfer. The whole process would be under acute schedule pressure and may require safety and operations waivers.

4. *Rescue Mission.* A rescue from the ISS, which has multiple hatches, airlocks, and at least one other egress vehicle available (Soyuz), is less complex and risky than a rescue from a stranded Space Shuttle by a second Space Shuttle. When NASA first evaluated free-space transfer of crew, which would be required to evacuate the Shuttle in an autonomous mission, a number of safety concerns were identified. This analysis has since been updated in support of the Hubble Servicing Mission feasibility assessment to identify all of the potential issues and safe solutions.

5. *TPS Repair.* NASA's nominal TPS repair method uses the ISS robotic arm to stabilize an extravehicular activity (EVA) crew person over the repair worksite. NASA has since developed an alternate method for stabilizing the crewmember on the Orbiter Boom Sensor System (OBSS) for repairs on missions where the ISS arm is unavailable.

NASA IMPLEMENTATION

Note: This section refers to inspection and repair during nominal Shuttle missions to the ISS.

NASA greatly expanded our capabilities to detect debris liberation during ascent, to identify locations where debris may have originated, and to identify impact sites on the Orbiter TPS for evaluation. The ability to see debris liberated during ascent through the addition of high-speed ground and vehicle-mounted cameras and radar, complemented by the impact detection sensor system and a suite of on-orbit inspection assets, aids in providing the data required to ensure an effective inspection and, if necessary, repair of the Orbiter TPS. These capabilities, paired with NASA's improved insight into the impact and damage tolerance of the Orbiter, allow the Mission Management Team (MMT) to make informed decisions about whether any impacts sustained by the Orbiter represent a threat to mission success or the safety of the crew and vehicle. The data also help determine whether any repairs attempted have successfully mitigated risk to a degree necessary for a safe entry.

On STS-114 and STS-121, NASA successfully used a combination of Space Shuttle and ISS assets to image the Shuttle TPS and identify and characterize any damage sustained during ascent and on orbit. These same assets will be used and improved on future missions. These inspection assets and methods include the OBSS, the Shuttle Remote Manipulator System (SRMS), the Space Station Remote Manipulator System (SSRMS), an experimental wing leading edge (WLE) impact sensor detection system, and the R-bar Pitch Maneuver (RPM). Each inspection method provides information to improve insight into the conditions of the Orbiter TPS.

Evaluation of the imagery and data collected during ascent and on orbit determines the need for further focused inspection. NASA has established criteria for focused on-orbit inspections to evaluate the length, width, and depth of potential critical damage sites. These criteria are based on our expanded understanding of debris transport mechanisms and the capabilities of the Orbiter TPS. Plans are in place for further inspection, evaluation, and repair for tile or Reinforced Carbon-Carbon (RCC) damage that exceeds the damage criteria. Appropriate risk assessment of each potential damage site that exceeds the damage criteria is conducted and presented to the MMT for evaluation.

During missions, NASA assesses the health of the TPS using the data collected through inspections and makes recommendations on whether a repair is required or whether the TPS can be used as is. If a repair is necessary, NASA determines which repair method is required to enable the Orbiter to withstand the aero-thermal environment of entry and landing. In the event a safe entry is not possible, NASA has also made plans to keep the Space Shuttle crew on the ISS and mount a rescue mission. However, the CSCS capability will not be used to justify flying an otherwise unsafe vehicle and will only be used in the direst of situations.

For STS-114 and STS-121, NASA's central mission objective was to verify the performance of the integrated Shuttle system. As a result, inspection was one of our operational priorities. On both RTF missions, NASA actively inspected the entire TPS, including the WLE, nose cap, chin panel, and tile. NASA also performed additional focused inspections as required with the OBSS, guided by the results of ascent imagery, the initial OBSS scans, and crew camera photos.

With the combination of resources that were available at RTF, NASA has the capability to inspect for and detect critical damage on nearly 100 percent of the Space Shuttle's TPS tile and RCC. However, on most missions, there is limited operational time available for inspections, and conditions during inspection may not always be optimal. Inspections that take place early in the mission will detect damage from ascent debris, but may not find damage sustained while on orbit, such as damage from a micrometeoroid or orbital debris strike. NASA balances these operational concerns to ensure that mission planning makes the best use of available crew time to minimize risk throughout the mission.

Detection/Inspection

In February 2004, the SSP established an Inspection Tiger Team to review all inspection capabilities and to develop a plan to integrate these capabilities before RTF. The Tiger Team succeeded in producing a comprehensive in-flight inspection, imagery analysis, and damage assessment strategy that is implemented through the existing flight planning process. The best available cameras and laser sensors suitable for detecting critical damage in each TPS zone were used in conjunction with digital still photographs taken from the ISS during the Orbiter's approach. The Tiger Team strategy also laid the foundation for a more refined impact sensor and imagery system following the first two successful flights. This plan was later enhanced to clearly establish criteria for transitioning from one suite of inspection capabilities to another and the timeline for these transitions.

Along with the work of the Tiger Team, the Shuttle Systems Engineering and Integration Office developed a TPS Readiness Determination Operations Concept, which is documented in the Operations Integration Plan (OIP) for TPS Assessment. This document specifies the process for collection, analysis, and integration of inspection data in a way that ensures effective and timely mission decision-making. The TPS assessment process begins with the activities leading up to launch and continues through postlanding. The prelaunch process includes an approved configuration for imagery. Any deviation from this configuration is presented at the Flight Readiness Review (FRR) and during the subsequent prelaunch MMT reviews. Additionally, the Ice/Debris Inspection Team performs a series of prelaunch walkdowns of the pad and vehicle to identify potential debris sources and provides this information to the TPS assessment process.

During the mission, the TPS assessment process is divided into three steps: data collection, data processing, and Orbiter damage assessment. The data collection sources provide information on debris, debris trajectory, impact locations, damage, or depth of damage. During the data processing step, this information is analyzed to determine the integrity of the TPS. The findings are provided to the Orbiter Damage Assessment Team (DAT) to determine where there is potential TPS damage. The team develops recommendations to the MMT on whether the damaged TPS is safe to fly as is or whether a repair is needed, as well as which type of repair is required. The definition of the DAT membership and the processes that the DAT uses to assess damages are documented in the Orbiter Damage Assessment Process Annex to the OIP for TPS Assessment.

Postlanding, the TPS assessment process continues with a walkdown of the Orbiter by the Ice/Debris Inspection Team, which photographs observed TPS defects. The TPS assessment process concept has been exercised in several simulations, and was successfully applied on the STS-114 and STS-121 missions.

Damage Threshold

NASA defined the critical damage threshold for TPS inspections as the ability to detect damage of 1 inch (longest dimension) for tile around main landing gear or ET umbilical doors and 3 inches for acreage tile, and to detect cracks 0.020 inch by 2 inches long and 0.020 inch deep for RCC. Through an extensive test program and the application of analytical models developed to predict the capabilities of damaged tile and RCC, NASA determined that damage smaller than this threshold should not result in increased risk to entry. With the combination of resources available at RTF, NASA had the capability to detect this damage.

OBSS

The OBSS consists of sensors on the end of a 50-ft boom structure. The system is installed on the starboard sill of the Orbiter payload bay (figure 6.4-1-1) and is used in conjunction with the Shuttle Remote Manipulator System (SRMS). It is the primary system used to inspect WLE RCC, and to measure the depth of damage sustained by the Orbiter TPS. The OBSS carries a laser camera system (LCS) and a laser dynamic range imager (LDRI), which provide three-dimensional views of TPS damage. Two-dimensional sensors on the OBSS include the Intensified Television Camera (ITVC) and the integrated sensor inspection system (ISIS) digital camera (IDC), which was added on STS-121. The video from the OBSS is recorded on board the Shuttle and downlinked via the Orbiter communications system. The data from all of these sources are processed and analyzed on the ground as part of the TPS assessment process. In addition, the OBSS has the capability to support an EVA crewmember in foot restraints for focused inspection and repair activities.

Upper Pedestal   Composite Sections from RMS Spares

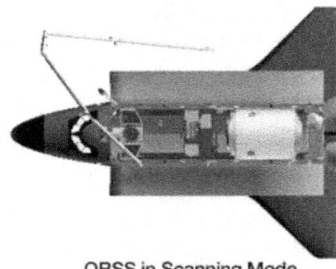

OBSS in Scanning Mode

Figure 6.4-1-1. Orbiter Boom Sensor System.

ISS Imagery During R-bar Pitch Maneuver

The primary method of inspecting the acreage tile across the bottom of the Orbiter is still photo imagery taken by the ISS crew as the Orbiter approaches for docking. This maneuver, the RPM, was practiced by Shuttle flight crews in the simulator prior to STS-114 (figure 6.4-1-2). During STS-114 and STS-121, the Orbiter paused its approach to the ISS when it was 600 ft away and pitched over to present its underside to the ISS. The ISS crew took overlapping high-resolution digital images of the Orbiter's TPS tile and downlinked them to the ground. Areas of concern identified by the RPM photos were re-inspected in greater detail while the Orbiter was docked to the ISS.

ISS cameras used to photograph the Orbiter on approach and during the RPM have the capability to detect critical damage in all areas of the Orbiter TPS tile. NASA's analysis indicates that, for flat surfaces perpendicular to the camera, the 400mm photos have an analytical resolution of 0.3 inch and the 800mm photos have a 0.15-inch analytical resolution.

Other Imagery Assets

Other imagery assets include the SRMS, the Space Station Remote Manipulator System (SSRMS), and other digital camera assets on board the Shuttle and the ISS. The SRMS and SSRMS can inspect areas of the Orbiter TPS within their operational reach, such as the crew cabin area, forward lower surface, and vertical tail, using their closed circuit television camera systems. Other digital assets include the still cameras available to EVA crewmembers in the event an EVA inspection is required to do focused inspection of the TPS that may have suspected damage.

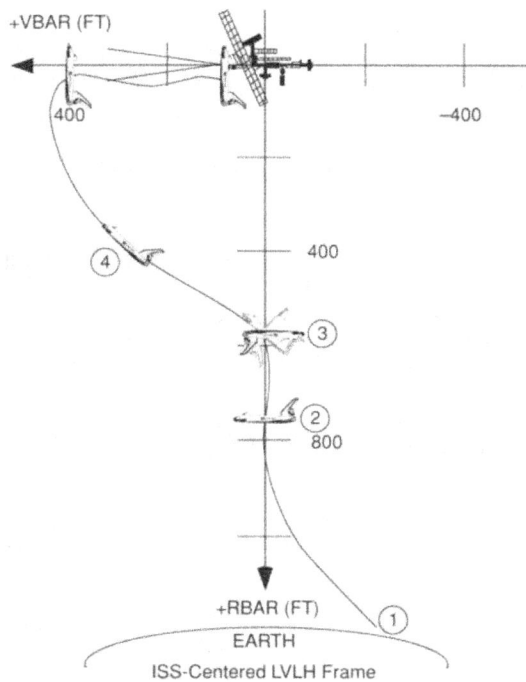

Figure 6.4-1-2. Orbiter RPM for inspection and approach to ISS.

WLE Impact Detection System

The WLE Impact Detection System was developed from an existing technology that had been previously flown as an experiment on the Shuttle. Initially, NASA hoped to include WLE sensors as a key element of our ability to detect damage. Today, these sensors are used primarily as a "pointing" device to cue TPS areas needing further inspection by the OBSS.

The WLE sensor system is composed of accelerometer and temperature sensors in both of the wing cavities and attached to the wing spar behind the RCC. The WLE sensor system data are collected during ascent and while on orbit and are downlinked to the ground via the Orbiter communications system. These data help to identify possible debris impact areas in the vicinity of the WLE RCC panels. If an impact is detected, engineers can determine the location of the sensor(s) that measured the impact and, through the TPS assessment process, recommend a more focused inspection of the suspect area later in the mission. Due to the limited battery life, there is a finite period of time for impact detection using this system. However, on STS-114, the batteries lasted far longer than predicted. NASA is implementing a change in the sensor boxes, beginning with STS-118, to extend the duration of on-orbit impact monitoring.

Repair

Despite comprehensive efforts to develop TPS repair materials and techniques, the state-of-the-art technology in this area has yielded only modest technology improvements. As a result, continued effort does not hold promise of significant capabilities beyond those that already exist.

Comprehensive, vehicle-wide TPS repair capability was not a constraint to RTF or to continued flight. A capability to effect emergency repairs to the widest range of damage practicable was introduced on STS-114, evolved, and improved for subsequent missions. The repair capabilities available for STS-114 included a limited capability to repair minor tile damage and small- to medium-sized RCC damage.

For tile, the repair capability available at RTF included an emittance wash used to enhance the thermal performance of damaged tile and an experimental, ablative fill material and applicator used to repair larger areas of tile damage – up to 10 inches by 20 inches, and the depth of the tile. STS-114 also carried an experimental carbon-silicon carbide (C-SiC) tile overlay that could be used to repair damage sites up to 10 inches by 20 inches. For RCC repair, NASA had plug repair tools and materials that would potentially allow us to fix holes up to 4 inches in diameter and crack repair material for small cracks and gouges in the wing leading edge RCC. The crack repair material and application techniques were successfully demonstrated during the STS-114 development test objective (DTO). The STS-114 crew also conducted a demonstration of the plug repair inside the crew cabin. Additional demonstrations of the crack repair system and test sample generation were performed during STS-121 DTOs to establish their functionality in the combined environments of space. Plug repair hardware is available in the event of a contingency.

NASA has developed a TPS readiness determination operations concept that is documented in the OIP for TPS Assessment and the OIP Damage Assessment Annex. These documents specify the process for collecting, analyzing, and applying the diverse inspection data in a way that ensures effective and timely mission decision-making. This process is used to determine whether damage sustained is safe for entry or requires repair, and whether an attempted repair will render the Orbiter safe for entry. Damage assessment tools used during the mission are the same as those used for preflight inspection criteria validation and include aero-heating environments, cavity heating augmentation factors, damaged tile thermal assessment tools, and structural analysis tools. There are two elements to determine whether damage sustained is safe for entry. The first is a use-as-is assessment to determine whether a repair should be attempted. The second is a follow-on assessment to determine whether any repairs attempted have made the Orbiter safe for entry.

The analytical and decision-making processes associated with TPS repair and inspection were exercised by the MMT through simulations and during the STS-114 and STS-121 missions.

TPS Repair Access

The EVA crew can use either the SRMS, SSRMS, or the OBSS to gain access to repair sites on the Orbiter. On STS-114, the Shuttle crew used the SSRMS to access the belly of the Orbiter to remove protruding gap fillers. Prior to the completion of the STS-121 assessment of OBSS stability to perform TPS repairs (for repair areas to which the SRMS or SSRMS cannot provide access), NASA developed a combined SRMS and SSRMS "flip around" operation, called the Orbiter Repair Maneuver (ORM), to allow TPS repairs while the Space Shuttle is docked to the ISS. The ORM involved turning the Shuttle into a belly-up position that provides SSRMS access to the repair site. The assessment performed during an STS-121 EVA demonstrated that the SRMS and OBSS can be used as a stable EVA worksite for TPS repair, and the ORM is not required for Orbiter repairs. Using OBSS for such operations entails far less operational risk than the ORM.

NASA has continued to develop EVA tools and techniques for TPS repair. NASA has already developed prototype specialized tools for applying and curing TPS repair materials. We also have developed new and innovative EVA techniques for working with the fragile Shuttle TPS system while ensuring that crew safety is maintained. EVAs for TPS repair represent a significant challenge; the experiences gained through the numerous complex ISS construction tasks performed over the past several years are contributing to our ability to meet this challenge.

RCC Repair

NASA has evaluated RCC repair concepts using the expertise of six NASA centers, 12 primary contractors, and the United States Air Force Research Laboratory. The main technical challenges to repairing RCC are maintaining a bond to the RCC coating during entry heating and meeting very small edge step requirements for repair patches and fills.

NASA has implemented two complementary repair concepts—plug repair and crack repair—that together will enable repair of limited RCC damage. Plug repair consists of a cover plate intended to repair medium-sized holes in the WLE from 1.1 inches to 4 inches in diameter. Crack repair uses a non-oxide experimental adhesive (NOAX) material intended to fill cracks and missing coating areas in the WLE. Both concepts are expected to have limitations in terms of damage characteristics, damage location, and testing/analysis.

RCC plug repair requires heavily bruised or through penetrations of the RCC; as a result, some damage types may require drilling to use the plug repair. Currently, RCC plug repair can be successfully installed on 62 percent of the WLE but has not achieved the required step and gap for all RCC surfaces. Additionally, plug repair does not fix damage larger than the plug area. Eighteen unique plug repair contours were delivered to ISS on STS-121 for contingency repair. Two full scale plug repair tests were conducted in the arc jet and passed.

Successful RCC crack repair requires a complex operation. Process controls during EVA in the challenging environment of space will determine the success of repair. The RCC crack repair has been successful in repairing cracks up to 0.065 inch wide, chipped coating, and gouges in ground tests; these repairs were also verified by arc-jet testing. Additional arc-jet and thermal-vacuum testing was conducted on uncured NOAX and samples of chipped coating and gouged RCC that were impacted prior to the repair. Results of all tests indicate the current NOAX formula can survive the entry environment without being actively cured by a heater; however, on-orbit passive curing of the repair is recommended prior to entry.

Prior to RTF, NASA also initiated a research and development effort to repair medium-sized holes with a flexible patch. The flexible patch would have been directly applied over holes and cracks found on RCC panels. The synergy of using the same repair concept for both holes and cracks will greatly reduce the total hardware required for each mission. However, due to relatively low technical maturity and a long development schedule, further development of flexible patch repair was halted.

A fourth repair concept, RCC rigid overwrap, encountered problems during development and was shown to be infeasible to implement in the near term; as a result, it was also deleted from consideration. Additionally, NASA evaluated several new overlay concepts for large area repairs of both tile and RCC using flexible carbon silicon carbide (CSiC). However, due to their low technical maturity level and the need for considerable additional research and development before the repair concepts could be applied, development of these concepts was also halted.

Tile Repair

A limited tile repair capability was developed for on-orbit testing for STS-114. This capability included an emittance wash application to repair shallow damage, a cure in place ablator (CIPA) repair material with CIPA applicator to repair larger damage (up to 10 inches by 20 inches and the depth of the tile), and a CSiC tile repair mechanical overlay designed to repair large tile damage areas and lost tiles.

On STS-114, NASA completed a demonstration of the tile repair emittance wash application during the first EVA. The emittance wash was applied directly and indirectly to portions of two damaged tile test articles. Two CIPA applicators and a mechanical overlay were flown but not tested on the first flight. With the exception of the overlay mechanical concept, all of these tile repair materials and tools were verified by ground testing and certified as safe to fly and safe to use on STS-114, if an emergency tile repair was required. The overlay mechanical hardware and EVA tools were certified as safe to fly only.

The emittance wash is a silicon carbide (SiC) material mixed with a carrier material. It provides an emissive coating to the damaged tiles that prevents small gouges in the tile from further erosion. This keeps the damage shallow and prevents cavity heating effects, preserving the insulating capability of the tile. The emittance wash could also be used to prime and seal CIPA repairs to the tiles.

The CIPA-dispensed STA-54 is a two-part room temperature vulcanizing-based material. Both the material and the applicator have encountered significant challenges during development. Most significant of these challenges was recurrent bubbling in the material. Ground tests of repaired tiles using STA-54 demonstrated the material's ability to provide thermal protection, even with a 1-G distribution of bubbles. Material property testing was unsuccessful in reducing areas of uncertainty about the material properties of STS-54, including the material's ability to cure during the thermal cycling of Earth orbit and its adhesion to tile during entry. Consequently, the material and applicator required extensive system-level testing in the combined space environment to understand their utility for on-orbit repairs. In addition, the level of toxicity of one of the STA-54 components prior to mixing and dispensing was determined to be high, which required a triple level of containment for STA-54 stowage. In December 2005, the Program terminated the CIPA/STA-54 project, including the STS-121 CIPA DTO, in favor of consolidating available tile repair resources on the emittance wash and mechanical overlay techniques.

The mechanical overlay repair consists of a thin ceramic cover plate, batting insulation, ceramic washers, and ceramic augers. The repair is performed by filling the damaged tile cavity with a Saffil batting insulation, then placing a thin CSiC cover plate and high-temperature gasket seal over the damaged tile area. Coated SiC ceramic augers (screws) with accompanying SiC ceramic washers are screwed into undamaged tiles to attach the overlay. The 12-inch by 25-inch overlay is capable of covering a 10-square inch by 20-square inch damage area.

STATUS

The following actions have been completed:

- Quantified SRMS, SSRMS, and ISS digital still camera inspection resolution
- Feasibility analyses for docked repair technique using SRMS and SSRMS
- Air-bearing floor test of overall boom to SRMS interface
- OBSS conceptual development, design requirements, and preliminary design review, systems design review, initial OV-103 vehicle integration testing at Kennedy Space Center with both sensors
- Engineering assessment for lower surface radio frequency communication during EVA repair
- Simplified Aid for EVA Rescue technique conceptual development and testing
- Feasibility testing on tile repair material
- Tile repair material transition from concept development to characterization and qualification tests
- 1-G suited tests on tile repair technique
- Initial KC-135 tile repair technique evaluations
- Vacuum dispense and cure of the tile repair material with key components of the EVA applicator
- Review of all Shuttle systems for compatibility with the docking repair scenario
- Inspection Tiger Team strategy formulated

- Down-selected to two complementary RCC repair techniques for further development (plug repair, crack repair), with the elimination of rigid wrap repair for RTF
- Developed the inspection and repair of the RCC and tile operations concept (figure 6.1-4-3)
- The digital cameras that ISS crew uses to photograph the Shuttle TPS were launched on a Russian Progress vehicle and are now on board the ISS

Transport, impact, and material analyses and tests have provided a clear enough picture of the WLE and RCC's characteristics to allow NASA to make an informed risk trade for a practicable inspection plan. This inspection plan is based on potential debris sources and impact likelihood, specific RCC panel capabilities, and LDRI capabilities that have been demonstrated beyond its certified performance.

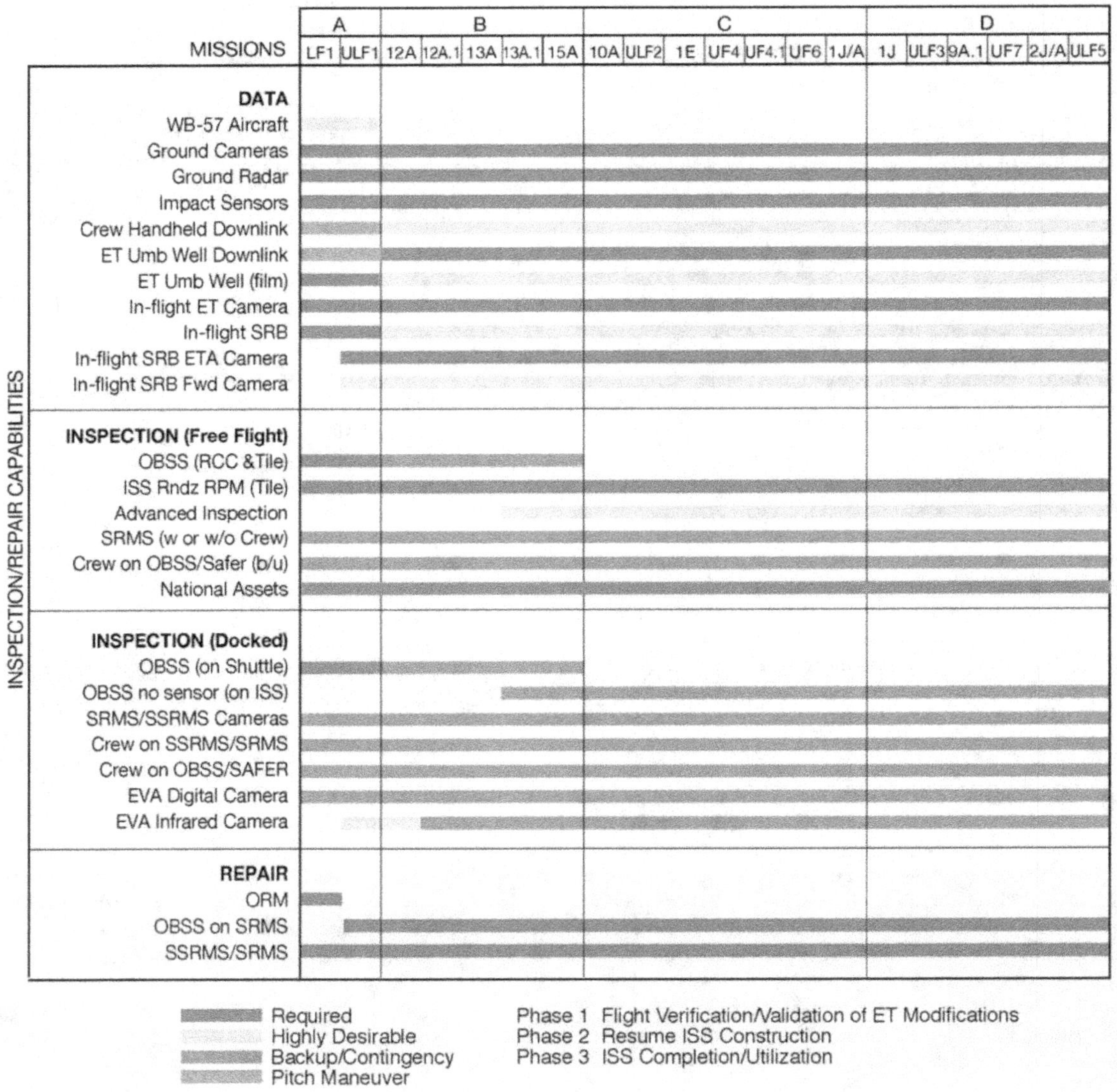

Figure 6.4-1-3. Integrated operations concepts for inspection and repair.

Individually, each warning device/inspection method described previously does not provide the total information needed to accurately determine the condition of the Orbiter prior to committing to entry. They are not redundant systems, per se, in that each provides a different piece of the puzzle, offering overlapping information to improve our knowledge of the Orbiter's condition. We can accept failure of one or more warning devices and have the confidence that we will be able to characterize potential debris liberations and possible damage to the TPS tile and RCC components.

FINAL UPDATE

TPS Inspection

During STS-114, NASA had unprecedented insight into the performance of the Space Shuttle system during ascent and the on-orbit status of the Orbiter. This new ability to completely image the Orbiter during the mission revealed some unexpected foam losses from the ET (see R3.2-1) and two on-orbit anomalies that were addressed by the MMT: protruding gap fillers and a damaged thermal blanket near one of the Orbiter windows. The quality of the images gained through inspection allowed the MMT to complete a thorough real-time analysis of the two anomalies to determine the risk posed by them to the safety of the Orbiter and the crew. As a result, the MMT directed the STS-114 crew to conduct an unplanned EVA task, which successfully removed the protruding gap fillers. In addition, NASA was able to detect and inspect minor damage sites on the acreage tile and clear them for entry.

The OBSS performed exceptionally well and produced better-than-expected results and resolution of the Orbiter TPS. The sensors were considered adequate compared to the Orbiter critical damage size. Issues with structural margins during ascent and landing were resolved before the STS-114 flight. On STS-121, OBSS Sensor Package 2 carried one additional high-resolution digital camera, the ISIS IDC to use for focused inspections. This camera improved NASA's ability to distinguish actual damage from false indicators at the 0.08-inch diameter level. This improved capability significantly reduces the amount of in-flight analytical time required to assess potential damage, and improves the MMT's ability to make critical in-flight decisions (figure 6.4-1-4).

Lessons learned from STS-114 and STS-121 have been documented in the OIP for TPS Assessment and OIP Damage Assessment Annex. The key lessons learned included: 1) the necessity for a uniform ascent reporting database to integrate the ascent debris observations from imagery, radar, and the WLE sensors, and 2) the need for a capability to provide imagery to operational users as well as analysts. NASA also updated the ascent debris detection process to account for the addition of the enhanced launch vehicle imaging system solid rocket booster (SRB) ET attach ring and SRB forward skirt aft pointing cameras. Updates were also made to the damage inspection process to account for the addition of the OBSS high-resolution ISIS digital camera, the new requirements for a gap filler inspection and an ET umbilical door closure inspection, and the availability of an EVA infrared camera for RCC inspection. The OIP Annex was updated to include the latest inspection criteria and the late MMOD inspection and assessment process. The EVA infrared camera was flown as a DTO for STS-121. A simulation plan, similar to that developed in support of STS-114, was executed to demonstrate the ability to assess TPS for STS-121.

For STS-114 and STS-121, NASA performed OBSS operations the second and fourth crew flight days. On the second flight day, prior to docking with the ISS, the crew used the OBSS to inspect the nose cap and the underside and apex of the 22 leading edge RCC panels on each wing. The data from these inspections were fed into the TPS assessment process for Orbiter damage assessment. OBSS cameras were used during Flight day 4 (and Flight day 5 on STS-114) for focused inspections on areas determined likely to have experienced debris hits.

TPS Late Inspection

The STS-121 crew performed the first late flight day inspection to look for MMOD impacts. Although the Orbiter was inspected early in the flight to check for ascent damage, this late inspection reduces the risk of entering the Earth's atmosphere with damage to RCC caused by MMOD impacts while on orbit. On Flight Day 11, while docked to ISS, the port WLE RCC was inspected with the OBSS. After undocking, the OBSS was used on Flight Day 12 to inspect the starboard wing and nose cap RCC. The digital data from the inspection was downlinked to Earth for analysis by the DAT, and no problem areas were identified. In the unlikely event this late inspection had identified damage unacceptable for reentry, NASA could have attempted repairs or invoked CSCS.

Although late inspection of RCC reduces entry risk, it introduces other operational risks that must be considered. NASA has weighed these risks and decided to baseline late inspection for all future flights.

TPS Repair

On STS-114, NASA flew simulated tile and RCC damage test articles in the payload bay to enable the crew to practice tile and RCC repair techniques. Similar tests were conducted

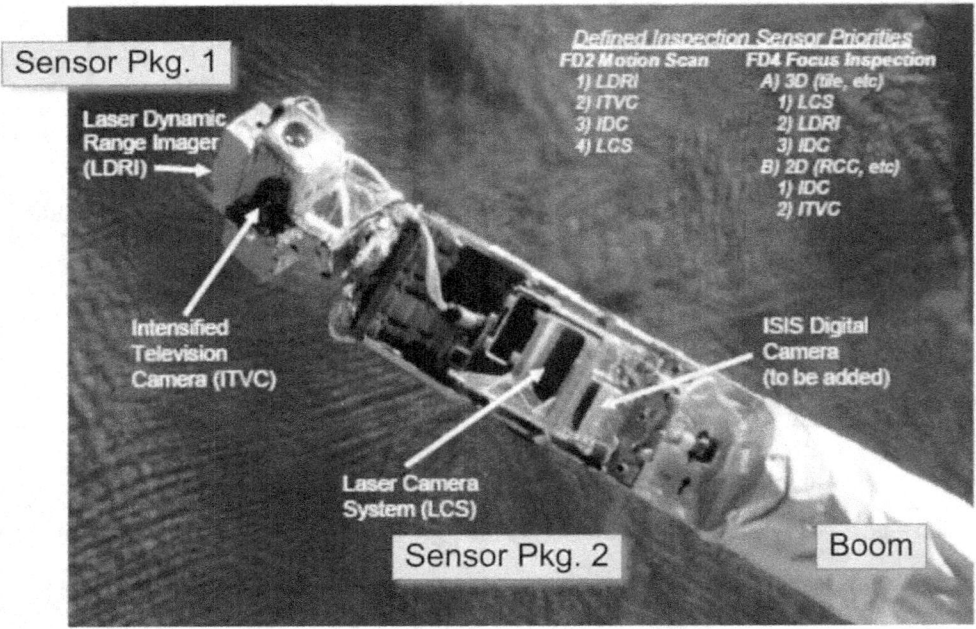

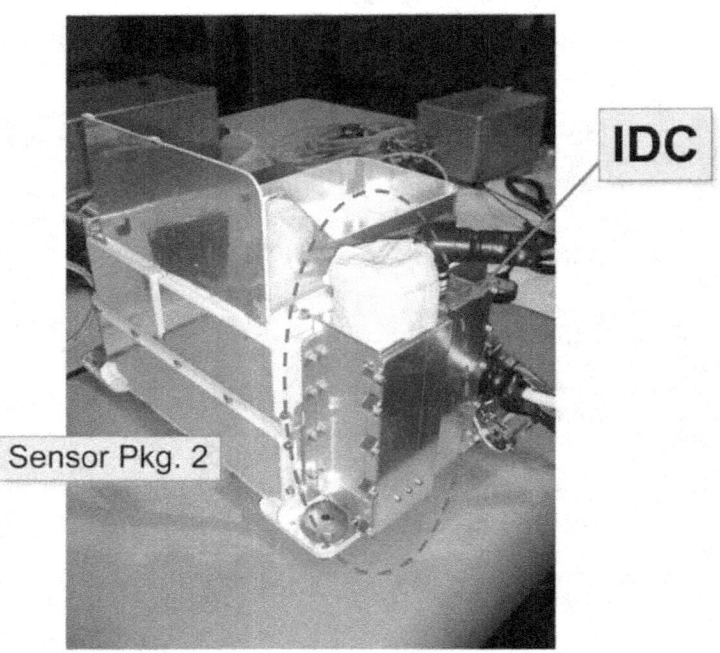

Figure 6.4-1-4. OBSS (sensor package 1) and ISIS digital camera (IDC) (sensor package 2).

on STS-121. During STS-114, tile repair emittance wash application and RCC NOAX crack repair material evaluation were demonstrated during an EVA: Preliminary results showed that both techniques are viable. Also during STS-114, an intravehicular activity demonstration of the mechanical aspects of the RCC plug repair was successfully conducted.

In addition to planned TPS repair capability, a test was performed during STS-121 to further evaluate RCC crack repair material, tools, and techniques and generate test samples. The prepared test samples, once returned to ground, were subjected to arc-jet testing to assess their performance in a flight-like entry environment.

NASA's long-term TPS repair strategy is to focus on on-orbit verifiable repair techniques that show the most promise for an on-orbit repair, as well as being capable of flight certification to specific design reference cases. To carry out this strategy, the Program had previously terminated the flexible patch and rigid overwrap projects for RCC in order to focus on NOAX crack repair and plug repair techniques. NASA pursued development of six additional plug designs to cover most of Panels 8-10, and defined the architecture for an analytical model of the NOAX crack repair material. RCC crack repair material will be certified to the fullest extent possible for small cracks and gouges, and verification will be completed for the plug repair system.

In December 2005, the Program also stopped work on other RCC large area repair techniques. For tile repair, the Program consolidated available resources on emittance wash and mechanical overlay techniques and terminated the CIPA/STA-54 project, including the STS-121 CIPA DTO Now that the most promising TPS repair techniques have been selected, the Program has focused on testing, verifying, and validating those repair techniques.

In April 2006, the Program authorized the Tile Repair Ablator Dispenser (T-RAD) project which uses the STA-54 material developed for the CIPA project. T-RAD is a simplified version of the CIPA concept. The Program plans to manifest T-RAD hardware and STA-54 material on every flight, making it available for on-orbit tile repair if the need arises. A DTO to demonstrate STA-54 material and T-RAD on-orbit performance will be performed on a subsequent flight (tentatively scheduled for STS-119). Additional STA-54 material toxicity assessment has concluded that the toxicity level is lower than the previously determined level, thereby eliminating the operational constraint of triple containment required for crew module stowage.

Numerous improvements have been made to the first generation overlay repair hardware since STS-114. Improvements include incorporation of a better auger material, a larger cover-plate size (15 inches x 25 inches), an improved batting dispenser design and improved EVA tools. The first-generation hardware, currently on ISS, will be replaced with the improved second generation hardware on STS-118.

Hubble Space Telescope Servicing Mission 4 Decision

On October 31, 2006, The NASA Administrator announced that NASA would conduct a fifth servicing mission of the Hubble Space Telescope (HST), referred to as Servicing Mission 4 (SM-4). The decision to proceed to conduct a SM-4 mission followed an extensive review by the relevant NASA mission directorates – the Science Mission Directorate (SMD) and the Space Operations Mission Directorate (SOMD) – of all safety and technical issues associated with conducting such a mission. The SM-4 mission will replace the Wide Field Camera, add the Cosmic Origins Spectrograph instrument, and bring critical battery and attitude control systems on the HST to full functionality. An unprecedented repair of the Space Telescope Imaging Spectrograph will also be performed, which will restore the full set of spectroscopic tools making them available to astronomers worldwide.

The overall risks associated with the SM-4 mission are comparable to risks associated with a mission to the ISS, except that repair of any potential damage is more challenging on an autonomous mission. To reduce the risks of the ascent debris environment, the SM-4 mission will be performed after introducing the final redesign for the ET liquid hydrogen ice/frost ramps. The first ET with this configuration is ET-128, which will be available for flight in spring 2008. The SM-4 mission also will be conducted after the Space Shuttle Program has the opportunity to evaluate the effectiveness of the redesign on ET-128 and use ET-129 or a later ET for SM4. The Hubble Team demonstrated that the MMOD environment for the HST mission is no different than that experienced in the four preceding servicing missions. Although the risk of MMOD damage in this environment is slightly higher than that in which the ISS operates, NASA has determined that the risk to the Orbiter and crew is acceptable.

NASA has made great progress in developing on-orbit inspection and repair techniques of the Orbiter TPS tile and RCC panels to assure that they are acceptable for entry. Although the RPM is not available on the HST mission, the OBSS and a special inspection using the SRMS can effectively be used to inspect and identify critical impact damage to tile and RCC panels. This inspection approach is sufficiently accounted for in the SM-4 mission timeline as outlined by NASA. As in any other mission, there would be impacts to mission content if critical damage to the TPS were identified during the SM-4 mission; repair of damaged

TPS tiles and focused detailed inspection would override any mission objective. In addition to repair, to further mitigate risk of ascent or on-orbit damage to the Orbiter, NASA is planning a crew rescue capability for the SM-4 mission in the unlikely event that TPS damage is too great to be adequately repaired for reentry. The Launch-On-Need capability instituted by NASA following the *Columbia* accident is more difficult with the SM-4 mission than for an ISS mission simply because we do not have a CSCS capability to secure the crew while another Orbiter is prepared to fly the rescue mission. However, the Hubble Team has prepared an acceptable plan to ready a second Orbiter for launch within the limited time available to rescue the crew. The initial planning will use two launch pads simultaneously to provide this capability.

This approach may impact the SM-4 launch date, but will have a minimal impact on the completion of the ISS and Constellation program activities.

Following the *Columbia* accident, there was no certainty that the SM-4 mission could be safely undertaken. Through the successful implementation and execution of the RTF requirements and continued effort and dedication of the NASA team, what were once considered insurmountable obstacles have been leveled. It is now possible to proceed with this much needed servicing mission to extend the life of the HST and its scientific achievements into the next decade.

SCHEDULE

Responsibility	Due Date	Activity/Deliverable
SSP	Jul 03 (Completed)	1-G suited and vacuum testing begins on tile repair technique
SSP	Aug 03 (Completed)	Generic crew and flight controller training begins on inspection maneuver during approach to ISS
SSP	Aug 03 (Completed)	KC-135 testing of tile repair technique
SSP	Oct 03 (Completed)	Start of RCC repair concept screening tests
SSP	Dec 03 (Completed)	Tile repair material selection
SSP	Jun 04 (Completed)	Baseline in-flight repair technique requirements and damage criteria
SSP	Sep 04 (Completed)	Initial human thermal-vacuum, end-to-end tile repair tests
JSC/Mission Operations Directorate	Oct 04 (Completed)	Formal procedure development complete for inspection and repair
SSP	Jan 05 (Completed)	RCC repair concept downselect
SSP and ISS Program	Jun 05 (Completed)	All modeling and systems analyses complete for damage assessment
SSP	Jun 05 (Completed)	Tile repair materials and tools delivery
SSP	STS-114 (Completed), STS-121 (Completed)	On-orbit test of TPS repair tools and process

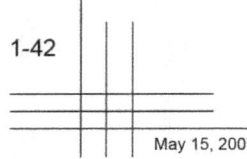

1-42

NASA's Implementation Plan for Space Shuttle Return to Flight and Beyond

May 15, 2007

Columbia Accident Investigation Board
Recommendation 3.3-3

To the extent possible, increase the Orbiter's ability to successfully re-enter Earth's atmosphere with minor leading edge structural sub-system damage.

Note: NASA has closed this recommendation through the formal Program Requirements Control Board process. The following summary details NASA's response to the recommendation and any additional work NASA performed beyond the *Columbia* Accident Investigation Board recommendation.

BACKGROUND

The STS-107 accident demonstrated that the Space Shuttle Leading Edge Structural Subsystem (LESS) is vulnerable, and damage to the LESS can cause the loss of the Orbiter. The Space Shuttle Program (SSP) conducted a comprehensive test and analysis program to redefine the maximum survivable LESS damage for entry. This information supports the requirements for inspection and ultimately the boundaries within which a Thermal Protection System (TPS) repair can be performed. In addition, the SSP has pursued LESS improvements that will increase the Orbiter's capability to enter the Earth's atmosphere with "minor" damage to the LESS. These improvements and NASA's efforts to define minor and critical damage using foam impact tests, arc jet tests, and wind tunnel tests are only mentioned here, since they are covered in recommendations R3.3-1, R3.3-2, R3.3-4, and R6.4-1.

NASA IMPLEMENTATION

The SSP has evaluated operational adjustments in vehicle and trajectory design for reducing thermal effects on the LESS during entry. Possibilities included weight reduction by cargo jettison, cold-soaking the damaged area of the Orbiter by shading it from direct sunlight, lowering the orbit to reduce maximum heat loads during deorbit, selecting entry opportunities that minimize structural loading and entry trajectory shaping. Additionally, NASA considered modifying the angle-of-attack profile.

STATUS

Evaluations in each of the above areas are complete. These evaluations were conducted within existing certification limits for entry trajectory conditions experienced during Shuttle missions to the International Space Station. The results showed only minor improvements in the entry thermal environment for Reinforced Carbon-Carbon (RCC). These results were presented to the SSP in July 2004. At that time, the SSP directed Mission Operations to conduct further evaluations that were not constrained by existing certification limits. The evaluations were conducted to determine whether major improvements in reducing thermal effects could be attained by exceeding certification limits for entry trajectory and angle of attack and, if so, by how much.

FINAL UPDATE

Evaluations of operational adjustments in vehicle and trajectory design, including utilization of a modified angle of attack profile, showed only minor improvements could be made in the entry thermal environment for RCC. Only with a substantially increased risk of guidance, navigation, and control uncertainties could noticeable improvements be obtained. However, these analysis results provided a clearer understanding of the relationship between all entry parameters and the associated risks to the Orbiter. These evaluations resulted in flight rules and operational procedures that establish the framework within which NASA can consider certified and uncertified options to provide additional thermal margin in the event of minor leading edge damage. These flight rules and procedures were presented and accepted by the SSP through the Program Requirements Control Board process in March 2005 in preparation for the STS-114 Return to Flight mission.

SCHEDULE

Responsibility	Due Date	Activity/Deliverable
SSP	Jul 04 (Completed)	Vehicle/trajectory design operational adjustment recommendation

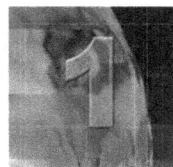

Columbia Accident Investigation Board
Recommendation 3.3-4

In order to understand the true material characteristics of Reinforced Carbon-Carbon components, develop a comprehensive database of flown Reinforced Carbon-Carbon material characteristics by destructive testing and evaluation.

Note: NASA has closed this recommendation through the formal Program Requirements Control Board process. The following summary details NASA's response to the recommendation and any additional work NASA performed beyond the *Columbia* Accident Investigation Board recommendation.

BACKGROUND

The only material property data initially available for flown Reinforced Carbon-Carbon (RCC) components were removed from Orbiter Vehicle (OV)-102 and destructively tested by the Space Shuttle Program (SSP). To obtain these data, material specimens were cut and tested from the lower surface of Panel 10 left (10L) after 19 flights and Panel 12 right (12R) after 15 flights. The results from these tests were compared to the analytical model and indicated that the model was conservative.

NASA IMPLEMENTATION

An RCC material characterization program has been implemented using existing flight assets to obtain additional data on strength, stiffness, stress-strain response, strain-rate sensitivity, and fracture properties of RCC for comparison to earlier test data. The SSP established a plan to determine the impact resistance of RCC in its current configuration using previously flown Panels 9L and 16R. In addition, tension, compression, in-plane shear, interlaminar shear, and interlaminar tension (coating adherence) properties were developed. Data on the attachment lug mechanical properties, corner mechanical properties, and coating adherence were also obtained. NASA maintains a comprehensive database developed with the information from these evaluations and characterization programs.

Mechanical property specimens excised from the upper surface, apex region, and lower surface of Panel 8L (OV-104 with 26 flights) have been tested, along with additional specimens taken from the apex region of Panels 10L and 12R. The data from these tests were distributed to teams to perform material property and impact analyses. As expected, the results showed slightly degraded properties, when compared with new material, but are still well above the conservative design allowables used in the mission life models for RCC. Panel 6L (OV-103 with 30 flights) has been used to perform thermal and mechanical testing to determine material susceptibility to crack propagation during the flight envelope. Panel 9L (OV-103 with 27 flights) was severely cracked during a series of full-scale, damage threshold determination impact tests. Specimens from the damaged region have been excised for damage tolerance assessment in the arc jet facility. In addition, mechanical property specimens adjacent to the damage zone were used to determine strength properties for use in the impact analysis correlation effort. Panel 16L was also subjected to repeated impacts until notable damage was observed in the RCC (cracking and delamination) to provide additional impact analysis correlation and determination of the damage threshold.

Three new Panel 9Ls were subjected to impact testing for further damage model correlation. Mechanical property specimens from Panel 9R (with 30 flights) from OV-103 were tested using methods similar to those used on Panels 10L and 12R, to compare its material properties to the analytical model and to add to the database.

FINAL UPDATE

The study of materials and processes is central to understanding and cataloging the material properties and their relation to the overall health of the wing leading edge subsystem. Materialography and material characteristics (porosity, coating/substrate composition, etc.) for RCC panels have been evaluated with the objective of correlating mechanical property degradation to microstructural/chemical changes and nondestructive inspection results. Since its development, the database could be used to direct future design upgrades and mission/life adjustments. The long-term plan will include additional RCC assets, as required to ensure that the database is fully populated (ref. R3.8-1).

SCHEDULE

Responsibility	Due Date	Activity/Deliverable
SSP	Sep 03 (Completed)	Selection of Panel 8L test specimens for material property testing
SSP	Sep 03 (Completed)	Panel 9L impact test number 1
SSP	Sep/Oct 03 (Completed)	Material property testing of Panel 8L specimens
SSP	Oct 03 (Completed)	Panel 9L impact test number 2 and 3
SSP	Jun 04 (Completed)	Panels 10L and 12R apex mechanical property testing
SSP	Aug 04 (Completed)	Panel 16R impact testing

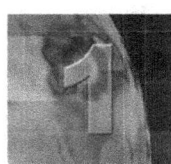

Columbia Accident Investigation Board
Recommendation 3.3-5

Improve the maintenance of launch pad structures to minimize the leaching of zinc primer onto Reinforced Carbon-Carbon components.

Note: NASA has closed this recommendation through the formal Program Requirements Control Board (PRCB) process. The following summary details NASA's response to the recommendation and any additional work NASA performed beyond the Columbia Accident Investigation Board recommendation.

BACKGROUND

Zinc coating is used on launch pad structures to protect against environmental corrosion. "Craze cracks" in the Reinforced Carbon-Carbon (RCC) panels allow rainwater and leached zinc to penetrate the panels and cause pinholes.

NASA IMPLEMENTATION

Before Return to Flight (RTF), Kennedy Space Center (KSC) had enhanced the launch pad structural maintenance program to reduce RCC zinc oxide exposure to prevent zinc-induced pinhole formation in the RCC (figure 3.3-5-1). The enhanced program has four key elements. KSC enhanced the postlaunch inspection and maintenance of the structural coating system, particularly on the rotating service structure. Exposed zinc primer has been recoated to prevent liberation and rainwater transport of zinc-rich compounds. Additionally, postlaunch pad structural wash-downs have been assessed to minimize the corrosive effects of acidic residue on the pad structure. This helps prevent corrosion-induced damage to the topcoat and prevents exposure of the zinc primer. Enhancements of the automatic wash down system were also implemented on Pad A in fiscal year 2005 to help prevent exposure of the primer. NASA investigated options to improve the physical protection of Orbiter RCC hardware and determined that the existing protection is sufficient. NASA has also implemented a sampling program to monitor the effectiveness of efforts to inhibit zinc oxide migration on all areas of the pad structure.

FINAL UPDATE

NASA has enhanced the inspections and structural maintenance of the launch pads to reduce zinc leaching. The options developed for zinc minimization were presented to the Space Shuttle PRCB in April 2004 and approved for implementation. A new enhanced automatic wash down system was approved, and has been designed and installed on Launch Complex 39A. All applicable work authorization documents were incorporated prior to RTF including the sampling of rainwater runoff. The RCC Problem Resolution Team will continue to identify and assess potential mechanisms for RCC pinhole formation.

SCHEDULE

Responsibility	Due Date	Activity/Deliverable
Space Shuttle Program (SSP)	Dec 03 (Completed)	Complete enhanced inspection, maintenance, wash-down, and sampling plan
SSP	Apr 04 (Completed)	Present to the PRCB

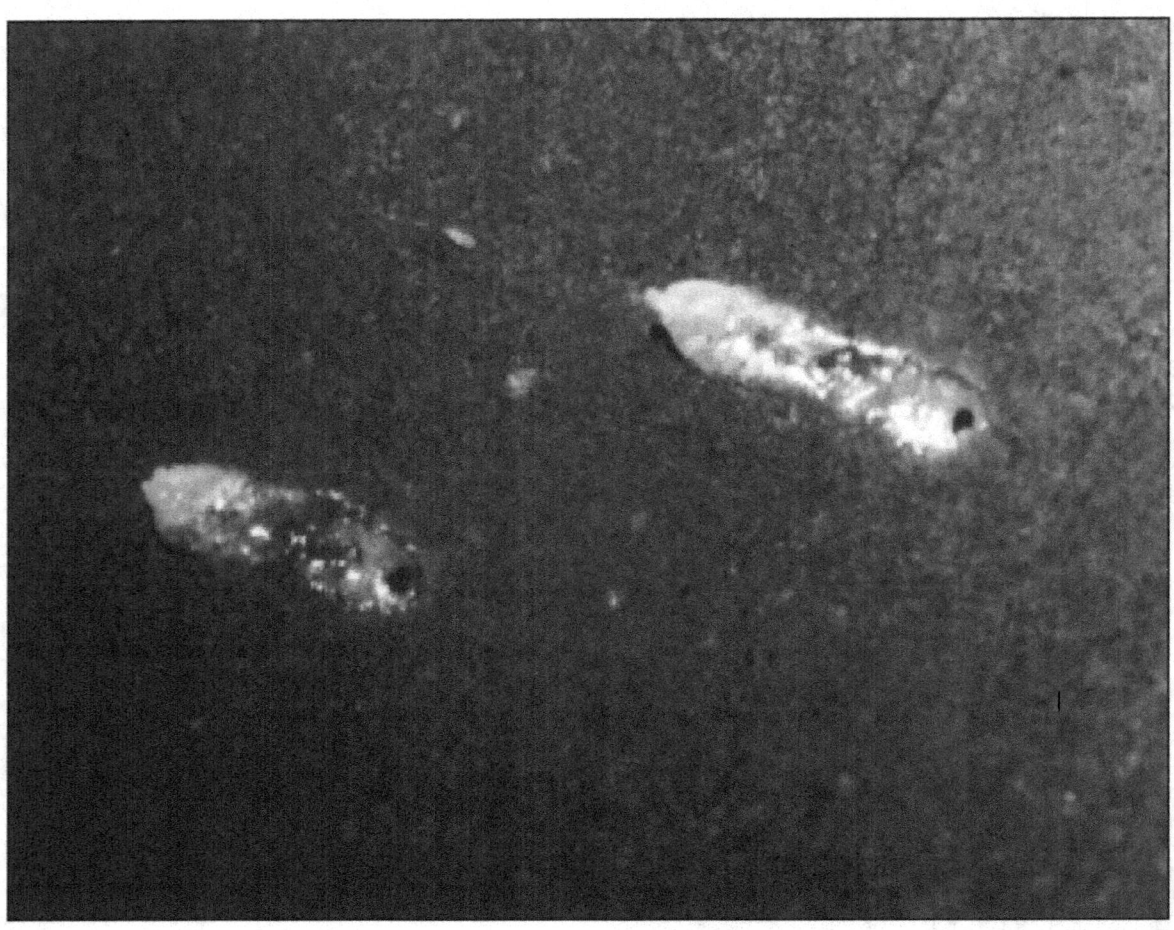

Figure 3.3-5-1. RCC pinholes.

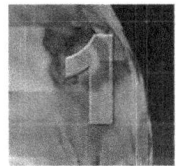

Columbia Accident Investigation Board
Recommendation 3.8-1

Obtain sufficient spare Reinforced Carbon-Carbon panel assemblies and associated support components to ensure that decisions on Reinforced Carbon-Carbon maintenance are made on the basis of component specifications, free of external pressures relating to schedules, costs, or other considerations.

Note: NASA has closed this recommendation through the formal Program Requirements Control Board (PRCB) process. The following summary details NASA's response to the recommendation and any additional work NASA performed beyond the *Columbia* Accident Investigation Board (CAIB) recommendation.

BACKGROUND

There are 44 wing leading edge (WLE) panels installed on an Orbiter. All of these components are made of Reinforced Carbon-Carbon (RCC). The panels in the hotter areas, panels 6 through 17, have a useful mission life of 50 flights or more. The panels in the cooler areas, panels 1 through 5 and 18 through 22, have longer lives, as high as 100 flights depending on the specific location. The "hot" panels (6 through 17) are removed from the vehicle every other Orbiter maintenance down period and are shipped to the original equipment manufacturer, Lockheed-Martin, for refurbishment. Because these panels have a long life span, we have determined that a minimum of one spare ship-set is sufficient for flight requirements.

Since few panels have required replacement, few new panels have been produced since the delivery of Orbiter Vehicle (OV)-105. Currently, Lockheed-Martin is the only manufacturer of these panels.

NASA IMPLEMENTATION

NASA's goal is to maintain a minimum of one spare ship-set of RCC WLE panel assemblies. To achieve this goal, six additional panel assemblies were procured to complete the spare ship-set.

STATUS

In addition to the six panels needed to complete one spare ship-set, NASA has procured enough raw materials to build up to four additional ship-sets of RCC panels. The Space Shuttle Program Leading Edge Subsystem Prevention/Resolution Team has developed a prioritized list of additional spare panels over and above the one ship-set of spare panels currently required to support the Program. The prioritization of the list was based on the requirements for the spare ship-set, impact tolerance testing, and development of damage repair techniques.

FINAL UPDATE

During Return to Flight prior to STS-114, OV-103 panel 8L was removed and installed on OV-104 because OV-104's panel 8L was removed for impact testing. Panel 8L from the spare ship-set was installed on OV-103. The replacement panel 8L has been in fabrication and is scheduled to be delivered to replenish the spare ship-set in May 2007.

Panel 8R from the spare ship-set was used to replace panel 8R removed from OV-103 during STS-121 turn-around processing. Panel 8R removal was due to a suspect discrepancy detected during post STS-114 flight processing. The replacement panel 8R to replenish the spare ship-set is scheduled to be delivered in February 2008.

SCHEDULE

Responsibility	Due Date	Activity/Deliverable
Space Shuttle Program (SSP)	Jun 03 (Completed)	Authorization to build six panels to complete ship-set
SSP	Mar 05 (Completed)	Delivery of six additional panels

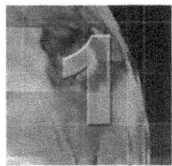

Columbia Accident Investigation Board
Recommendation 3.8-2

Develop, validate, and maintain physics-based computer models to evaluate Thermal Protection System damage from debris impacts. These tools should provide realistic and timely estimates of any impact damage from possible debris from any source that may ultimately impact the Orbiter. Establish impact damage thresholds that trigger responsive corrective action, such as on-orbit inspection and repair, when indicated.

Note: NASA has closed this recommendation through the formal Program Requirements Control Board process. The following summary details NASA's response to the recommendation and any additional work NASA performed beyond the Columbia Accident Investigation Board recommendations.

BACKGROUND

Foam impact testing, sponsored by the *Columbia* Accident Investigation Board (CAIB), proved that engineering analysis capabilities at the time of *Columbia* required upgrades and improvements to adequately predict vehicle response during certain events. In particular, the CAIB found that NASA's impact analysis software tool at the time, Crater, failed to correctly predict the level of damage to the Thermal Protection System (TPS) due to the External Tank foam impact to *Columbia* during STS-107 ascent and contributed to an inadequate debris impact assessment.

NASA IMPLEMENTATION

In addition to improving Crater and other predictive impact models, the Space Shuttle Program (SSP) assigned an action to all Program elements to evaluate the adequacy of all preflight and in-flight engineering analysis tools.

The SSP elements have investigated the adequacy of existing analysis tools to ensure that limitations or constraints in use are defined and documented, and formal configuration management control is maintained. Additionally, tools that are used to clear mission anomalies have undergone a more detailed assessment that included a review of the requirements and verification activities. Results of these element reviews have been briefed in detail at the SSP Integration Control Board (ICB) prior to briefing the specific findings and recommendations to the SSP Manager at the Program Requirements Control Board (PRCB). From these efforts, NASA has created a set of validated physics-based computer models for assessing items such as damage from debris impacts.

STATUS

The SSP worked with the Boeing Company, Southwest Research Institute, Glenn Research Center, Langley Research Center, Johnson Space Center Engineering Directorate, and other organizations to develop and validate potential replacement tools for Crater. Each model offers unique strengths and is a significant improvement beyond the current analytical capability. The pre-STS-107 damage estimation tools, such as Crater, are no longer in use.

An integrated analysis and testing approach was used to develop the models for Reinforced Carbon-Carbon (RCC) components. The analysis is based on comprehensive dynamic impact modeling. Testing was performed on RCC coupons, subcomponents, and wing leading edge panels to provide basic inputs to and validation of these models. Testing to characterize various debris materials were performed as part of model development. An extensive TPS tile impact testing program was performed to increase this knowledge base. Over 1,000 total impact tests were conducted on TPS tile and RCC materials.

In parallel with the model development and its supporting testing, an integrated analysis was developed involving debris source identification, transport, and impact damage, and resulting vehicle temperatures and margins. Through this integrated analysis, impact damage thresholds to determine whether the Orbiter can safely withstand observed damage without requiring on-orbit repair have been established. Insight from this work has been used to identify Space Shuttle modifications (e.g., TPS hardening, trajectory changes) to eliminate unsafe conditions. In addition, the integrated analytical tools are used to support real-time assessment of RCC and tile damage detected during the TPS inspection activities. Actual tile damage sites are assessed against mission-specific entry conditions allowing a risk/benefit trade among return, repair, and rescue. Key integrated analysis tools for damage assessment have undergone review by the NASA Engineering and Safety Center (NESC) prior to the configuration baseline at the Orbiter Configuration Control Board. These tools have been validated against wind tunnel testing at Langley Research Center and arc-jet testing at Ames Research Center and Johnson Space Center. Analytical tools to assess repair with ablator material remain uncorrelated at this time.

All SSP elements presented initial findings and plans for completing their assessments to the ICB in July 2003 and have now completed their assessments. Most SSP models and tools have been reviewed for accuracy and completeness. The remaining reviews will be completed before Return to Flight.

FINAL UPDATE

The SSP system engineering and integration technical areas have successfully evaluated the adequacy of their math models and tools. The NESC has also assessed the adequacy of Bumper (ref. R4.2-4) to perform risk management associated with micrometeoroid and orbital debris (MMOD).

Foam impact tests have provided empirical data to allow test validation of the analytical models and to define the limits of the models' applicability.

Verification and validation of the analytical models was completed in May 2005. This included a detailed review of the models by the NESC. The models were baselined in May 2005.

SCHEDULE

Responsibility	Due Date	Activity/Deliverable
SSP	Jul 03 (Completed)	Report math models and tools assessment initial findings and plans to ICB and PRCB
SSP	Sep 03 (Completed)	Integrated plan for debris transport, impact assessment, and TPS damage modeling
SSP	Dec 03 (Completed)	Reverification/validation of MMOD risk models
NESC	Dec 04 (Completed)	Independent technical assessment of the BUMPER software tool
SSP	May 05 (Completed)	Report math models and tools assessment final findings and recommendations to ICB and PRCB
SSP	May 05 (Completed)	TPS impact testing and model development
SSP	May 05 (Completed)	Verification/validation of new impact analysis tools
SSP	May 05 (Completed)	Verification/validation of new tile damage assessment tools

Columbia Accident Investigation Board
Recommendation 3.4-1

Upgrade the imaging system to be capable of providing a minimum of three useful views of the Space Shuttle from liftoff to at least Solid Rocket Booster separation, along any expected ascent azimuth. The operational status of these assets should be included in the Launch Commit Criteria for future launches. Consider using ships or aircraft to provide additional views of the Shuttle during ascent. [RTF]

Note: The Stafford-Covey Return to Flight Task Group held a plenary session on June 8, 2005, and NASA's progress toward answering this recommendation was reviewed. The Task Group agreed the actions taken were sufficient to fully close this recommendation.

BACKGROUND

NASA's evaluation of the STS-107 ascent debris impact was hampered by the lack of high-resolution, high-speed ground cameras. In response to this, tracking camera assets at the Kennedy Space Center (KSC) (figure 3.4-1-1) and on the Air Force Eastern Range needed to be improved to provide upgraded data during Shuttle ascent.

Multiple views of the Shuttle's ascent from varying angles and ranges provide important data for engineering assessment and discovery of unexpected anomalies. These data are important for validating and improving Shuttle performance, but less useful for pinpointing the exact location of potential damage.

Ground cameras provide visual data suitable for detailed analysis of vehicle performance and configuration from prelaunch through Solid Rocket Booster separation. Images can be used to assess debris shed in flight, including origin, size, and trajectory. In addition to providing information about debris, the images would provide detailed information on the Shuttle systems used for trend analysis that would allow us to further improve the Shuttle. Together, these help to identify unknown environments or technical anomalies that might pose a risk to the Shuttle.

NASA IMPLEMENTATION

NASA has developed a suite of improved ground and airborne cameras that fully satisfies this Recommendation. This improved suite of ground cameras maximizes our ability to capture three complementary views of the Shuttle and provide the Space Shuttle Program (SSP) with engineering data to give us a better and continuing understanding of the ascent environment and the performance of the Shuttle hardware elements within this environment. Ground imagery also helps us to detect ascent debris and identify potential damage to the Orbiter for on-orbit assessment. There are four types of imagery that NASA acquires from the ground cameras: primary imagery—film images used as the primary analysis tools for launch and ascent operations; fallback imagery—backup imagery for use when the primary imagery is unavailable; quick-look imagery—imagery provided to the Image Analysis labs shortly after launch for initial assessments; and tracker imagery—images used to guide the camera tracking mounts and for analysis when needed. Any anomalous situations identified in the post-ascent "quick-look" assessments are used to optimize the on-orbit inspections described in Recommendation 6.4-1.

Figure 3.4-1-1. Typical KSC long-range tracker.

NASA has increased the total number of ground cameras and added additional short-, medium-, and long-range camera sites, including nine new quick-look locations. Since all but one future Shuttle missions are planned to the International Space Station, the locations of the new cameras and trackers are optimized for 51.6-degree-inclination launches. However, our improved ground-based camera system will also provide good coverage and imagery for a mission to 28-degree inclination (such as to the Hubble Space Telescope). Previously, camera coverage was limited by a generic configuration originally designed for the full range of possible launch inclinations and ascent tracks. NASA has also added High-Definition Television (HDTV) digital cameras and 35mm/16mm motion picture cameras for quick-look/fallback and primary imagery, respectively. In addition, NASA has taken steps to improve the underlying infrastructure for distributing and analyzing the additional photo imagery obtained from ground cameras.

System Configuration

NASA divides the Shuttle ascent into three overlapping periods with different imaging requirements. These time periods provide for steps in lens focal lengths to improve image resolution as the vehicle moves away from each camera location:

- Short-range images (T-10 seconds through T+57 seconds)
- Medium-range images (T-7 seconds through T+100 seconds)
- Long-range trackers (T-7 or vehicle acquisition through T+165 seconds)

For short-range imaging, NASA has two Photographic Optic Control Systems (POCS), a primary and a backup, to control the fixed-film cameras at the launch pad, the Shuttle Landing Facility, and the remote areas of KSC where manned cameras are not allowed. There is significant redundancy in this system: Each POCS has the capability of controlling up to 512 individual cameras at a rate of 400 frames per second. There are approximately 75 cameras positioned for launch photography. POCS redundancy is also provided by multiple sets of command and control hardware and by multiple overlapping views, rather than through backup cameras. The POCS are a part of the Expanded Photographic Optic Control Center (EPOCC). EPOCC is the hub for the ground camera system.

The medium- and long-range tracking devices are on mobile platforms (e.g., Kineto Tracking Mount (KTM)), allowing them to be positioned optimally for each flight. The three trackers on the launch pad are controlled with the Pad Tracker System (PTS). PTS is a KSC-designed and -built system that provides both film and video imagery. It has multiple sets of command and control hardware to provide system redundancy. Each of the medium- and long-range tracking cameras is independent, assuring that no single failure can disable all of the trackers. Further, each of the film cameras on the trackers uses HDTV as a backup. For each flight, NASA can optimize the camera configuration, evaluating the locations of the cameras to ensure that the images provide the necessary resolution and coverage.

The planned locations at Launch Complex 39-B for short-, medium-, and long-range tracking cameras are as shown in figures 3.4-1-2, 3.4-1-3, and 3.4-1-4, respectively. As studies improve the understanding of vehicle coverage during ascent, these positions may change. Existing cameras

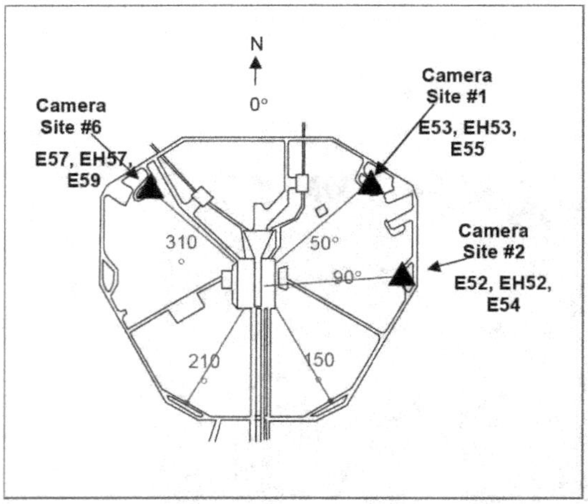

Figure 3.4-1-2. Short-range camera sites.

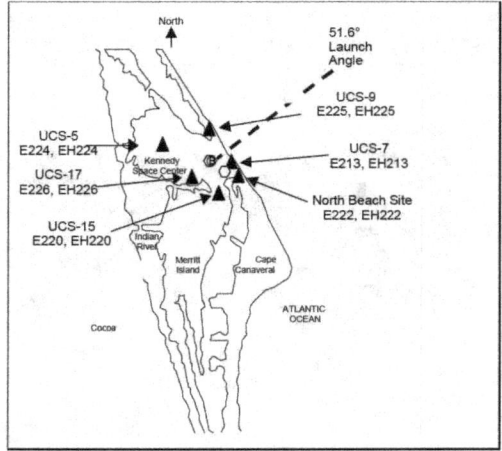

Figure 3.4-1-3. Medium- and long-range tracker sites.

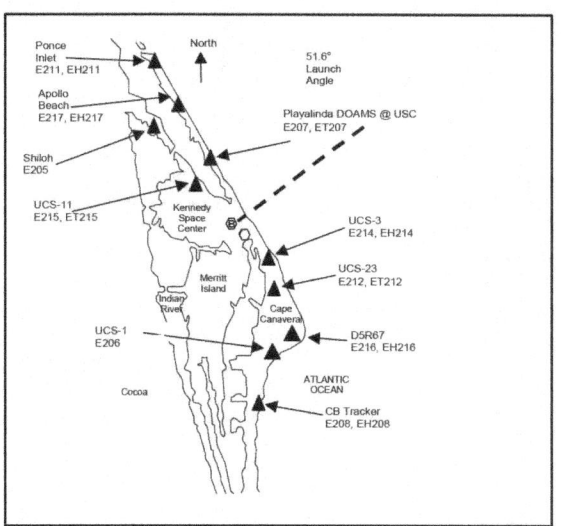

Figure 3.4-1-4. Medium- and long-range tracker sites.

Figure 3.4-1-5. WB-57 aircraft.

can be moved, modernized, and augmented to comply with new requirements.

In addition to ground cameras, NASA approved the development and implementation of an aircraft-based imaging system known as the WB-57 Ascent Video Experiment (WAVE) to provide both ascent and entry imagery. The use of an airborne imaging system provides opportunities to better observe the vehicle during days of heavier cloud cover and in areas obscured from ground cameras by the exhaust plume following launch.

The primary hardware for the WAVE consists of a 32-in. ball turret system mounted on the nose of two WB-57 aircraft (figure 3.4-1-5). The use of two aircraft flying at an altitude of 60,000 ft allows a wide range of coverage with each airplane providing imagery over a 400-mi path. The WAVE ball turret houses an optical bench that provides a location for installation of multiple camera systems (HDTV, infrared). The optics consist of a 4.2-m fixed focal length lens with an 11-in. aperture, and the system can be operated in both auto track and manual modes.

Although the ground cameras provide important engineering data for the Shuttle, they cannot have the resolution and coverage necessary to definitively establish that the Orbiter has suffered no ascent debris damage. No real-time decisions are based on ground imagery data. Rather, the comprehensive assessments of Orbiter impacts and damage necessary to ensure the safety of the vehicle and crew is conducted using on-orbit inspection and analysis.

NASA's analysis suggests that this upgraded suite of ground and airborne cameras significantly improves NASA's ability to obtain three useful views of each Shuttle launch, particularly in conditions of limited cloud cover.

Launch Requirements

NASA has optimized our launch requirements and procedures to support our ability to capture three useful views of the Shuttle, allowing us to conduct engineering analysis of the ascent environment. In accordance with NASA's Return to Flight (RTF) test flight plan, STS-114 and STS-121 were launched in daylight to maximize our ability to capture the most useful ground ascent imagery. For those and for subsequent flights, camera and tracker operability and readiness to support launch are ensured by an updated set of prelaunch equipment and data system checks that are conducted in the days prior to liftoff. These checkouts are documented in the system setup work documents and reported at the Flight Readiness Review. A final system status is reported to the Launch Director at T-20 minutes. In addition, specific launch commit criteria (LCC) have been added for those critical control systems and data collection nodes for which a power failure would prevent the operation of multiple cameras or disrupt our ability to collect and analyze the data in a timely fashion. The camera LCC is tracked to the T-9 minute milestone, and the countdown is not continued if the criteria are not satisfied.

With the additional cameras and trackers that were available at RTF, NASA provided sufficient redundancy in the system to allow us to gather ample data and maintain three useful views—even with the loss of an individual camera or tracker. As a result, it was not necessary to track the status of each individual camera and tracker after the final operability checks. This enhanced overall Shuttle safety by removing an unnecessary item for status tracking during the critical terminal countdown, allowing the Launch Control Team to concentrate on the many remaining key safety parameters. The LCCs remaining until the T-9 minute milestone protect the critical control systems and data collection nodes whose failure might prevent us from obtaining the engineering data necessary to assess vehicle health and function during the initial moments of ascent. For instance, the LCC requires that at least one POCS be functional at T-9 minutes, and that the overall system be stable and operating.

NASA has also confirmed that the existing LCCs related to weather constraints dictated by Eastern Range Safety satisfy the camera coverage requirements. NASA conducted detailed meteorological studies using Cape weather histories, which concluded that current Shuttle launch weather requirements, coupled with the wide geographic area covered by the ground camera suite and the airborne assets, adequately protect our ability to capture sufficient views of the Shuttle during ascent. The weather LCCs balance launch probability, including the need to avoid potentially dangerous launch aborts, against the need to have adequate camera coverage of ascent. The extensive revitalization of the ground camera system accomplished since the *Columbia* accident provides the redundancy that makes such an approach viable and appropriate.

STATUS

The Program Requirements Control Board (PRCB) approved an integrated suite of imagery assets that provides the SSP with the engineering data necessary to validate the performance of the External Tank (ET) and other Shuttle systems, detect ascent debris, and identify and characterize damage to the Orbiter. On August 12, 2004, the PRCB approved funding for the camera suite, to include procurement and sustaining operations. The decision package included the deletion of several long- and medium-range cameras after the first two re-flights, contingent on clearing the ET and understanding the ascent debris environment.

The 14 existing trackers are undergoing refurbishment at White Sands Missile Range, and total refurbishment completion is scheduled for 2008. Trackers and optics were borrowed from other ranges to support the first few launches. NASA also approved funding to procure additional spare mounts, as well as to fund studies on additional capability in the areas of infrared and ultraviolet imagery, adaptive optics, and high-speed digital video, and in the rapid transmission of large data files for engineering analysis.

For RTF, NASA doubled the total number of camera sites from 10 to 20, each with two or more cameras. At RTF, NASA had three short-range camera sites around the perimeter of the launch pad; six medium-range camera sites, one at the Shuttle Launch Facility; and 10 long-range camera sites. To accommodate the enhanced imagery, we installed high-volume data lines for rapid image distribution and improved KSC's image analysis capabilities.

NASA also procured additional cameras to provide increased redundancy and refurbished existing cameras. NASA ordered 35 camera lenses to supplement the existing inventory and purchased two KTM Digital Control Chassis to improve KTM reliability and performance. In addition, NASA procured 36 HDTV cameras to improve our quick-look capabilities.

The U.S. Air Force (USAF)-owned optics for the Cocoa Beach, Florida, camera (the "fuzzy camera" on STS-107) were returned to the vendor for repair. We have completed an evaluation on current and additional camera locations, and refined the requirements for camera sites. Additional sites were picked and are documented in the Launch and Landing Program Requirements Document 2000, sections 2800 and 3120. Additional operator training was provided to improve tracking, especially in difficult weather conditions.

NASA's plan for use of ground-based wideband radar and ship-based Doppler radar to track ascent debris is addressed in Part 2 of this document under item SSP-12, Radar Coverage Capabilities and Requirements.

NASA added redundant power sources to the command and control facility as part of our ground camera upgrade to ensure greater redundancy in the fixed medium/long-range camera system. NASA also added a third short-range tracker site.

The SSP has addressed hardware upgrades, operator training, and quality assurance of ground-based cameras according to the integrated imagery requirements assessment.

FINAL UPDATE

The overall result of the STS-114 Ground Camera Plan was successful. The planned and actual set-ups for film camera focus, exposure, and field of view provided acceptable data products. Three complementary views in each direction were obtained throughout ascent. All imagery products from the ground cameras were delivered to KSC, Johnson Space Center, and Marshall Space Flight Center within the timeline identified in National Space Transportation System Document 60540. The high-definition television cameras and resulting video data added for STS-114 RTF were also successful. No other deviations from the STS-114 Implementation Plan are necessary for ground cameras at this time.

WAVE was used on an experimental basis during the first two Space Shuttle RTF missions. The WAVE met many of its flight objectives concerning operations and flight capability; however, the operation of the camera equipment and the quality of data were compromised by a "jitter" problem.

The ground- and ship-based radars used during STS-114 developed operational experience with using high-resolution radars for debris tracking purposes, and successfully completed their most important first flight objective of gathering baseline data on the Shuttle and debris radar signature. For the first time in Program history, the C-band radar successfully imaged the full extent of expected debris performance during second stage.

Whereas STS-114 collected data largely with a temporary system borrowed from the U.S. Navy, STS-121 saw the first Shuttle use of the permanent configuration for both the C- and X-band radars. Systemic improvements in this configuration included moving from a 30- to a 50-foot-diameter main radar dish—with greater sensitivity and dynamic range than the system used during STS-114. The X-band radars also benefited from sea-based antenna pedestals and improved support locations off shore.

Following STS-121, NASA revisited its commitment to daylight launches and ET separation in lighted conditions. The first two RTF flights developed a benchmark using all sensor assets to validate RTF design modifications to the ET, characterization of the debris environment during liftoff and ascent, and determination of effectiveness of ground radars, wing leading edge impact sensors, and vehicle-mounted cameras. This benchmark was used to validate the debris detection capability of sensor assets that are successful at night and allow correlation of debris detection events. This validation, along with Ice Frost Ramp debris characterization resulting from the STS-115 flight, assured NASA that the combined capability of sensor assets to detect critical debris were sufficient to allow the night launch of the STS-116 mission in December 2006. For remaining Space Shuttle flights, day versus night launch opportunities will be reviewed on a flight-by-flight basis based on mission requirements.

The short-range trackers and one medium-range tracker are now remotely operated from within the Expanded Photographic Optic Control Center using a new control system. The KTM trackers continue to be refurbished with a total system upgrade completion date in mid-fiscal year 2008. The imagery data products from two trackers were eliminated due to budgetary constraints. The Cocoa Beach Tracker was moved about 3 miles north due to limited visibility early in the 51.6-degree ascent trajectory. The USAF-operated Eastern Test Range assets are being upgraded to replace the standard definition cameras with high-definition cameras.

SCHEDULE

Responsibility	Due Date	Activity/Deliverable
SSP	Aug 03 (Completed)	Program Approval of Ground Camera Upgrade Plan
SSP	Sep 03 (Completed)	Program Approval of funding for Ground Camera Upgrade Plan
SSP	Feb 04 (Completed)	Baseline Program Requirements/Document Requirements for additional camera locations
SSP	May 04 (Completed)	Begin refurbishment of 14 existing trackers. Will be ongoing until refurbishment of all trackers is complete (expected 2008). Trackers and Optics will be borrowed from other ranges to support launch until the assets are delivered
SSP	Jul 04 (Completed)	Critical Design Review for WAVE airborne imaging system
SSP	Mar 05 (Completed)	Install new optics and cameras
SSP	May 05 (Completed)	Baseline revised Launch Commit Criteria

Columbia Accident Investigation Board
Recommendation 3.4-2

Provide a capability to obtain and downlink high-resolution images of the External Tank after it separates. [RTF]

Note: The Stafford-Covey Return to Flight Task Group held a plenary session on December 15, 2004, and NASA's progress toward answering this recommendation was reviewed. The Task Group agreed the actions taken were sufficient to fully close this recommendation.

BACKGROUND

NASA agrees that it is critical to verify the performance of the External Tank (ET) modifications to eliminate ascent debris. Real-time downlink of this information may help in the early identification of some risks to flight. Prior to Return to Flight, the Space Shuttle had two on-board high-resolution cameras that photographed the ET after separation; however, the images from these cameras were available only postflight and were not downlinked to the Mission Control Center during the mission. Therefore, no real-time imaging of the ET was available to provide engineering insight into potential debris during the mission.

NASA IMPLEMENTATION

To provide the capability to downlink images of the ET after separation for analysis, NASA replaced the 35mm film camera in the Orbiter umbilical well with a high-resolution digital camera and equipped the flight crew with a handheld digital still camera with a telephoto lens. Umbilical and handheld camera images are downlinked after safe orbit operations are established. These images are used for quick-look analysis by the Mission Management Team to determine if any ET anomalies exist that require additional on-orbit inspections (see Recommendation 6.4-1).

STATUS

Fabrication, certification, and installation of the digital umbilical camera system were completed prior to STS-114. Installation of data transfer cabling is complete in all three Orbiters (figure 3.4-2-1).

FINAL UPDATE

The digital umbilical well camera worked extremely well during the STS-114 and STS-121 launches, and all images of the ET were successfully downlinked during the missions. The data from these cameras were critical in assessing the status of the ET post-separation; these cameras will be flown on all future Space Shuttle flights. The digital hand-held pictures also were of exceptional quality, and the crew was commended for its outstanding efforts. A change to add a flash to the digital umbilical well camera is planned for STS-123 (which is scheduled to launch in fiscal year 2008). This will provide a capability to view the external tank for night launches.

SCHEDULE

Responsibility	Due Date	Activity/Deliverable
Space Shuttle Program (SSP)	Nov 03 (Completed)	Orbiter umbilical well digital camera feasibility study
SSP	Apr 04 (Completed)	Preliminary design review/critical design review
SSP	Jan 05 (Completed)	OV-103 and OV-104 Orbiter umbilical well camera wiring and support structure installation
SSP	Sep 05 (Completed)	OV-105 Orbiter umbilical well camera wiring and support structure installation
SSP	Mar 05 (Completed)	Camera system functional testing
SSP	Launch −6 weeks	Installation of digital umbilical well camera prior to each flight

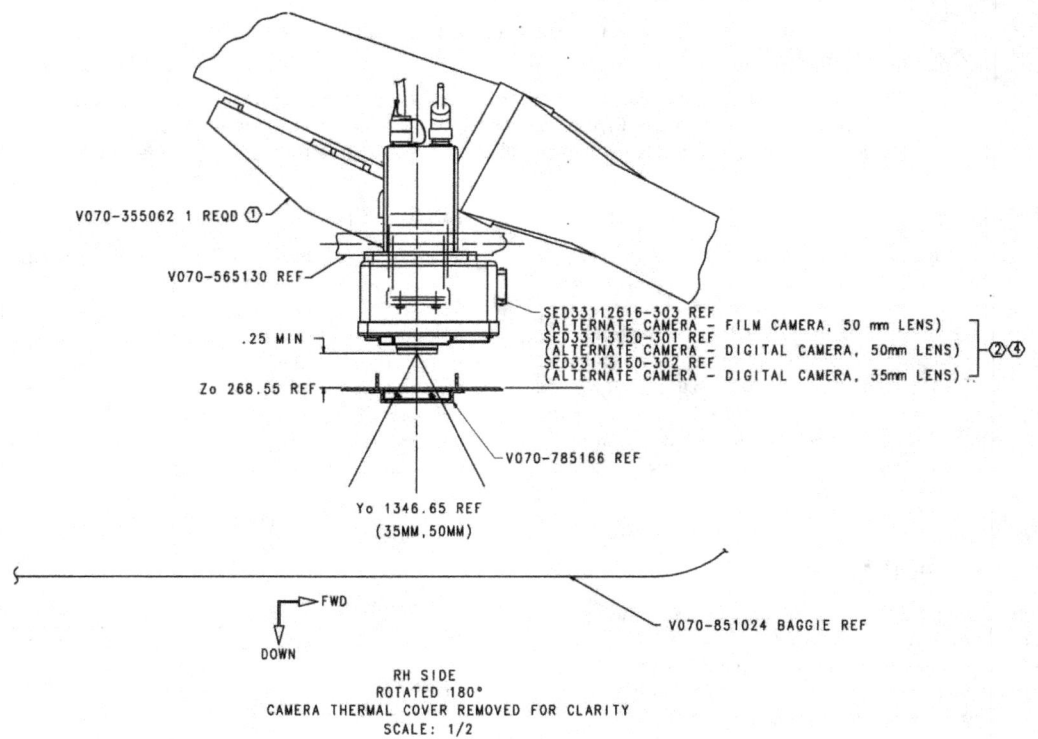

Figure 3.4-2-1. Schematic of umbilical well camera.

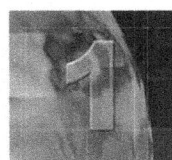

Columbia Accident Investigation Board
Recommendation 3.4-3

Provide a capability to obtain and downlink high-resolution images of the underside of the Orbiter wing leading edge and forward section of both wings' Thermal Protection System. [RTF]

Note: The Stafford-Covey Return to Flight Task Group held a plenary session on June 8, 2005, and NASA's progress toward answering this recommendation was reviewed. The Task Group agreed the actions taken were sufficient to fully close this recommendation.

BACKGROUND

The damage to the left wing of *Columbia* occurred shortly after liftoff, but went undetected for the entire mission. Although there was ground photographic evidence of debris impact, we were unaware of the extent of the damage. Therefore, NASA is adding on-vehicle cameras and sensors that will help to detect and assess damage.

NASA IMPLEMENTATION

For the first few missions after Return to Flight, NASA used primarily on-orbit inspections to meet the requirement to assess the health and status of the Orbiter's Thermal Protection System (TPS). (Details of our on-orbit inspections can be found in Recommendation 6.4-1.) This is because the on-vehicle ascent imagery suite does not provide complete imagery of the underside of the Orbiter or guarantee detection of all potential impacts to the Orbiter. However, on-vehicle ascent imagery is a valuable source of engineering, performance, and environments data and is useful for understanding in-flight anomalies. NASA's long-term strategy includes improving on-vehicle ascent imagery.

For STS-114, NASA had cameras on the External Tank (ET) liquid oxygen (LO_2) feedline fairing and the Solid Rocket Booster (SRB) forward skirt (figure 3.4-3-1). The ET LO_2 feedline fairing camera took images of the ET bipod areas and the underside of the Space Shuttle fuselage and the right wing from liftoff through the first 10 minutes of flight. The new location of the ET camera eliminated the likelihood that its views would be obscured by the booster separation motor (BSM) plume, a discrepancy observed on STS-112. These images were transmitted real time to ground stations.

The SRB forward skirt cameras took images from three seconds to 350 seconds after liftoff. These two cameras looked sideways at the ET intertank. The images from this location were stored on the SRBs and available after the SRBs were recovered, approximately three days after launch.

Beginning with STS-121, NASA introduced an additional complement of cameras on the SRBs: aft-looking cameras located on the SRB forward skirt and forward-looking cameras located on the SRB External Tank Attachment (ETA) Ring (figure 3.4-3-2). Together, these additional cameras provided comprehensive views of the Orbiter's underside during prelaunch and ascent.

STATUS

The Program Requirements Control Board approved the Level II requirements for the on-vehicle ascent camera system that were implemented for Return to Flight.

Because both on-vehicle cameras during ascent and on-orbit inspection are required to provide a complete picture of the status of the Orbiter's TPS, this recommendation was considered for closure by the Stafford-Covey Return to Flight Task Group in conjunction with Recommendation 6.4-1, Thermal Protection System On-Orbit Inspection and Repair.

FINAL UPDATE

All of the new vehicle-mounted cameras worked as expected during the STS-114 and STS-121 launches. The additional cameras flown on STS-121 provided astounding views of ascent, the Orbiter, the ET, and SRB separation. The final flight complement of vehicle-mounted cameras is in place.

NASA continues to research options to improve camera resolution, functionality in reduced lighting conditions, and alternate camera mounting configurations within schedule and budgeted constraints.

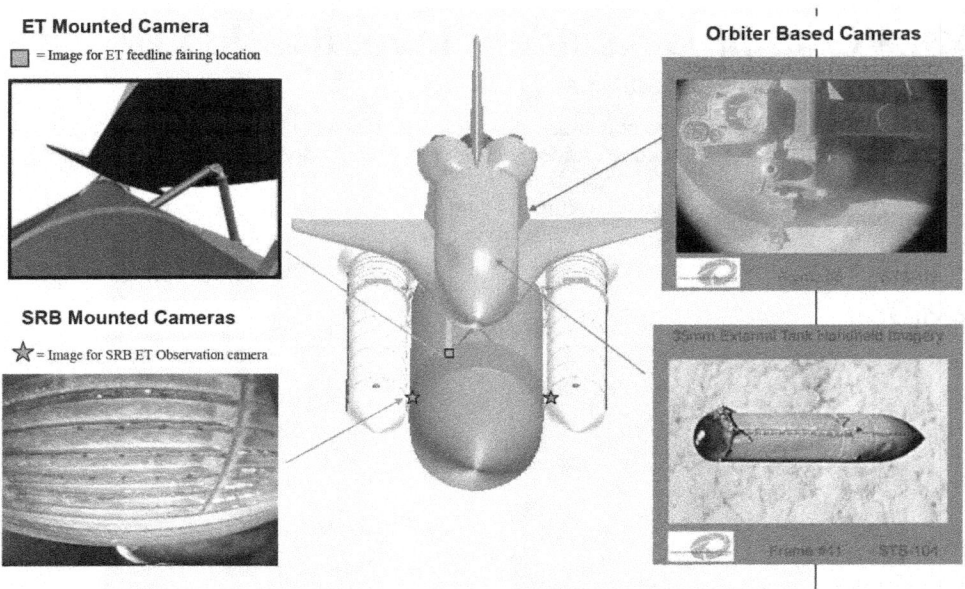

Figure 3.4-3-1. ET flight cameras (STS-114 configuration).

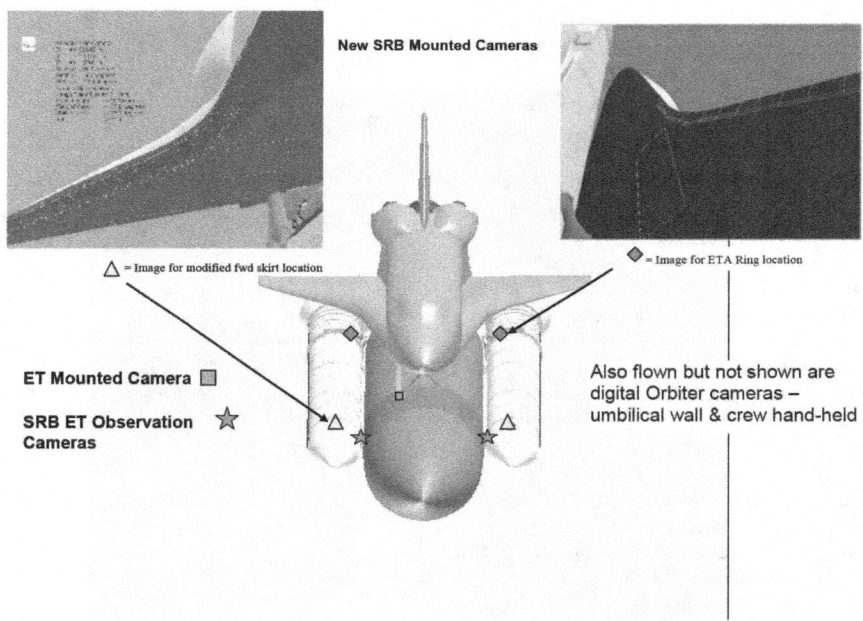

Figure 3.4-3-2. ET flight cameras (STS-121 configuration).

SCHEDULE

Responsibility	Due Date	Activity/Deliverable
Space Shuttle Program (SSP)	May 03 (Completed)	Start ET hardware modifications
SSP	Jul 03 (Completed)	Authority to proceed with ET LO_2 feedline and SRB forward skirt locations; implementation approval for ET camera
SSP	Mar 04 (Completed)	Systems Requirements Review
SSP	Jun 04 (Completed)	Begin ET camera installations
SSP	Oct 04 (Completed)	Begin SRB "ET Observation" camera installation
SSP	Sep 05 (Completed)	Review SRB camera enhancements for mission effectivity

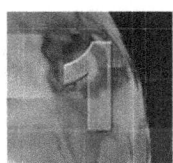

Columbia Accident Investigation Board
Recommendation 6.3-2

Modify the Memorandum of Agreement with the National Imagery and Mapping Agency (NIMA) to make the imaging of each Shuttle flight while on orbit a standard requirement. [RTF]

Note: The Stafford-Covey Return to Flight Task Group held a plenary session on December 15, 2004, and NASA's progress toward answering this recommendation was reviewed. The Task Group agreed the actions taken were sufficient to fully close this recommendation.

BACKGROUND

The *Columbia* Accident Investigation Board (CAIB) found, and NASA concurs, that the full capabilities of the United States to assess the condition of the *Columbia* during STS-107 should have been used but were not.

NASA IMPLEMENTATION

NASA has concluded a Memorandum of Agreement (MOA) with the National Imagery and Mapping Agency (subsequently renamed the National Geospatial-Intelligence Agency [NGA]) that provides for on-orbit assessment of the condition of each Orbiter vehicle as a standard requirement. In addition, NASA has completed discussions with other agencies to explore the use of appropriate national assets to evaluate the condition of the Orbiter vehicle. Additional agreements have also been signed. The operational teams have developed standard operating procedures to implement agreements with the appropriate government agencies at the Headquarters level.

NASA has determined which positions/personnel will require access to data obtained from external sources. NASA has ensured that all personnel are familiar with the general capabilities available for on-orbit assessment and that the appropriate personnel have the appropriate clearances and have access to that information. Testing and validation of these new processes and procedures was accomplished through simulations conducted prior to the first Return to Flight mission, STS-114. Since these actions involved receipt and handling of classified information, the appropriate security safeguards were observed during its implementation.

In April 2004, the Stafford-Covey Return to Flight Task Group reviewed NASA's progress and agreed to conditionally close this recommendation. The full intent of CAIB Recommendation 6.3-2 has been met and full closure of this recommendation was achieved in December 2004.

FINAL UPDATE

The processes were fully exercised from STS-114 onward, and the results were excellent. The security clearances were adequate and are being adjusted as people change positions.

SCHEDULE

An internal NASA process is being used to track clearances, training of personnel, and the process validation.

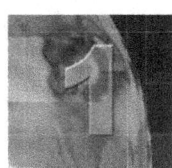

Columbia Accident Investigation Board
Recommendation 3.6-1

The Modular Auxiliary Data System instrumentation and sensor suite on each Orbiter should be maintained and updated to include current sensor and data acquisition technologies.

Note: NASA has closed this recommendation through the formal Program Requirements Control Board process. The following summary details NASA's response to the recommendation and any additional work NASA performed beyond the *Columbia* Accident Investigation Board (CAIB) recommendation.

BACKGROUND

The Modular Auxiliary Data System (MADS), which is also referred to in the CAIB Report as the "OEX recorder," is a platform for collecting engineering performance data. The MADS records data that provide the engineering community with information on the environment experienced by the Orbiter during ascent and entry, and with information on how the structures and systems responded to this environment. The Space Shuttle Program (SSP) requirements have been updated to include MADS.

The MADS hardware is 1970's technology and is difficult to maintain. NASA has recognized the problem with its sustainability for some time. The available instrumentation hardware assets can only support the existing sensor suite in each Orbiter. If any additional sensors are required, their associated hardware must be procured.

NASA IMPLEMENTATION

The SSP agrees that MADS needs to be maintained. The SSP approved the incorporation of the MADS subsystem into the Program requirements documentation. The Instrumentation Problem Resolution Team (PRT) reviews sensor requirements and post-flight data for various Orbiter systems to determine appropriate action for sensors. The PRT also ensures proper maintenance of the current MADS hardware. NASA has acquired MADS wideband instrumentation tape and certified it for flight. This will extend the operational availability of the MADS recorder. NASA has also extended the recorder maintenance and skills retention contract with the MADS vendor, Sypris.

FINAL UPDATE

The SSP maintains and updates as necessary the current MADS, including flight hardware, ground support equipment, and sensor and data acquisition components.

SCHEDULE

Complete.

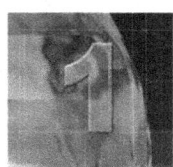

Columbia Accident Investigation Board
Recommendation 3.6-2

The Modular Auxiliary Data System should be redesigned to include engineering performance and vehicle health information and have the ability to be reconfigured during flight in order to allow certain data to be recorded, telemetered, or both, as needs change.

Note: NASA has closed this recommendation through the formal Program Requirements Control Board (PRCB) process. The following summary details NASA's response to the recommendation and any additional work NASA performed beyond the *Columbia* Accident Investigation Board (CAIB) recommendation.

BACKGROUND

The Modular Auxiliary Data System (MADS)* provides limited engineering performance and vehicle health information postflight. There are two aspects to this recommendation: (1) redesign for additional sensor information, and (2) redesign to provide the ability to select certain data to be recorded and/or telemetered to the ground during the mission. To meet these recommendations, a new system would have to be developed to replace MADS. This replacement would address system obsolescence issues and also provide additional capability.

The Space Shuttle Program (SSP) also baselined a requirement to add additional vehicle health monitoring capability. These capabilities increase the insight into the Orbiter's Thermal Protection System.

NASA IMPLEMENTATION

Initially, NASA planned to address the enhanced requirements for MADS through a new Vehicle Health Maintenance System (VHMS), which was part of the suite of upgrades comprising the Shuttle Service Life Extension Program. In January 2004, the Vision for Space Exploration was announced. The Vision refocused the mission of the SSP on support for and assembly of the International Space Station (ISS), and called for the retirement of the Space Shuttle following ISS assembly complete at the end of the decade. As a result of this Program reorientation and the focus on returning safely to flight following the loss of the *Columbia* and her crew, the SSP reevaluated its Program priorities. As a part of this reevaluation, the Shuttle Program reviewed its commitment to the VHMS upgrade and determined that it was not a high-priority investment. VHMS would have expanded the Shuttle's capability to monitor new instrumentation and telemeter the resulting data, but did not address a specific safety concern. Rather it was designed to improve engineering insight into the Space Shuttle's condition during a mission.

Instead of developing and installing a new VHMS system, the Orbiters have been modified to provide low-rate MADS digital data that are available for downlink during on-orbit operations. These low-rate data include temperature, strain gauge, and pressure sensors already installed in unique locations specific to each Orbiter. In addition, there are other non-MADS instrumentation systems that collect more vehicle health data. For instance, the Wing Leading Edge Impact Detection System (WLEIDS) collects temperature and possible impact data along the Orbiter's right and left wing leading edge structure. Data from the WLEIDS are downlinked during on-orbit operations.

FINAL UPDATE

The low-rate MADS digital data modification and WLEIDS is installed on all Orbiter vehicles. The WLEIDS, in particular, has enabled engineers to successfully monitor for possible impacts to the leading edge during ascent and while on-orbit for all missions since Return to Flight. The SSP will continue to assess the data collection requirements for the integrated vehicle and the Orbiter, and will provide status updates to the PRCB.

SCHEDULE

Complete.

---

*Note that the CAIB Report alternately refers to this as the OEX Recorder.

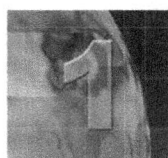
Columbia Accident Investigation Board
Recommendation 4.2-2

As part of the Shuttle Service Life Extension Program and potential 40-year service life, develop a state-of-the-art means to inspect all Orbiter wiring, including that which is inaccessible.

Note: With the establishment of a new national policy for U.S. space exploration in January 2004, the planned service life of the Space Shuttle was shortened. The Space Shuttle will be used to conduct a fifth servicing mission to the Hubble Space Telescope and complete assembly of the International Space Station (planned for the end of the decade), and then the Shuttle will be retired. Due to the reduced service life, NASA's approach to complying with this recommendation has been appropriately adjusted. These actions were closed through the formal Program Requirements Control Board (PRCB) process. The following summary details NASA's response to the recommendation and any additional work NASA performed beyond the *Columbia* Accident Investigation Board (CAIB) recommendation.

BACKGROUND

A significant amount of Orbiter wiring is insulated with Kapton, a polyimide film used as electrical insulation. Kapton-insulated wire has many advantages; however, over the years several concerns have been identified and addressed by the Space Shuttle Program (SSP) through both remedial and corrective actions.

One of these ongoing concerns, arc tracking, was highlighted during STS-93 as a result of a short circuit in the wiring powering one of the channels of the Space Shuttle Main Engine controllers. Arc tracking is a known failure mode of Kapton wiring in which the electrical short can propagate along the wire and to adjacent wiring. Following STS-93, NASA initiated an extensive wiring investigation program to identify and repair/replace discrepant wiring. NASA also initiated a program of Critical Wire Separation efforts. This program separated redundant critical function wires that were collocated in a single wire bundle into separate wire bundles to mitigate the risk of an electrical short on one wire arc tracking to an adjacent wire and resulting in the total loss of a system. In areas where complete separation was not possible, inspections are being performed to identify discrepant wire, repair/replace it, and to protect against damage that may lead to arc tracking. In addition, abrasion protection (convoluted tubing or Teflon wrap) is being added to wire bundles that carry circuits of specific concern and/or are routed through areas of known high damage potential.

The STS-93 wiring investigation also led to improvements in the requirements for wiring inspections, wiring inspection techniques, and wire awareness training of personnel working in the vehicle. Wiring was inspected, separated, and protected in the accessible areas during the general flight-to-flight Operations and Maintenance Requirements Specification Document (OMRSD) process. The wiring that was inaccessible during the OMRSD process was inspected, separated, and protected during the Orbiter Maintenance Down Period.

Currently, visual inspection is the most effective means of detecting wire damage. Technology-assisted techniques such as Hipot, a high-potential dielectric verification test, and time domain reflectometry (TDR), a test that identifies changes in the impedance between conductors, are rarely effective for detecting damage that does not expose the conductor or where a subtle impedance change is present. Neither is an effective method for detecting subtle damage to wiring insulation. However, for some areas, visual inspection is impractical. The Orbiters contain some wire runs, such as those installed beneath the crew module, that are completely inaccessible to inspectors during routine ground processing. Even where wire is installed in accessible areas, not every wire segment is available for inspection due to bundling and routing techniques. However, the results of wire inspections, particularly since STS-93, have shown that the vast majority of wire damage is caused by maintenance workers accessing and working in areas where wire bundles are present. Areas that must be accessed for normal flight-to-flight processing, such as the payload bay or the environmental control systems bay, are particularly vulnerable.

NASA IMPLEMENTATION

NASA initially took a broad approach to mitigating Orbiter wiring concerns by evaluating promising new technologies for nondestructive evaluation (NDE) of wires, benchmarking with the practices of other government agencies, improving its visual wire inspection techniques, and creating a study group to recommend improvements to wiring issues.

NASA's initial work on NDE involved the Ames Research Center (ARC), where engineers were developing a proposed Hybrid Reflectometer, a TDR derivative, to detect defects in wiring. At the Langley Research Center (LaRC) engineers were developing a wire insulation age-life tester and an ultrasonic crimp joint tool to measure the integrity of wire crimps as they are made. At the Johnson Space Center (JSC) engineers were evaluating a destructive age-life test capability.

Prior to the articulation of the Vision for Space Exploration, NASA was particularly interested in the issue of aging wiring as a part of the Shuttle Service Life Extension Program to the year 2020 and potential 40-year service life of the Orbiters. Military and civilian aircraft are also frequently flown beyond their original design lives. NASA began an effort to benchmark with industry, academia, and other government agencies to find the most effective means to address the aging wiring concerns. Examples are NASA's participation on the Joint Council for Aging Aircraft and its collaboration with the Air Force Research Laboratory.

To improve inspection techniques, the SSP more clearly defined requirements for Category I Inspections (cutting the minimum wire ties needed to perform repair/replacement, opening up bundles, and spreading out and inspecting the additional wires made available) and Category II Inspections (inspecting bundle periphery with 10× magnification, and opening bundles if damage was noted). The Program also planned to update a previous Boeing study that evaluated types of wire insulation other than Kapton, planned to identify and map "inaccessible" wiring, and considered potential wire replacement.

Finally, the SSP assigned an action to the Orbiter Project Office to research, evaluate, and present a comprehensive list of options to address the wiring issue in general and CAIB Recommendation 4.2-2 specifically. An Orbiter Wiring Working Group composed of engineers from SSP, JSC, and Kennedy Space Center (KSC) Engineering, United Space Alliance, and Boeing began this evaluation in 2003.

STATUS

In January 2004, a new national policy for U.S. Space Exploration was established and the planned life of the Space Shuttle was shortened. Following its Return to Flight, the Space Shuttle will be used to complete assembly of the International Space Station, planned for the end of the decade, and then the Space Shuttle will be retired. Due to this reduced service life, NASA's approach to complying with CAIB Recommendation 4.2-2 was appropriately adjusted.

On June 17, 2004, the Orbiter Wiring Working Group presented to the PRCB a four-prong, two-phase approach to address wiring issues and respond to CAIB Recommendation 4.2-2. The four prongs or options were: (1) inspect and protect, by continuing to improve upon current wiring inspections and activities at KSC; (2) invest in the development of NDE, including a wire insulation tester, a wire age life tester, and an ultrasonic wire crimp tool; (3) perform destructive evaluations to determine whether the Orbiter wiring does, in fact, show aging effects that are of concern; and (4) evaluate wire replacement for the Orbiters. The two phases related to NDE were Phase I – Proof of Concept and Phase II – Delivery of a Working Unit.

In light of the reduced service life of the Orbiter, the PRCB approved option (1), inspect and protect, and option (3), perform destructive evaluations. Options (2) and (4) were not approved and, as a consequence, further NDE work at the ARC and LaRC is no longer being funded by the SSP. The investment in NDE in option (2) was felt to offer little return on investment considering the relatively low technology readiness level of wiring NDE techniques. Also, few remaining flights could make use of the new NDE due to the time required to develop, test, and field operational units. In view of the planned retirement of the Space Shuttles in 2010, replacing Orbiter wiring was assessed as not cost effective.

In contrast, the inspect and protect approach continues with wiring damage corrective actions that have been in place since the post-STS-93 wiring efforts, including lessons learned to date. NASA also chartered the Orbiter Wiring Team to evaluate a wiring destructive testing program to better characterize the specific vulnerabilities of Orbiter wiring to aging and damage, and to predict future wiring damage, particularly in inaccessible areas.

The Aging Orbiter Working Group (AOWG) Wiring Subteam presented their findings to the SSP in November 2005. It concluded that current research suggests that aromatic polyimide does not degrade severely due to "aging time" in benign environments, there is no evidence that the wire insulation will undergo significant degradation in Orbiter-controlled service environments, and there are no proven destructive testing methods that would guarantee a complete resolution of the aging issue. The AOWG recommended that any wire that was removed from a vehicle through normal rework be routed to the AOWG chairperson to be sent to a materials and processes lab for failure analysis. The SSP agreed and directed the contractor to do so.

To formalize wiring inspection improvements, NASA revised Specification ML0303-0014, "Installation Requirements for Electrical Wire Harnesses and Coaxial Cables," with improved guidelines for wire inspection procedures and protection protocols. A new Avionics Damage Database was implemented to capture statistical data to NASA's ability to analyze and predict wiring damage trends. NASA also initiated an aggressive wire damage awareness program that limits the number of people given access to Orbiter areas where wiring can be damaged. In addition, specific training is now given to personnel who require entry to areas that have a high potential for wiring damage. This training has already helped raise awareness and reduce unintended processing damage.

FINAL UPDATE

To continually improve our understanding of wiring issues for the remaining service life of the Space Shuttle, information and technical exchanges dealing with aging wiring issues will continue between the SSP, NASA research centers, and other agencies such as the Federal Aviation Administration and the Department of Defense.

The SSP will continue to evaluate the risk of aging/damaged wiring against the other major risk drivers in the Program, within the constraints of current technical capabilities, and given the Shuttle's planned retirement at the end of the decade.

SCHEDULE

Responsibility	Due Date	Activity/Deliverable
SSP	Apr 04 (Completed)	Present project plan to the PRCB
Orbiter Project Office	Apr 05 (Completed)	Present findings of Orbiter Wiring Working Group

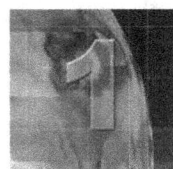

Columbia Accident Investigation Board
Recommendation 4.2-1

Test and qualify the flight hardware bolt catchers. [RTF]

Note: The Stafford-Covey Return to Flight Task Group held a plenary session on December 15, 2004, and NASA's progress toward answering this recommendation was reviewed. The Task Group agreed the actions taken were sufficient to fully close this recommendation.

BACKGROUND

The External Tank (ET) is attached to the Solid Rocket Boosters (SRBs) at the forward skirt thrust fitting by the forward separation bolt. The pyrotechnic bolt is actuated at SRB separation by fracturing the bolt in half at a predetermined groove, releasing the SRBs from the ET thrust fittings. The bolt catcher attached to the ET fitting retains the forward half of the separation bolt. The other half of the separation bolt is retained within a cavity in the forward skirt thrust post (figure 4.2-1-1).

The STS-107 bolt catcher design consisted of an aluminum dome welded to a machined aluminum base bolted to both the left- and right-hand ET fittings. The inside of the bolt catcher was filled with a honeycomb energy absorber to decelerate the ET half of the separation bolt (figure 4.2-1-2).

Static and dynamic testing demonstrated that the manufactured lot of bolt catchers that flew on STS-107 had a factor of safety of approximately 1. The factor of safety for the bolt catcher assembly should be 1.4.

NASA IMPLEMENTATION

NASA determined that the bolt catcher assembly and related hardware needed to be redesigned and qualified by testing as a complete system to demonstrate compliance with factor-of-safety requirements.

NASA completed the redesign of the bolt catcher assembly, the redesign and resizing of the ET attachment bolts and inserts, the testing to characterize the energy absorber material, and the testing to determine the design loads.

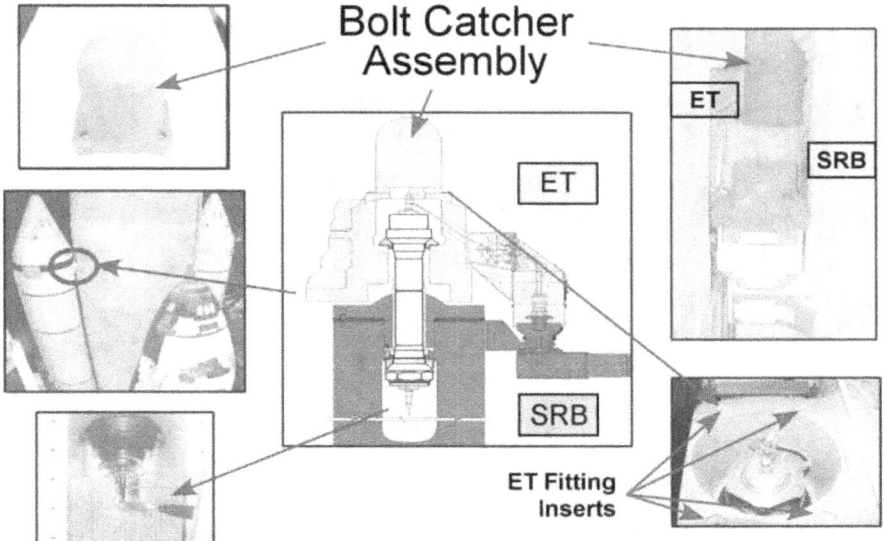

Figure 4.2-1-1. SRB/ET forward attach area.

Bolt catcher energy absorber

Bolt catcher energy absorber after bolt impact

Figure 4.2-1-2. Bolt catcher impact testing.

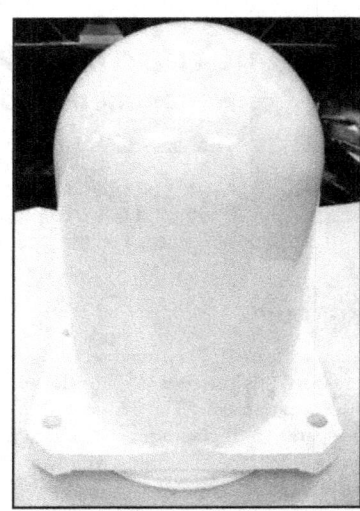

Figure 4.2-1-3. New one-piece forging design.

The redesigned bolt catcher housing is fabricated from a single piece of aluminum forging (figure 4.2-1-3) that removes the weld from the original design (figure 4.2-1-4). Further, new energy-absorbing material and thermal protection material was used (figure 4.2-1-4), and the ET attachment bolts and inserts (figure 4.2-1-5) were redesigned and resized.

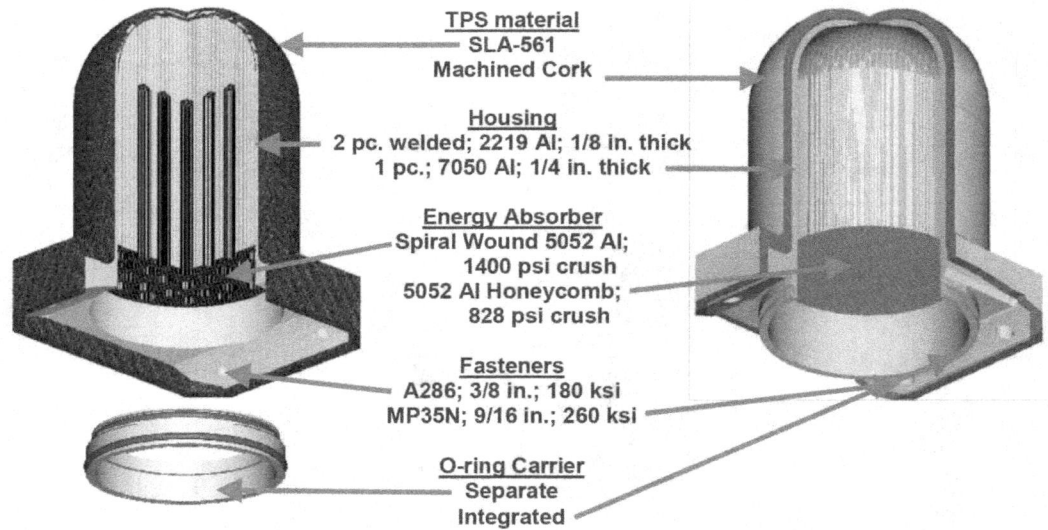

STS-107 Bolt Catcher Design | Final Bolt Catcher Redesign

TPS material
SLA-561
Machined Cork

Housing
2 pc. welded; 2219 Al; 1/8 in. thick
1 pc.; 7050 Al; 1/4 in. thick

Energy Absorber
Spiral Wound 5052 Al;
1400 psi crush
5052 Al Honeycomb;
828 psi crush

Fasteners
A286; 3/8 in.; 180 ksi
MP35N; 9/16 in.; 260 ksi

O-ring Carrier
Separate
Integrated

Figure 4.2-1-4. Old and new bolt catcher design comparison.

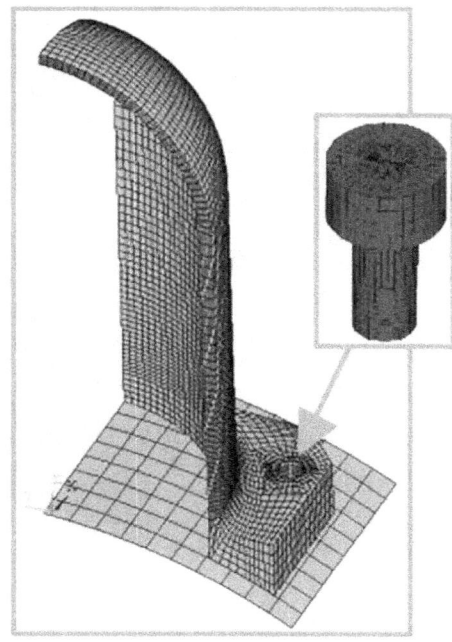

Figure 4.2-1-5. ET bolt/insert finite element model.

STATUS

Structural qualification to demonstrate that the assembly complies with the 1.4 factor-of-safety requirement is complete. Cork has been selected as the Thermal Protection System (TPS) material for the bolt catcher. TPS qualification testing is complete including weather exposure followed by combined environment testing, which includes vibration, acoustic, thermal, and pyrotechnic shock testing.

FINAL UPDATE

Postflight evaluation of the redesigned SRB bolt catcher was performed through review of ascent imagery (both visual and radar) recorded at the time of, and following, SRB separation. The ascent imagery that was analyzed for this evaluation resides on the Marshall Space Flight Center Engineering Photographic Analysis Web site, allowing NASA-wide access to these data.

Photographic evaluation revealed that the integrity of the bolt catcher dome machined cork TPS exhibited no signs of damage, which would appear as a color contrast on the video imagery if ablation or spalling of the white Hypalon topcoat revealed the underlying machined cork. Although some expected discoloration of the Hypalon paint was indicated due to aeroheating, there was no indication of loss of TPS.

Imagery analysis of the bolt catchers indicates that the SRB and ET forward separation bolt was retained in the SRB bolt catcher, and there was no indication of unexpected debris generated. This evaluation indicates that the redesigned SRB bolt catchers performed successfully on the STS-114 and STS-121 missions.

The redesigned SRB bolt catcher has performed successfully since its introduction on STS-114.

SCHEDULE

Responsibility	Due Date	Activity/Deliverable
Space Shuttle Program (SSP)	May 04 (Completed)	Complete Critical Design Review
SSP	Oct 04 (Completed)	Complete Qualification
SSP	Feb 05 (Completed)	First Flight Article Delivered

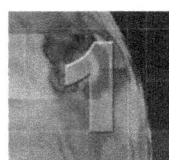

Columbia Accident Investigation Board
Recommendation 4.2-3

Require that at least two employees attend all final closeouts and intertank area hand-spraying procedures. [RTF]

Note: The Stafford-Covey Return to Flight Task Group held a plenary session on December 15, 2004, and NASA's progress toward answering this recommendation was reviewed. The Task Group agreed the actions taken were sufficient to fully close this recommendation.

BACKGROUND

External Tank (ET) final closeouts and intertank area hand-spraying processes typically require more than one person in attendance to execute procedures. At the time of *Columbia*, those closeout processes that were performed by a single person did not necessarily specify an independent witness or verification.

NASA IMPLEMENTATION

NASA has established a Thermal Protection System (TPS) verification team to verify, validate, and certify all future foam processes. The verification team assessed and improved the TPS applications and manual spray processes. Included with this assessment was a review and an update of the process controls applied to foam applications, especially the manual spray applications. Spray schedules, acceptance criteria, quality, and data requirements were established for all processes during verification using a Material Processing Plan (MPP). The plan defined how each specific part closeout will be processed. Numerous TPS processing parameters and requirements were enhanced, including additional requirements for observation and documentation of processes. In addition, a review was conducted to ensure the appropriate quality coverage based on process enhancements and critical application characteristics.

The MPPs were revised to require, at a minimum, that all ET critical hardware processes, including all final closeouts and intertank area hand-spray procedures, are performed in the presence of two certified Production Operations employees. The MPPs were also revised to include a step that requires technicians to stamp the build paper to verify their presence, and to validate that work was performed according to plan. Additionally, quality control personnel witness and accept each manual spray TPS application. Government oversight of TPS applications is determined upon completion of the revised designs and the identification of critical process parameters.

In addition to these specific corrective measures taken by the ET Project, in March 2004 the Space Shuttle Program (SSP) widened the scope of this corrective action in response to a recommendation from the Return to Flight Task Group (RTFTG). The scope was widened to include all flight hardware projects. An audit of all final closeouts is performed to ensure compliance with the existing guidelines that a minimum of two persons witness final flight hardware closures for flight for both quality assurance and security purposes.

The audits included participation from Project engineers, technicians, and managers. The following were used to complete the audit: comprehensive processing and manufacturing reviews, which included detailed work authorization and manufacturing document appraisals, and on-scene checks.

STATUS

The revised TPS verification activities are complete, and specific applicable ET processing procedures have been changed.

All major flight hardware elements (Orbiter, ET, Solid Rocket Booster, Solid Rocket Motor, extravehicular activity, vehicle processing, and main engine) have concluded their respective audits as directed by the March 2004 SSP initiative. The results of the audits were presented to the Program Manager on May 26, 2004. The two-person closeout guideline was previously well-established in the SSP and largely enforced by multiple overlapping quality assurance and safety requirements.

In April 2004, the Stafford-Covey Return to Flight Task Group reviewed NASA's progress and agreed to conditionally close this recommendation. The full intent of CAIB Recommendation 4.2-3 has been met and full closure of this recommendation was achieved in December 2004.

FINAL UPDATE

As a result of the audits, several SSP projects identified and addressed specific processing or manufacturing steps to extend this guideline beyond current implementation and also identified areas where rigorous satisfaction of this guideline could be better documented. Those changes are now complete. Changes to Program-level documents are also complete, including the requirement for the projects and elements to have a minimum of two people witness final closeouts of major flight hardware elements.

SCHEDULE

Responsibility	Due Date	Activity/Deliverable
ET	Dec 03 (Completed)	Review revised processes with RTFTG
All flight hardware elements	May 04 (Completed)	Audit results of all SSP elements due
ET	May 04 (Completed)	Assessment of Audit Results
SSP	May 04 (Completed)	SSP element audit findings presented to SSP Manager
SSP	Jun 04 (Completed)	Responses due; PRCB action closed
SSP	Jan 05 (Completed)	Revised requirements formally documented

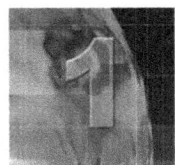

Columbia Accident Investigation Board
Recommendation 4.2-4

Require the Space Shuttle to be operated with the same degree of safety for micrometeoroid and orbital debris as the degree of safety calculated for the International Space Station. Change the micrometeoroid and orbital debris safety criteria from guidelines to requirements.

Note: NASA has closed this recommendation through the formal Program Requirements Control Board process. The following summary details NASA's response to the recommendation and any additional work NASA performed beyond the *Columbia* Accident Investigation Board recommendation.

BACKGROUND

Micrometeoroid and orbital debris (MMOD) is a continuing concern. The current differences between the International Space Station (ISS) and Orbiter MMOD risk allowances for a critical debris impact are based on the original design specifications for each of the vehicles. The ISS was designed for long-term MMOD exposure, whereas the Orbiter was designed for short-term MMOD exposure.

The debris impact factors that are considered when determining the MMOD risks for a spacecraft are mission duration, attitude(s), altitude, inclination, year, damage limits, damage tolerance, and on-board payloads. Current Orbiter impact damage guidelines dictate that there will be no more than a 1 in 200 risk for loss of vehicle from MMOD impact for any single mission. This guideline is comparable to the ISS MMOD risk assessments of 0.7 percent annual catastrophic risk from MMOD debris impact. Assuming five Space Shuttle flights per year, complying with the ISS standard would require an Orbiter annual average MMOD damage risk for loss of vehicle of 1 in 700 per mission.

NASA uses hypervelocity impact tests on representative Shuttle Thermal Protection System (TPS) and structure samples, and a computer modeling tool called the BUMPER Code to assess the risk from MMOD impact to the Orbiter during each flight. The risk assessments take into account mission duration, attitude variations, altitude, recent debris-generating breakups, meteor showers and storms, and other mitigating factors such as on-orbit inspection of TPS for potentially critical damage and the capability to repair TPS if critical damage is detected. Mitigation techniques to reduce MMOD risks are identified and evaluated based on risk assessment results.

NASA IMPLEMENTATION

Because the Space Shuttle will be retired in 2010, the short time remaining in the Program limits NASA's ability to implement the wide range of mitigations necessary to fully comply with this recommendation and reduce the risk of loss due to MMOD damage to 1 in 700 per year. However, NASA has implemented a number of changes to the MMOD risk assessment process to make it more realistic, and to Shuttle operations to reduce MMOD risk.

NASA made a significant improvement to the accuracy of Shuttle MMOD risk assessments by reducing the allowable limits of damage to the wing leading edge (WLE) and nose cap. The new WLE and nose cap damage limits for MMOD damage are shown in figure 4.2-4-1. This figure shows the damage limit as a function of location on the WLE and nose cap, the maximum size of damage that is allowed, and typical orbital debris particle diameter that would cause the damage. The WLE/nose cap damage limit was determined based on arc-jet tests of hypervelocity-impacted WLE/nose cap samples, analysis of damage growth during entry, internal wing heating models including plasma ingestion, and criteria establishing over-temperature limits for internal insulation-protecting critical structures. The figure below indicates the undersides of the WLE and nose cap are vulnerable to small damage that can lead to burn-through during entry and over-temperature of critical structure. The failure criteria prior to the *Columbia* accident allowed a greater amount of damage to the underside of the WLE and nose cap.

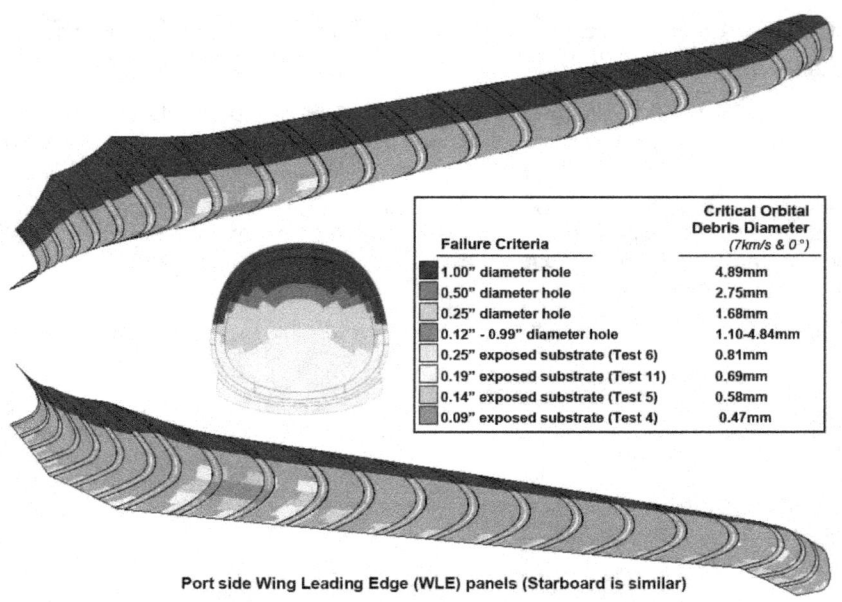

Figure 4.2-4-1. WLE and nose cap damage limit from MMOD impact to ensure safe entry.

In addition to improving the accuracy of its risk assessments, NASA implemented two changes to vehicle operations that reduce the risk from MMOD impact. First, NASA changed the orientation of Orbiter while docked to the ISS to reduce exposure to MMOD. Second, NASA implemented a late inspection regime during missions to identify MMOD damage sustained during docked operations. These two changes are described in detail below.

Just after Shuttle docks to ISS, the ISS-Shuttle stack is maneuvered such that ISS leads and Shuttle trails; i.e., a 180-degree yaw is executed after dock (figure 4.2-4-2). Because significantly more MMOD impacts occur on leading surfaces than trailing surfaces, the Shuttle Orbiter vehicle, and particularly the lower surface of the WLE and nose cap sensitive to MMOD damage, is better protected from MMOD impact in the flight attitude after the 180-degree yaw. This change reduced MMOD risk by a factor of five compared to the previously used ISS-Shuttle mated attitude (where the 180-degree yaw maneuver was not performed after dock).

A late inspection is performed of the WLE and nose cap to evaluate if damage exceeding allowable limits occurred during the mission. The inspection is performed prior to entry, just after undock from ISS. If the late inspection finds damage that exceeds allowable limits, TPS repair procedures and/or contingency Shuttle crew support (CSCS) procedures can be implemented. MMOD risks are decreased by a factor of two with late inspection and repair or CSCS as necessary.

The STS-116 MMOD risk assessment, including the effects of the change in Shuttle-ISS docked attitude and late inspection, indicated that there was a 1 in 309 chance of MMOD impact exceeding damage limits, which is better than the 1-in-200 guideline.

Further, to comply with the recommendation to change the MMOD safety criteria from guidelines to requirements, NASA is updating its meteoroid and orbital debris risk requirements.

FINAL UPDATE

NASA continues to emphasize MMOD risk reduction during flight planning and mission execution using vehicle attitudes that minimize MMOD impacts on sensitive areas of the vehicle, as well as late inspection to evaluate the integrity of the TPS prior to entry. NASA also conducts inspections of the vehicle post flight to identify MMOD damage, trend damage over time, monitor changes in the MMOD environment, and assess vehicle design margins to actual MMOD impacts.

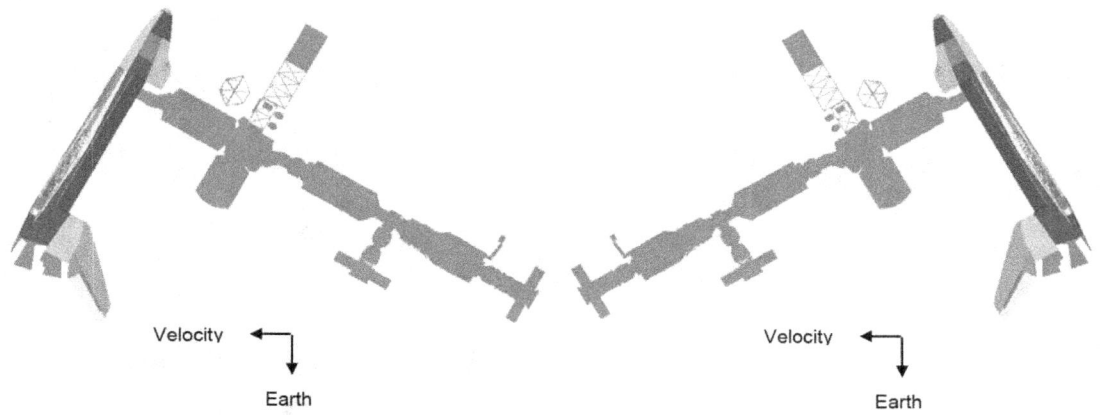

Figure 4.2-4-2. Left image illustrates initial attitude of ISS-Shuttle stack at dock. Right image shows ISS-Shuttle stack after 180-degree yaw maneuver, with ISS leading and Shuttle trailing.

SCHEDULE

Responsibility	Due Date	Activity/Deliverable
SSP	Dec 03 (Completed)	Assess adequacy of MMOD requirements
SSP	Apr 04 (Completed)	WLE Sensor System Critical Design Review
SSP	Nov 04 (Completed)	WLE Impact Detection System hardware delivery (OV-103)
SSP	Apr 05 – Phase 1 (Completed) Dec 05 – Phase 2	Assess WLE RCC impact damage tolerance – Perform hypervelocity and arc-jet tests
SSP	May 05 (Completed)	Flight-by-flight SSP review of forward work status and MMOD requirements

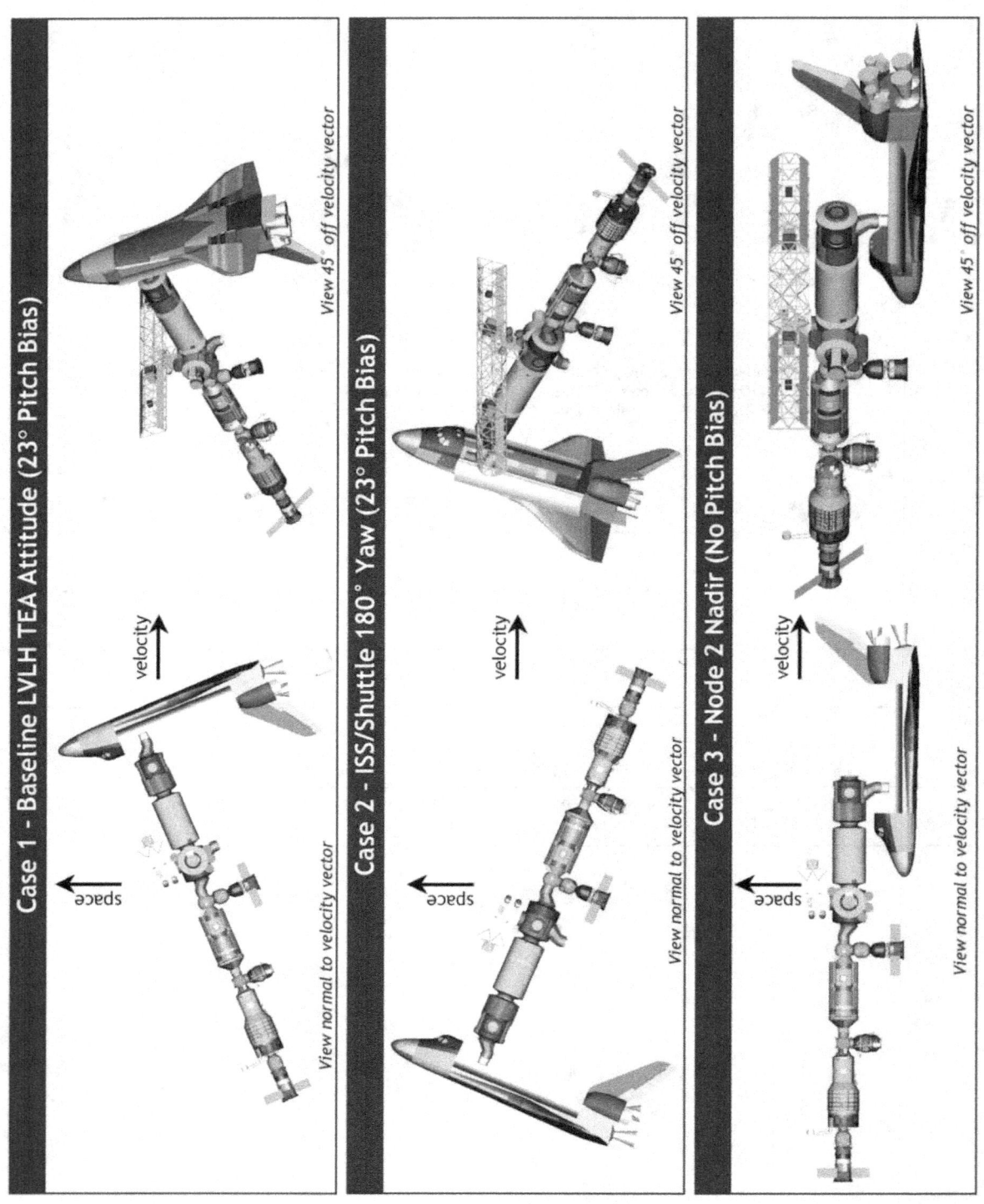

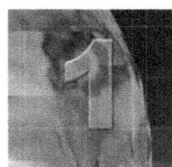

Columbia Accident Investigation Board
Recommendation 4.2-5

Kennedy Space Center Quality Assurance and United Space Alliance must return to the straightforward, industry-standard definition of "Foreign Object Debris," and eliminate any alternate or statistically deceptive definitions like "processing debris." [RTF]

Note: The Stafford-Covey Return to Flight Task Group held a plenary session on December 15, 2004, and NASA's progress toward answering this recommendation was reviewed. The Task Group agreed the actions taken were sufficient to fully close this recommendation.

BACKGROUND

Beginning in 2001, debris at Kennedy Space Center (KSC) was divided into two categories, "processing debris" and foreign object debris (FOD). FOD was defined as debris found during the final or flight-closeout inspection process. All other debris was labeled processing debris. The categorization and subsequent use of two different definitions of debris led to the perception that processing debris was not a concern.

NASA IMPLEMENTATION

NASA and United Space Alliance (USA) have changed work procedures to consider all debris equally important and preventable. Rigorous, industry-standard definitions of FOD have been adopted from National Aerospace FOD Prevention, Inc., and include guidelines and industry standards that address FOD, Foreign Object Damage, and Clean-As-You-Go. FOD is redefined as "a substance, debris or article alien to a vehicle or system which would potentially cause damage."

KSC chartered a multidiscipline NASA/USA team to respond to this recommendation. Team members were selected for their experience in important FOD-related disciplines including processing, quality, and corrective engineering; process analysis and integration; and operations management. The team began by fact-finding and benchmarking to better understand the industry standards and best practices for FOD prevention. They visited the Northrop Grumman facility at Lake Charles, La.; Boeing Aerospace at Kelly Air Force Base, Texas; Gulfstream Aerospace in Savannah, Ga.; and the Air Force's Air Logistics Center in Oklahoma City, Okla. At each site, the team studied the FOD prevention processes, documentation programs, and assurance practices.

Armed with this information, the NASA/USA team developed a more robust FOD prevention program that not only fully answered the *Columbia* Accident Investigation Board (CAIB) recommendation, but also raised the bar by instituting a myriad of additional improvements. The new FOD program is anchored in three fundamental areas of emphasis: First, it eliminates various categories of FOD, including "processing debris," and treats all FOD as preventable and with equal importance. Second, it reemphasizes the responsibility and authority for FOD prevention at the operations level. FOD prevention and elimination are stressed and the work force is encouraged to report any and all FOD found by entering the data in the FOD database. This activity is performed with the knowledge that finding and reporting FOD is the goal of the Program and employees will not be penalized for their findings. Third, it elevates the importance of comprehensive independent monitoring by both contractors and the Government.

USA has also developed and implemented new work practices and strengthened existing practices. This new rigor will reduce the possibility for temporary worksite items or debris to migrate to an out-of-sight or inaccessible area, and it serves an important psychological purpose in eliminating visible breaches in FOD prevention discipline.

FOD "walkdowns" have been a standard industry and KSC procedure for many years. These are dedicated periods during which all employees execute a prescribed search pattern throughout the work areas, picking up all debris. USA has increased the frequency and participation in walkdowns, and has also increased the number of areas that are regularly subject to them. USA has also improved walkdown effectiveness by segmenting FOD walkdown areas into zones. Red zones are all areas within three feet of flight hardware and all areas inside or immediately above or below flight hardware. Yellow zones are all areas within a designated flight hardware operational processing area. Blue zones are desk space and other administrative areas within designated flight hardware operational processing areas.

Additionally, both NASA and USA have increased their independent monitoring of the FOD prevention program. The USA Process Assurance Engineering organization randomly audits work areas for compliance with such work rules as removal of potential FOD items before entering work areas and tethering of those items that cannot be removed (e.g., glasses), tool control protocol, parts protection, and Clean-As-You-Go housekeeping procedures. NASA Quality personnel periodically participate in FOD walkdowns to assess their effectiveness and oversee contractor accomplishment of all FOD program requirements.

An important aspect of the FOD prevention program has been the planning and success of its rollout. USA assigned FOD Point of Contact duties to a senior employee who led the development of the training program from the very beginning of plan construction. This program included a rollout briefing followed by mandatory participation in a new FOD Prevention Program Course, distribution of an FOD awareness booklet, and hands-on training on a new FOD tracking database. Annual FOD Prevention training is required for all personnel with permanent access permissions to controlled Shuttle processing facilities at KSC. This is enforced through the KSC Personnel Access Security System. Another important piece of the rollout strategy was the strong support of senior NASA and USA management for the new FOD program and their insistence upon its comprehensive implementation. Managers at all levels have taken the FOD courses and periodically participate in FOD walkdowns.

The new FOD program has a meaningful set of metrics to measure effectiveness and to guide improvements. As of July 2004, FOD walkdown findings are tracked in the Integrated Quality Support Database. This database also tracks FOD found during closeouts, launch countdowns, post-launch pad turnarounds, landing operations, and NASA quality assurance audits. "Stumble-on" FOD findings are also tracked, as they offer an important metric of program effectiveness independent of planned FOD program activities. For all metrics, the types of FOD and their locations are recorded and analyzed for trends to identify particular areas for improvement. Monthly metrics reporting to management highlight the top five FOD types, locations, and observed workforce behaviors, along with the prior months' trends. Continual improvement is a hallmark of the revitalized FOD program.

STATUS

NASA and USA completed the initial benchmarking exercises, identified best practices, modified operating plans and database procedures, and conducted the rollout orientation and initial employee training. Official, full-up implementation began on July 1, 2004, although many aspects of the plan existed in the previous FOD prevention program in place at KSC. Assessment audits by NASA and USA were conducted beginning in October 2004. Corrective Action Plans were developed and worked. The findings and observations identified during the two audits were resolved. Schedules for the verification of the actions taken and for verifying the effectiveness of the corrective actions have been accomplished to ensure the ongoing effectiveness of the FOD prevention program. Audits of the program are required every 2 years per NSTS 60538. Continual improvement will be vigorously pursued for the remainder of the life of the Shuttle. In July 2004, the Stafford-Covey Return to Flight Task Group reviewed NASA's progress and agreed to conditionally close this recommendation. The full intent of CAIB Recommendation 4.2-5 has been met, and full closure of this recommendation was achieved in December 2004. NASA and USA have gone beyond the recommendation to implement a truly world-class FOD prevention program.

FINAL UPDATE

The revised FOD prevention program has been fully embraced by the Space Shuttle processing workforce. Work practices and methods that prevent FOD generation, monitor work area cleanliness, collect statistical data to monitor program effectiveness, and identify areas for process improvement have proven successful over the past year. Since the implementation of the revised FOD prevention program at KSC, USA-designated FOD monitors have performed and documented over 114,500 FOD walkdowns in flight hardware processing areas and zones where flight hardware has been present. During these activities, over 8,700 individual FOD finding entries have been recorded in the FOD database, which represents a rate of approximately 7.5 FOD findings per 100 walkdowns performed.

Specifically for STS-114 and STS-121, the efforts for FOD prevention at Launch Pad 39-B were exceptional. Although the Return to Flight criteria for debris elimination were quite stringent, the overall condition of the launch pad was subjectively described as "pristine" by personnel who have spent many years performing prelaunch operations at the pad. Attention to initiatives, such as "clean-as-you-go," ensured FOD on the launch pad was minimized during STS-114 and STS-121 prelaunch processing.

The attention to FOD prevention continues during processing activities for all subsequent missions.

SCHEDULE

Responsibility	Due Date	Activity/Deliverable
Space Shuttle Program (SSP)	Ongoing	Review and trend metrics
SSP	Oct 03 (Completed)	Initiate NASA Management walkdowns
SSP	Dec 03 (Completed)	FOD Control Program benchmarking
SSP	Jan 04 (Completed)	Revised FOD definition
SSP	Apr 04 (Completed)	Draft USA Operating Procedure released for review
SSP	Jul 04 (Completed)	Begin implementation of FOD surveillance
SSP	Oct 04 (Completed)	Baseline audit of implementation of FOD definition, training, and surveillance
SSP	Every 2 years	Periodic surveillance audits per NSTS 60538

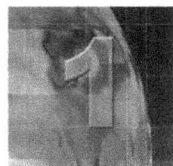

Columbia Accident Investigation Board
Recommendation 6.2-1

Adopt and maintain a Shuttle flight schedule that is consistent with available resources. Although schedule deadlines are an important management tool, those deadlines must be regularly evaluated to ensure that any additional risk incurred to meet the schedule is recognized, understood, and acceptable. [RTF]

Note: The Stafford-Covey Return to Flight Task Group held a plenary session on June 8, 2005, and NASA's progress toward answering this recommendation was reviewed. The Task Group agreed the actions taken were sufficient to fully close this recommendation.

BACKGROUND

NASA has enhanced and strengthened our risk management system that balances technical, schedule, and resource risks to successfully achieve safe and reliable operations. Safe and reliable operations are assured by first focusing on the technical risks and taking the needed time and financial resources to properly resolve technical issues. Once technical risks are eliminated or reduced to an acceptable level, program managers turn to the management of schedule and resource risks to preserve safety. Schedules are integral parts of program management and provide for the integration and optimization of resource investments across a wide range of connected systems. The Space Shuttle Program (SSP) must have a visible schedule with clear milestones to effectively achieve its mission. Schedules associated with all activities generate very specific milestones that must be completed for mission success. Nonetheless, schedules of milestone-driven activities will be extended when necessary to ensure safety. NASA will not compromise safe and reliable operations in our effort to optimize schedules.

NASA IMPLEMENTATION

NASA's priorities will always be operating safely and accomplishing our missions successfully. NASA will adopt and maintain a Shuttle flight schedule that is consistent with available resources. Schedule threats are regularly assessed and unacceptable risk will be mitigated. In support of the Program Operating Plan (POP) process, NASA Shuttle Processing and United Space Alliance (USA) Ground Operations management use the Equivalent Flow Model (EFM) to plan resources that are consistent with the Shuttle flight schedule provided in the POP guidelines. The EFM is a computerized tool that uses a planned manifest and past performance to calculate processing resource requirements. The EFM concept was partnered among USA and NASA Shuttle Processing in fiscal year 2002 and is based on the total flight and ground workforce. The workforce, a primary input to the EFM tool, comprises fixed resources, supporting core daily operations, and variable resources that fluctuate depending on the manifest. Using past mission timelines and actual hours worked, an "equivalent flow" is developed to establish the required processing hours for a baseline processing flow. The baseline "equivalent flow" content is adjusted to reflect the work content in the planned manifest (i.e., Orbiter Major Modifications, Operations and Maintenance Requirements Specification interval requirements, mini-mods, etc.) to arrive at the total equivalent flows in the year for all vehicles in processing. This in turn drives the resource requirement to process those equivalent flows. The result is a definition of an achievable schedule that is consistent with the available workforce needed to meet the technical requirements. If the achievable schedule exceeds the schedule provided in the POP guidelines, one of three actions is available:

- The workforce needed to meet the requirements is identified as an over-guide threat and is accommodated within the budget,
- The schedule is adjusted to meet the available workforce, or
- The technical requirements are adjusted.

The result is an achievable schedule that is consistent with the available resource for processing the Space Shuttle at the Kennedy Space Center (KSC).

To assess and manage the manifest, NASA has developed a process, called the Manifest Assessment System (MAS), for Space Shuttle launch schedules that incorporates all manifest constraints and influences and allows adequate margin to accommodate a normalized amount of changes. This process entails building in launch margin, cargo and logistics margin, and crew timeline margin while preserving the technical element needed for safe and reliable operations. MAS simulates the Space Shuttle flight production process of all flights in the manifest, considering resource sharing (facilities and equipment) in its multi-flow environment. MAS is a custom software application powered by the Extend™ simulation software package and the Efficiency Quotient, Inc (EQI) Scheduling Engine; data supporting the application software is prepared in Oracle database tables. USA Ground Operations is using MAS to assess the feasibility of proposed technical and manifest changes to determine how changes to facility availability, the schedule, or duration of flight production activities effect the overall manifest schedule. Figure 6.2-1-1 illustrates an example of the Space Shuttle manifest.

The Extend™ simulation engine uses EQI custom model blocks to simulate the flight production process for every flight in the scenario as a multi-flow process. A simulation "item," representing each payload and each flight, passes through the activities of the template specified for the flight. The process model attempts to adhere to the schedule provided. However, delays may occur along the way due to constraints to launch, including lighting, orbit thermal constraints, Russian launch vehicle constraints, and facility or vehicle availability. The ability to define and analyze the effects of Orbiter Maintenance Down Period (OMDP) variations and facility utilization are also part of the system.

MAS results are presented through graphical depictions and summary reports. Figure 6.2-1-2 illustrates the simulation results overlaid on the display of the Space Shuttle manifest. "Drill-down" features allow the user to investigate why the results are as presented and enable modifications to mitigate conflicts. Subsequent runs can then validate proposed changes to resolve conflicts.

Scenario datasets can be saved and shared among users in different locations to communicate the complex details of different manifest options under assessment. Coordinated results can then be presented to senior management for their consideration.

By sharing information with the Program-level scheduling tools, MAS can provide integrated analysis of current schedules and projected schedules in the same simulation. This capability enables a more useful way to implement realistic, achievable schedules while successfully balancing technical, schedule, and resource risks to maintain safe and reliable operations.

Schedule deadlines and milestones are regularly evaluated so that added technical requirements and workload changes can be adjusted based on the available resources. New requirements technically required to maintain safe and reliable operations become mandatory, and a NASA KSC Shuttle Processing and USA Ground Operations assessment concerning impacts to accomplish this added work is made. The results of this assessment are presented to Program Management, and schedule milestones and launch dates are adjusted when the necessary resources to accomplish the new requirements are not available. New technical requirements that are highly desirable or can be implemented on an as-available basis are deferred; schedule and resource risks would be incurred. There are numerous forums held as needed (daily/weekly/monthly) in which the SSP management is provided status from each of the Program Elements on current technical requirements, operational requirements, and reasons for necessary adjustments to schedules.

Policies are in place to assure the workforce health in the face of schedule deadlines. The NASA Maximum Work Time Policy, found in KSC Safety Practices Handbook (KHB 1710.2, section 3.4) includes daily, weekly, monthly, yearly, and consecutive hours worked limitations. Deviations require senior management approval up to the KSC Center Director and independent of the Space Shuttle Program. KSC work time safeguards assure that when available resource capacity is approached, the schedule is adjusted to safely accommodate the added work. When possible, launches are planned on Wednesdays or Thursdays to minimize weekend hours and associated costs; repeated launch attempts are delayed to reduce crew and test team fatigue. Overtime hours and safety hazard data are continually monitored by KSC and Space Shuttle Program management for indications of workforce stress, and when management and/or an employee deem it appropriate time-outs are called.

Robust processes are in place to assess and adjust schedules to prevent an excessive workload and maintain safe and reliable operations. These processes maintain a Shuttle flight milestone schedule that is consistent with available resources. Evidence of this practice is demonstrated by the SSP's willingness to judiciously move milestones, which was done repeatedly in the Return to Flight (RTF) effort. During this effort, manifest owners initiated work to identify their requirements. SSP coordinated with the International Space Station (ISS) Program to create an RTF integrated schedule. The SSP Systems Engineering and Integration Office reported the RTF Integrated Schedule weekly to the SSP Program Requirements Control Board. Summary briefs were provided at each Space Flight Leadership Council meeting.

In 2002, NASA made some management changes in key human space flight programs to ensure that Shuttle flight schedules were appropriately maintained and amended to be consistent with available resources. The Office of Space Operations established the position of Deputy Associate Administrator for International Space Station and Space Shuttle Programs (DAA for ISS/SSPs) to manage and direct both programs. This transferred the overall program management of the ISS and SSP from Johnson Space Center to Headquarters, as illustrated in figure 6.2-1-3. The DAA for

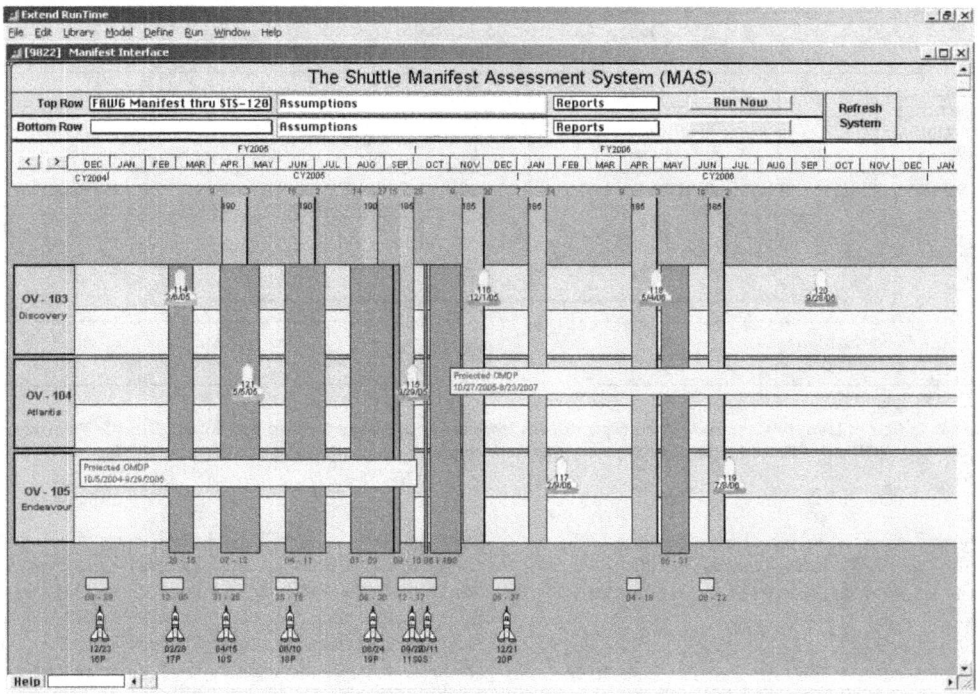

Figure 6.2-1-1. Space Shuttle manifest.

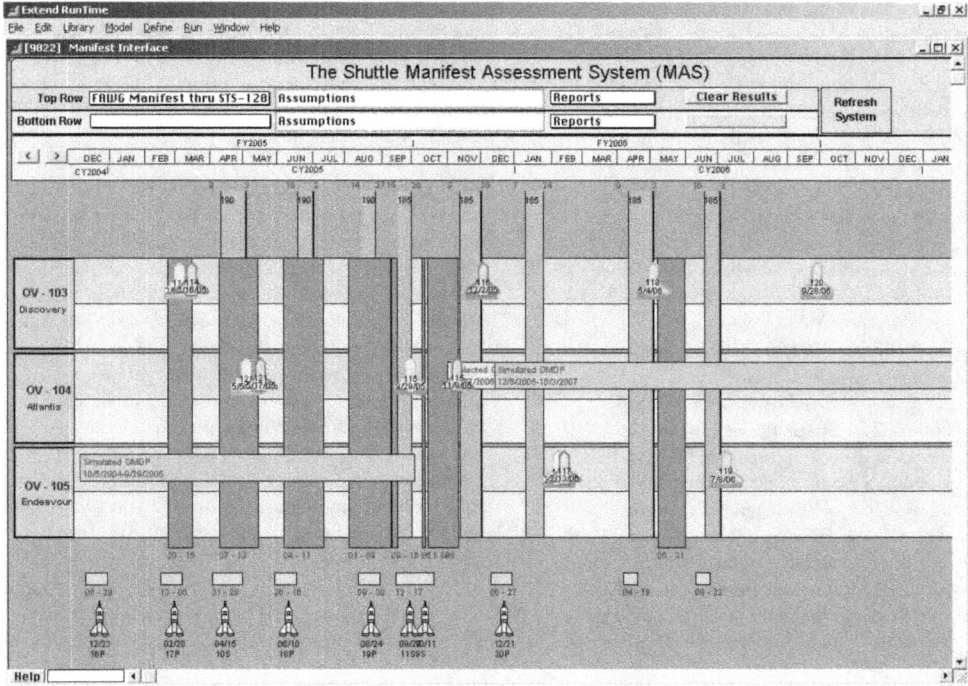

Figure 6.2-1-2. Space Shuttle manifest with simulation results.

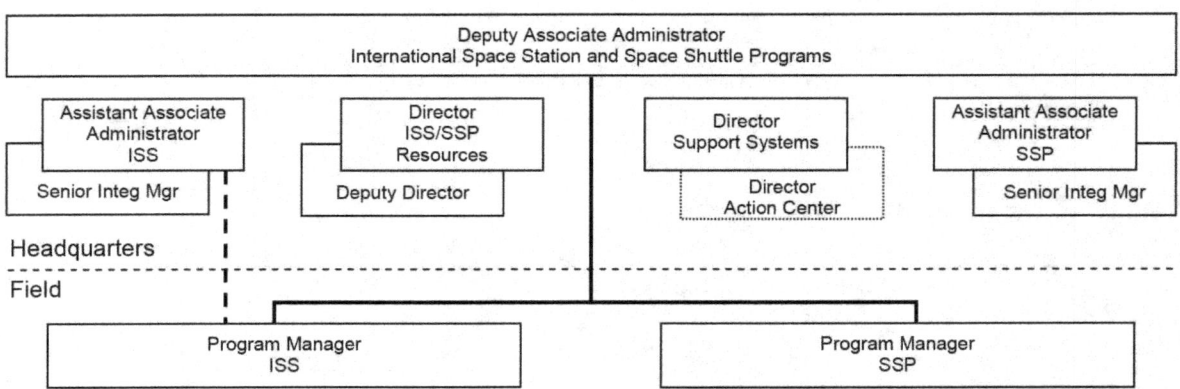

Figure 6.2-1-3. Office of Deputy Associate Administrator for International Space Station and Space Shuttle Programs (Office of Space Operations) is organized to maximize performance oversight.

ISS/SSP was accountable for the execution of the ISS and SSP, and the authority to establish requirements, direct program milestones, and assign resources, contract awards, and contract fees.

As illustrated in figure 6.2-1-4, the Office of DAA for ISS/SSP employed an integrated resource evaluation process to ensure the effectiveness of both programs. Initial resource allocations are made through our annual budget formulation process. At any given time, there are three fiscal year budgets in work: the current fiscal year budget, the presentation of the next fiscal year Presidential budget to Congress, and preparation of budget guidelines and evaluation of budget proposals for the follow-on year. This overlapping budget process, illustrated in figure 6.2-1-5, provides the means for reviewing and adjusting resources to accomplish an ongoing schedule of activities with acceptable risk. Quarterly Program Management Reviews have begun in fiscal year 2005 to assess program and project technical, schedule, and cost performance against an established baseline. These reviews will continue as another tool to assure that the SSP is executed within available resources.

Defined mission requirements, policy direction, and resource allocations were provided to the ISS and SSP Managers for execution. The Space Flight Leadership Council provided specific direction during RTF efforts. The Office of DAA for ISS/SSP continually evaluated the execution of both programs as policy and mission requirements are implemented with the assigned resources. Resource and milestone concerns are identified through this evaluation process. Continued safe operation of the ISS and SSP is the primary objective of program execution; technical and safety issues are evaluated by the Headquarters DAA staff in preparation for each ISS and SSP mission and continuously as NASA prepares for RTF. As demonstrated in actions before the *Columbia* accident and continually during the RTF process, adjustments are made to program milestones, such as launch windows, to assure safe and successful operations. Mission anomalies, as well as overall mission performance, are fed back into each program and adjustments are made to benefit future flights.

In June 2005, prior to the STS-114 Flight Readiness Review, the Space Flight Leadership Council was decommissioned since it had successfully completed all work outlined by its charter. Subsequent actions relating to RTF were conducted by the Space Operations Mission Directorate (SOMD) and the Space Shuttle Program.

In August 2005, following the first RTF mission, NASA and the SOMD made several changes to roles and responsibilities that were reflected in organizational changes. NASA changed the reporting structure of the centers, removing them from the Mission Directorates' chain of command and placing them instead under the NASA Associate Administrator, a new position. Under this new governance structure, the Mission Directorates provide strategic direction and oversight of missions in their area of responsibility. As a result of the mission focus, the position of DAA for ISS/SSP was eliminated, and the Program Managers for SSP and ISS began reporting directly to the Associate Administrator for Space Operations. Similarly, the ISS and SSP headquarters staff was placed under the Associate Administrator for Space Operations.

The SOMD ISS and SSP staff reviews and assesses the status of both programs daily. The staff continues to evolve

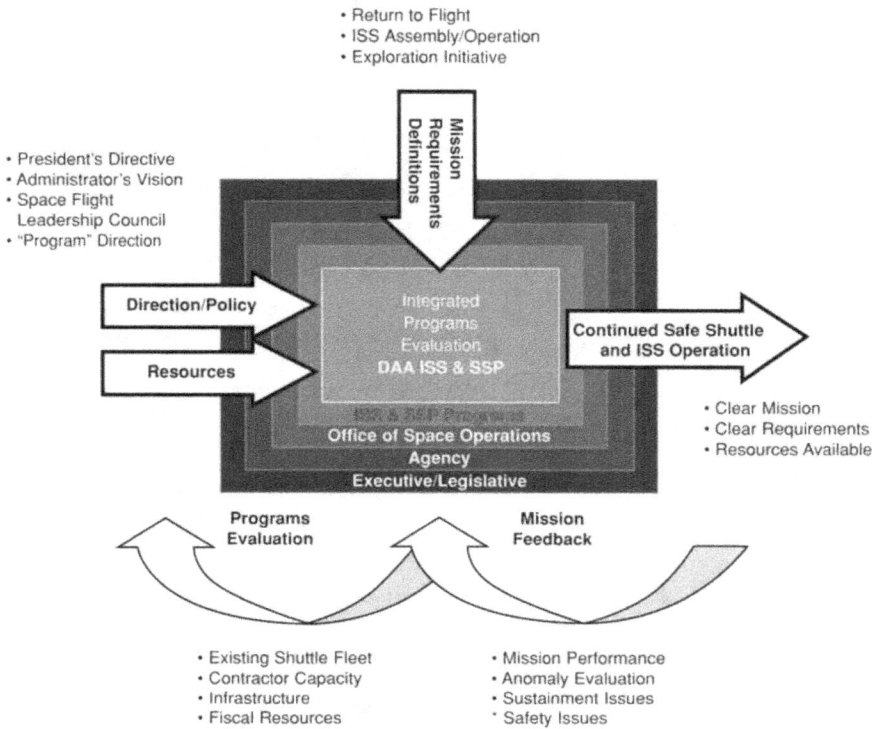

Figure 6.2-1-4. Integrated Resource Evaluation Process is employed by NASA Headquarters, Office of Space Operations.

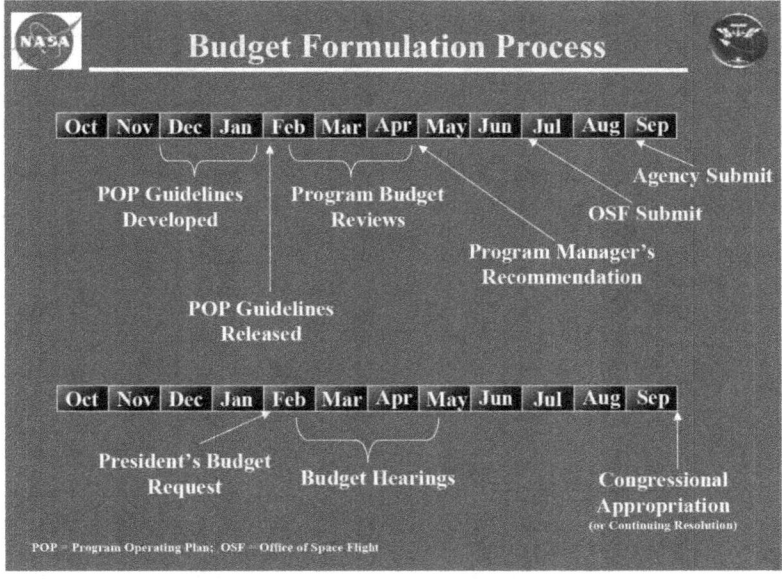

Figure 6.2-1-5. ISS and SSP annual budget formulation process.

the NASA Management Information System (MIS). The One-NASA MIS provides NASA senior management with access to non-time-critical program data and offers a portal to a significant number of NASA Center and program management information systems and Web sites. The extensive information on the One-NASA MIS includes the Key Program Performance Indicators (KPPIs). They use the KPPIs to present required information to the SOMD Program Management Council and the Agency Program Management Council.

Overall, SOMD has implemented a comprehensive process for continually evaluating the effectiveness of the SSP. This process allows the SOMD staff to recognize and rapidly respond to changes in status and to act transparently to elevate issues such as schedule changes that may require decisions from the appropriate leaderships. NASA senior leadership and the SOMD staff have repeatedly demonstrated an understanding of acceptable risk, and have responded by changing milestones to assure continued safe and reliable operations.

STATUS

NASA has repeatedly demonstrated its willingness and ability to make changes in the manifest to account for resource constraints to maintain safe and reliable operations. The *Columbia* accident has resulted in new requirements that must be factored into the manifest. The ISS and SSP are working together to incorporate RTF changes into the ISS assembly sequence. A system review of currently planned flights is constantly being performed. After all of the requirements have been analyzed and identified, a launch schedule and ISS manifest will be established. NASA will continue to add margin that allows some changes while not causing downstream delays in the manifest.

SSP Flight Operations and Integration has scheduling and manifesting responsibility for the Program, working both the short-term and long-term manifest options. The current proposed manifest launch dates are all "no earlier than" (NET) dates, and are contingent upon establishing a launch date for a given mission. A computerized manifesting capability, called the MAS, is being used to more effectively manage the schedule margin, launch constraints, and manifest flexibility. The primary constraints to launch, including lighting (for STS-121), orbit thermal constraints, and Russian Launch Vehicle constraints, have been incorporated into MAS and tested to ensure proper effects on simulation results. The ability to define and analyze the effects of OMDP variations and facility utilization are also now part of the system. The system will be improved in the future to include increased flexibility in resource loading enhancements. Development will continue on the computer-aided tools to manage the manifest schedule margin, launch constraints, and manifest flexibility.

Prior to RTF, the launch constraints were well understood and their schedules were incorporated into the manifest. The launch window established for STS-114 was based on the External Tank separation lighting and launch-on-need Shuttle rescue constraints. This same philosophy is being used for the STS-121 launch date.

POST STS-114 UPDATE

In August 2005, after the successful completion of the first RTF mission, STS-114, NASA removed institutional responsibilities, such as center management, from the Mission Directorates. Under this new governance structure, the Mission Directorates provide strategic direction and oversight of missions in their area of responsibility. As a result of the mission focus, the position of DAA for ISS/SSP was eliminated, and the Program Managers for SSP and ISS began reporting directly to the Associate Administrator for Space Operations.

The SSP projects will continue to provide information on their status and constraints to give senior managers the ability to make informed decisions on the Space Shuttle manifest. The STS-121 and subsequent launch windows will be established based on ability of the SSP to meet the requirements established for that mission.

SCHEDULE

Responsibility	Due Date	Activity/Deliverable
SSP	Aug 03 (Completed)	Baseline the RTF constraints schedule
SSP	Jun 05 (Completed)	Establish STS-114 baseline schedule

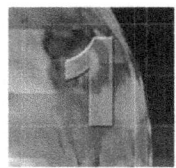

Columbia Accident Investigation Board
Recommendation 6.3-1

Implement an expanded training program in which the Mission Management Team faces potential crew and vehicle safety contingencies beyond launch and ascent. These contingencies should involve potential loss of Shuttle or crew, contain numerous uncertainties and unknowns, and require the Mission Management Team to assemble and interact with support organizations across NASA/Contractor lines and in various locations. [RTF]

Note: The Stafford-Covey Return to Flight Task Group held a plenary session on June 8, 2005, and NASA's progress toward answering this recommendation was reviewed. The Task Group agreed the actions taken were sufficient to fully close this recommendation.

BACKGROUND

During Space Shuttle missions, the Shuttle Mission Management Team (MMT) is responsible for oversight of the operations teams during both prelaunch and in-flight activities. The countdown and flight operations are conducted by the operations teams according to rules and procedures approved by Space Shuttle Program (SSP) Management well prior to operations and documented in NSTS 07700. The Mission Operations Team leads all nominal mission operations. The MMT provides guidance to the operations teams for situations that fall outside normal operations; the MMT also redefines programmatic priority when in-flight anomalies or off-nominal conditions result in conflicting priorities.

In the past, the MMT's work was focused on the immediate decision-making required of in-flight operations. In this environment, the concerns of individual engineers, the quality of risk assessments, and the pedigree of engineering assessments were sometimes poorly understood. Training for managers on the MMT reflected this bias and consisted primarily of briefings and simulations that focused on the prelaunch and launch phases, including launch aborts. In retrospect, this approach did not adequately sensitize NASA management in general, and the MMT specifically, to actively seek out potential concerns and issues raised by support teams and working groups. As a result, the MMT training has changed to focus on: clearer communication processes centered on bringing out differing opinions; maintaining awareness of decisions that impact the remainder of the flight; and ensuring an understanding of the roles and responsibilities of team members and supporting working groups and teams, including data-sharing processes and required milestones to support real-time operations. All of these changes will improve the ability of the MMT to identify and address in-flight problems effectively.

The MMT is responsible for making SSP decisions regarding preflight and in-flight activities and operations that exceed the authority of the launch director or the flight director.

The MMT's responsibilities for a specific Space Shuttle mission start with the first scheduled meeting two days prior to a scheduled launch (L-2). Kennedy Space Center (KSC) prelaunch activities continue through launch and terminate when the Shuttle has achieved safe orbit. At that time, MMT activities transfer to the Johnson Space Center (JSC). The flight MMT meets daily during the subsequent on-orbit, entry, and landing phases and terminates with crew egress from the vehicle. When the flight MMT is not in session, all MMT members are on-call and required to support emergency MMTs convened because of anomalies or changing flight conditions.

NASA IMPLEMENTATION

NASA's response to this recommendation was implemented in two phases: (1) review and revise MMT processes and procedures; and (2) develop and implement a training program consistent with those process revisions.

Processes and Procedures

NASA contracted with the Behavioral Excellence Strategic Team and several past flight directors, including Gene Kranz and Glynn Lunney, to study the MMT processes and make recommendations to improve communications, decision-making, and operational processes. The result was a demonstrable improvement in critical decision-making efficiency and more open communication among MMT members to resolve critical issues.

NASA established a process for the review and resolution of off-nominal mission events to assure that all such issues are identified to and resolved by the flight MMT. The Space Shuttle Systems Engineering and Integration Office maintains and provides an integrated anomaly list at each MMT meeting. As appropriate, anomalies are assigned a formal office of primary responsibility (OPR) for technical evaluation and will be subject to an independent risk assessment by Safety and Mission Assurance (SMA). The MMT has one SSP SMA core member and three institutional SMA advisory members from JSC, KSC, and Marshall Space Flight Center. In addition, the program has added the independent Engineering Technical Authority and Safety Technical Authority to the MMT as advisors.

The MMT secretary maintains and displays an action tracking log to ensure all members are adequately informed of the anomaly's status. Closure of actions associated with the anomaly, as determined by the MMT Chair, is tracked and documented on the MMT website. Response to actions must include a description of the issue (observation and potential consequences), technical analysis details (including databases, employed models, and methodologies), recommended actions, associated mission impacts, and flight closure rationale, if applicable. These steps are all designed to eliminate the possibility of critical missteps by the MMT due to incomplete or un-communicated information. These steps also make the process of anomaly resolution more transparent to all stakeholders. NASA documents these changes in the Mission Evaluation Room console handbook that includes MMT reporting requirements, a flight MMT reporting process for on-orbit vehicle inspection findings, and MMT meeting support procedures. The MMT secretary also maintains and displays an MMT Decisions Log that tracks and documents MMT decisions that fall outside the nominal mission objectives or changes in programmatic decisions.

Additional improvements were made to MMT internal processes and procedures, including more clearly defining requirements for the frequency of MMT meetings and the process for requesting an emergency MMT meeting. NASA now conducts daily MMT meetings beginning with the L-2 day MMT. Membership, organization, and chairmanship of the preflight and in-flight MMTs are standardized. The Manager, KSC Launch Integration, chairs the preflight MMTs. The SSP Deputy Manager chairs the on-orbit MMTs.

MMT membership has been expanded and augmented with core and/or advisory members from SMA, the NASA Engineering and Safety Center, and engineering and program management disciplines. The MMT membership's responsibilities have been clearly defined, and MMT membership and training status for each mission is established by each participating organization in writing at the Flight Readiness Review (FRR). Each MMT member also has clearly defined processes for MMT support and problem reporting.

Procedures for Flight MMT meetings are standardized through the use of predefined templates for agenda formats, presentations, action item assignments, decisions logs, and readiness polls. This ensures that the communication and resolution of issues are performed in a consistent, rigorous manner. Existing SSP meeting support infrastructure and a collaboration tool are used to ensure that critical data are distributed before scheduled meetings and that MMT meeting minutes are distributed the following morning after each meeting. In addition, NASA established formal processes for the review of findings from ascent and on-orbit imagery analyses, postlaunch hardware inspections, ascent reconstruction, and all other flight data reviews to ensure timely, effective reviews of key data by the MMT.

Using research on improving communications for critical decision making, NASA refurbished the Mission Control Center and redesigned the MMT's working space to provide increased seating capacity and communications improvements. Improvements include a video-teleconferencing capability, a multi-user collaboration tool, and a larger room to allow more subject matter experts and MMT members. A large "C"-shaped table now seats all members of the MMT and encourages open communication by eliminating a hierarchical seating arrangement. The MMT Command Center has been operational since the November 2004.

Training

All MMT members, except those serving exclusively in an advisory capacity or as a Department of Defense Mission Support representative, are required to complete a minimum set of training requirements to attain initial qualification prior to performing MMT responsibilities. MMT members must also participate in an ongoing training program to maintain qualification status, which is renewed annually. Training records are maintained to ensure compliance with those requirements. NASA has employed an independent training consultant to assist in developing and refining these training activities to support a flying manifest and to evaluate overall training effectiveness.

In addition, to ensure adequate backups are available, at least two people are trained to fill each MMT core position prior to each flight. This protects the integrity of the integrated MMT process against individuals' inability to perform their role for any reason.

STATUS

The SSP published a formal MMT training plan (NSTS 07700, Volume II, Program Structure and Responsibilities, Book 2 – Space Shuttle Program Directives, Space Shuttle Program Directive 150) that defines the generic training requirements for MMT certification. This plan is comprised of three basic types of training: courses and workshops, MMT simulations, and self-instruction. Courses, workshops, and self-instruction materials were selected to strengthen individual expertise in human factors, critical decision making, and risk management of high-reliability systems. On-line courses are geared to develop a basic understanding of mission operations and concepts to allow efficiencies and team cohesiveness during MMT meetings. MMT training activities are now standard practice throughout each year. Prior to December 2006, a total of 18 simulations were completed, including end-to-end contingency simulations conducted every

18 months and simulations to address MMT actions related to Contingency Shuttle Crew Support (CSCS), as well as other critical skills and decision-making scenarios. These simulations bring together the flight crew, flight control team, launch control team, engineering staff, outside agencies, and MMT members to improve communication and teach better problem-recognition and decision-making skills. All MMT members complete the requisite training prior to each flight.

As a result of the significant changes made to the MMT processes and structure, the MMT is a strong, operations-oriented, problem-solving, and critical decision-making management body. The MMT's technical engineering sub-teams perform the engineering root-cause analysis and technical problem solving, and identify options and make recommendations to the MMT. This paradigm shift has resulted in more focused decision-making by the MMT, making better use of the unique expertise of the MMT membership.

Risk management is now a major consideration at each MMT meeting. Each identified hazard is required to have a clear risk assessment performed and presented to the MMT so the appropriate risk versus risk tradeoffs can be discussed and decided upon. Supporting analyses, assumptions, issues, and ramifications are a part of this discussion.

Communication between the MMT, its various sub-teams, and remote locations has improved tremendously, in part due to the physical layout of the new MMT Command Center and the ability to tie in remote sites through a video-teleconference link. As a result, the MMT and sub-team members receive data at the required times, members are more aware of mission timelines, and all stakeholders can view actions in real time since these actions are posted electronically in the MMT Command Center and on the video-teleconference circuits. This improved communication results in an improved state of situational awareness and ability to make informed, appropriate decisions more quickly.

The MMT completed one final simulation prior to Return to Flight. This simulation addressed all MMT actions related to a CSCS capability decision. (See SSP-3 for details on CSCS.)

Revisions to project and element processes, which were established consistent with the new MMT requirements, follow formal Program approval processes. All SSP projects completed a flight MMT reporting process for launch imagery analysis and on-orbit vehicle inspection findings.

FINAL UPDATE

The MMT's performance during STS-114 bore out the success of the improved training program. The MMT successfully dealt with several in-flight anomalies, including protruding gap fillers and damage to a thermal blanket. The team was able to use effectively the new imagery and data available from the mission, and draw on the technical expertise of a wide range of people from various centers to make informed decisions to preserve the safety of the crew and vehicle on orbit and for entry.

The SSP began preparations for continued MMT training post-STS-114. The MMT yearly training calendar is posted and includes the classes and simulations available to MMT members to meet their training requirements.

The MMT will conduct at least one mission simulation each year. Additionally, a prelaunch contingency simulation will be conducted every 18 months. Members now receive training credit for supporting missions during MMTs. Prebriefs to MMT members are conducted 1 week prior to each flight or prior to a simulation. The MMT team reviews all pertinent programmatic decisions made for the flight or simulation; this includes a review of the mission objectives and priorities as well as a review of Lessons Learned. This process ensures confidence in the ability of MMT members to know and perform their tasks and ensures mission cognizance and team cohesiveness prior to each flight or simulation. The simulations exercise the ability of the MMT to perform integrated risk assessments through technical presentations. The Engineering teams provide the technical presentations to the MMT. The presentations include discussion of uncertainties, assumptions, and dissenting opinions during technical assessments. In addition, SMA has implemented an anonymous reporting process to ensure dissenting opinions are brought to the MMT's attention during missions and simulations. Debriefs are also conducted approximately 2 weeks post flight as well as immediately following simulations. Lessons Learned are captured during each flight or simulation debrief and posted on the MMT training webpage.

NASA changed the management structure of the SSP and Space Operations Mission Directorate after STS-114 and it will continue to evolve to ensure that the MMT critical functions described above are maintained and enhanced. All MMT simulations include objectives to exercise the interfaces with advisory members, technical support personnel, and NASA Headquarters. The SSP will continue to evaluate and update the MMT training plan.

SCHEDULE

Responsibility	Due Date	Activity/Deliverable
SSP	Oct 03 (Completed)	MMT Interim training plan
SSP	Oct 03 (Completed)	MMT process changes to Program Requirements Change Board
SSP	Oct 03 (Completed)	Project/element process changes
SSP	Nov 03 – Return to Flight (Completed)	MMT training
SSP		**MMT Simulation Summary**
	Nov 03 (Completed)	MMT On-Orbit simulation
	Dec 03 (Completed)	MMT SSP/International Space Station (ISS) Joint On-Orbit simulation
	Feb 04 (Completed)	MMT On-Orbit simulation
	Apr 04 (Completed)	MMT Prelaunch simulation
	May 04 (Completed)	MMT On-Orbit simulation involving Thermal Protection System (TPS) inspection
	Jun 04 (Completed)	MMT Prelaunch simulation
	Jul 04 (Completed)	MMT On-Orbit simulation
	Sep 04 (Completed)	MMT On-Orbit simulation
	Nov 04 (Completed)	MMT SSP/ISS Joint On-Orbit simulation involving TPS inspection
	Dec 04 (Completed)	MMT Prelaunch simulation
	Jan 05 (Completed)	MMT Prelaunch Contingency simulation
	Feb 05 (Completed)	MMT Prelaunch/On-Orbit/Entry Integrated simulation involving TPS inspection
	May 05 (Completed)	MMT Contingency Crew Survival Capability Sim
SSP	Dec 03 (Completed)	Status to Space Flight Leadership Council and Stafford/Covey Task Group
SSP	Feb 04 (Completed)	MMT final training plan
SSP	Apr 04 (Completed)	Status to Stafford/Covey Task Group
SSP	Aug 04 (Completed)	Miscellaneous MMT process and training revisions to address simulations lessons learned
SSP	Sep 04 (Completed)	Status to Stafford/Covey Return to Flight Task Group
SSP	Nov 04 (Completed)	Complete refurbishment of MMT Command Center
SSP	Feb 05 (Completed)	Update MMT Training Plan

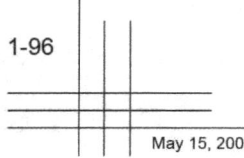

Columbia Accident Investigation Board
Recommendations 9.1-1, 7.5-1, 7.5-2, and 7.5-3

R9.1-1 Prepare a detailed plan for defining, establishing, transitioning, and implementing an independent Technical Engineering Authority, independent safety program, and a reorganized Space Shuttle Integration Office as described in R7.5-1, R7.5-2, and R7.5-3. In addition, NASA should submit annual reports to Congress, as part of the budget review process, on its implementation activities. [RTF]

R7.5-1 Establish an independent Technical Engineering Authority that is responsible for technical requirements and all waivers to them, and will build a disciplined, systematic approach to identifying, analyzing, and controlling hazards throughout the life cycle of the Shuttle System. The independent technical authority does the following as a minimum:

- Develop and maintain technical standards for all Space Shuttle Program projects and elements
- Be the sole waiver-granting authority for all technical standards
- Conduct trend and risk analysis at the sub-system, system, and enterprise levels
- Own the failure mode, effects analysis and hazard reporting systems
- Conduct integrated hazard analysis
- Decide what is and is not an anomalous event
- Independently verify launch readiness
- Approves the provisions of the recertification program called for in Recommendation [R9.2-1]

The Technical Engineering Authority should be funded directly from NASA Headquarters and should have no connection to or responsibility for schedule or program cost.

R7.5-2 NASA Headquarters Office of Safety and Mission Assurance should have direct line authority over the entire Space Shuttle Program safety organization and should be independently resourced.

R7.5-3 Reorganize the Space Shuttle Integration Office to make it capable of integrating all elements of the Space Shuttle Program, including the Orbiter.

Note: The Stafford-Covey Return to Flight Task Group held a plenary session on June 8, 2005, and NASA's progress toward answering this recommendation was reviewed. The Task Group agreed the actions taken were sufficient to fully close this recommendation.

Note: On March 6, 2007, NASA released an update to the "NASA Space Flight Program and Project Management Requirements," otherwise known as NASA Procedural Requirements (NPR) 7120.5D. At the time of the release of this Final Update to the Implementation Plan (April 2007), NPR 7120.5D applies to all current and future NASA space flight programs and projects, and lays out the governance model and management structure for ensuring that sufficient checks and balances are in place between key organizations to ensure that decisions have the benefit of different points of view. NPR 7120.5D incorporates lessons learned since Return to Flight, including recommendations given by the Stafford-Covey Return to Flight Task Group. What follows is a description of the governance structure that was in place as of October 2006 and in effect at the time of the second Return to Flight mission, STS-121. This information is provided for historical purposes only.

INTRODUCTION

NASA, under the leadership of the Office of Safety and Mission Assurance (OSMA) and the Office of the Chief Engineer, is implementing a plan addressing the Agency-wide response to Recommendation 9.1-1 – referred to as the "9.1-1 Plan" and titled "NASA's Plan for Implementing Safe and Reliable Operations." Although the *Columbia* Accident Investigation Board (CAIB) only recommended that NASA prepare a detailed plan for 9.1-1 prior to Return to Flight (RTF), NASA has begun the transformation called for in the three relevant Chapter 7 recommendations.

The CAIB's independent investigation revealed areas in NASA's organization and its operations that needed substantial improvement before returning the Space Shuttle to safe and reliable flight operations. This report addresses three fundamental changes that NASA is making to improve the safety and reliability of its operations:

- Restore specific engineering technical authority, independent of programmatic decision-making.

- Increase the authority, independence, and capability of the Agency Safety and Mission Assurance (SMA) organizations.

- Expand the role of the Space Shuttle Integration Office to address the entire Space Shuttle system.

These changes reflect careful and diligent review of the CAIB's investigation as a basis for implementation of their recommendations. Specifically, these changes address CAIB Recommendations R9.1-1 and its accompanying Recommendations R7.5-1, R7.5-2, and R7.5-3.

As a first necessary step to put the CAIB's recommendations regarding independent technical authority into practice, the NASA Administrator designated the Chief Engineer as the NASA Technical Authority (TA). The Chief Safety and Mission Assurance Officer provides leadership, policy direction, functional oversight, assessment, and coordination for the safety and quality assurance disciplines across the Agency. The role of the Shuttle Integration Office (now the Shuttle Systems Engineering and Integration Office) has been strengthened so that it integrates all of the elements of the Space Shuttle Program (SSP). These three organizational changes—an independent technical authority, a separate and distinct independent SMA, and a focused Program management structure—form a foundation for ensuring safe and reliable operations for NASA's Space Shuttle and other missions.

Section I of this report, the first change, was issued in November 2004 to provide NASA's plan to restore specific engineering technical authority, independent of programmatic decision-making, in all of NASA's missions. Section 4.5 provides NASA's progress on implementing technical authority. Section II describes the role of SMA and how the second change increases the authority, capability, and independence of the SMA community. Section III addresses how the third change expands the role of the new Space Shuttle Systems Engineering and Integration Office to address the entire Space Shuttle system. Section 4.3 addresses the relationship of the roles and responsibilities of the ITA and SMA organizations.

NASA IMPLEMENTATION

Independent Technical Authority (R7.5-1)

This plan answers the CAIB Recommendation 7.5-1 by aggressively implementing an independent technical authority at NASA that has the responsibility, authority, and accountability to establish, monitor, and approve technical requirements, processes, products, and policy.

Technical Authority

The NASA Chief Engineer, as the TA, governs and is accountable for technical decisions that affect safe and reliable operations and is using a warrant system to further delegate this technical authority. The TA provides technical decisions for safe and reliable operations in support of mission development activities and programs and projects that pose minimum reasonable risk to humans; i.e., astronauts, the NASA workforce, and the public. Sound technical requirements necessary for safe and reliable operations will not be compromised by programmatic constraints, including cost and schedule.

As the NASA TA, the NASA Chief Engineer is working to develop a technical conscience throughout the engineering community, that is, the personal responsibility to provide safe technical products coupled with an awareness of the avenues available to raise and resolve technical concerns. Technical authority and technical conscience represent a renewed culture in NASA governing and upholding sound technical decision-making by personnel who are independent of programmatic processes. This change affects how technical requirements are established and maintained as well as how technical decisions are made, safety considerations being first and foremost in technical decision-making. Five key principles govern the independent technical authority. This authority:

1. Resides in an individual, not an organization;
2. Is clear and unambiguous regarding authority, responsibility, and accountability;
3. Is independent of Program Management;
4. Is executed using credible personnel, technical requirements, and decision-making tools; and
5. Makes and influences technical decisions through prestige, visibility, and the strength of technical requirements and evaluations.

Warrant System

The Chief Engineer has put technical authority into practice through a system of governing warrants issued to individuals. These Technical Warrant Holders (TWHs) are proven subject matter experts with mature judgment who are operating with a technical authority budget that is independent from Program budgets and Program authority. This technical authority budget covers the cost of the TWHs and their agents as they execute their responsibility for establishing and maintaining technical requirements, reviewing technical products, and preparing and administering technical processes and policies for disciplines and systems under their purview.

The warrant system provides a disciplined formal procedure that is standardized across the Agency, and a process that is recognized inside and outside NASA in the execution of independent technical authority.

Technical Conscience

Technical conscience is personal ownership of the technical product by the individual who is responsible for that product. Committee reviews, supervisory initials, etc., do not relieve these individuals of their obligation for a safe and reliable mission operation if their technical requirements are followed. Technical conscience is also the personal principle for individuals to raise concerns regarding situations that do not "sit right" with NASA's mandate for safe and reliable systems and operations. With adoption of technical authority and the warrant system, technical personnel have the means to address and adjudicate technical concerns according to the requirements of the situation. TA and the TWHs provide the means for independent evaluation and adjudication of any concern raised in exercising technical conscience.

On November 23, 2004, the NASA Administrator issued the policy and requirements to implement technical authority through a technical warrant process. This policy was issued under NPD 1240.4 NASA Technical Authority (draft) and NPR 1240.1 Technical Warrant System (draft), and is in accordance with the plan. In December, NASA Chief Engineer Rex Geveden assigned Walter Hussey as Director of ITA Implementation to focus the Agency's internal efforts on this cultural transformation. The Chief Engineer has identified and selected TWHs and issued warrants for 26 critical areas, including all major systems for the Space Shuttle. After their selection and training, these newly assigned TWHs are now executing the responsibilities of their warrants. The Space Shuttle TWHs are making the technical decisions necessary for safe and reliable operations and are involved in RTF activities for the Space Shuttle.

NASA is selecting additional TWHs to span the full range of technical disciplines and systems needed across the Agency. The Chief Engineer issued several new warrants in March 2005, including one for Systems Safety Engineering that will help revitalize the conduct of safety analyses (failure mode and effects analysis (FMEA), hazards analysis, reliability engineering, etc.) as part of design and engineering, and will continue to issue warrants as required.

Independent Safety (R7.5-2)

This plan answers the CAIB Recommendation 7.5-2 by aggressively addressing the fundamental problems brought out by the CAIB in three categories: authority, independence, and capability.

SMA Authority

To address the authority issue raised by the CAIB, NASA has strengthened OSMA's traditional policy oversight over NASA programs and Center line organizations with the explicit authority of the Administrator through the Deputy Administrator/Chief Operating Officer (COO) to enforce those policies. The Chief Safety and Mission Assurance

Officer provides leadership, policy direction, functional oversight, assessment, and coordination for the safety, quality, and mission assurance disciplines across the Agency. Operational responsibility for the requirements of these disciplines rests with the Agency's program and line organizations as an integral part of the NASA mission. To increase OSMA's "line authority" over field SMA activities, NASA has taken four important steps:

1. The Chief Safety and Mission Assurance Officer now has explicit authority over selection, relief, and performance evaluation of all Center SMA Directors as well as the lead SMA managers for major programs, including Space Shuttle and International Space Station (ISS), as well as the Director of the Independent Verification and Validation (IV&V) Center.

2. The Chief, OSMA will provide a formal "functional performance evaluation" for each Center Director to their Headquarters Center Executive (HCE) each year.

3. "Suspension" authority is delegated to the Center Directors and their SMA Directors. This authority applies to any program, project, or operation conducted at the Center or under that center's SMA oversight regardless of whether the Center also has programmatic responsibility for that activity.

4. The SMA community, through their institutional chain of command up to the COO, now has the authority to decide the level of SMA support for the project/program.

NASA SMA support for the SSP consists of dedicated Program office staff, technical support from the Centers, and functional oversight from the Headquarters OSMA. A senior SMA professional heads the Program's SMA Office as the Space Shuttle SMA Manager. The SMA Manager reports directly to the Program Manager and is responsible for execution of the safety and quality assurance requirements within the Program. The Program SMA Office integrates the safety and quality assurance activities performed by all Space Operations Centers for the various projects and Program elements located at those Centers. The Center SMA Directorates provide technical support to the Program's SMA Manager. They also provide independent safety and quality assurance functions in the form of independent assessments, safety, and reliability panel reviews. Finally, they provide a cadre of personnel dedicated to OSMA's Independent Assessment of compliance function.

SMA Independence

The CAIB recommendation requires that the OSMA be independently funded. After the Report of the Presidential Commission on the Space Shuttle *Challenger* Accident, also known as the Rogers Commission Report, NASA created the Office of Safety, Reliability and Quality Assurance, later renamed OSMA, and specifically set up its reporting and funding to be separate from that of the Chief Engineer's office and any of the Program Enterprises. At the time of *Columbia*, all funding for OSMA was in the corporate General and Administrative (G&A) line, separate from all other program, institutional, and mission support and functional support office funding. As for personnel, all permanent OSMA personnel are dedicated to OSMA and, therefore, are independent of program or other mission support and functional support offices. This plan retains that independent reporting and funding approach consistent with the CAIB recommendation.

With respect to Center-based civil servants and their support contractors performing safety, reliability, and quality assurance tasks, this plan calls for significant change. This plan establishes that the institution, not the program, decides SMA resource levels. Under the oversight of the Headquarters HCEs, Centers will set up SMA-"directed" service pools to allow SMA labor to be applied to programs and projects in the areas and at the levels deemed necessary by the SMA Directors and their institutional chain of authority. SMA will pre-coordinate the use of their resources with the programs to foster understanding of how SMA labor will be used. This approach will guarantee both organizational and funding independence from the programs in a way that fully addresses the CAIB's findings. Finally, the Headquarters OSMA will, for the first time, be a voting member of the Institutional Committee wherein institutional (including SMA service pool) budget decisions are made for the Agency. To aid OSMA in its resource oversight and approval responsibilities, each center SMA Directorate will develop an Annual Operating Agreement that calls out all SMA activities at the center, industrial, program support, and independent assessment.

Under NASA's old definition of independence, which focused on organizational independence, the SSP Program and Project Managers had funding approval authority for about 99% (based on fiscal year (FY) 03 estimates) of the total SMA funding level for Shuttle (includes all contractor and Center NASA and support contractor SMA resources). The remaining 1% consisted of Center SMA supervisor time (paid by Center General and Administrative funds) and approximately $2M per year of Space Shuttle Independent

Assessment (IA) activity (paid for by OSMA). Under NASA's new definition of independence, which now includes the directed service pool, the SSP has funding approval authority for only about 70% of the total SMA funding level. This funding pays for Shuttle prime and subcontractor SMA and for the small civil service SMA Management Office in the Program. The remaining funding approval is accomplished through the directed service pool. This accounts for all Center SMA Civil Service (CS), all SMA support contractors, and OSMA's IV&V and IA that supports Space Shuttle.

SMA Capability

To address SMA capability, all of the Centers have reviewed their SMA skills and resources for adequacy and added positions as required. In particular, the Space Operations Centers have all addressed staffing deficiencies as part of Shuttle RTF, and they have already begun hiring to fill vacancies. Headquarters OSMA has increased significantly its ability to provide functional oversight of all NASA SMA programs. Staffing has been increased in the Headquarters office from 48 to 51 people, partly to accommodate increased liaison needs created by addition of NASA Engineering and Safety Center (NESC), IV&V, and new assurance programs. At the time of *Columbia*, OSMA had a budget of $6M per year for IA, its primary corporate assurance tool. OSMA will continue to send IA funding to the Space Flight Centers for use by SMA Directorates in performing Center audits and supporting OSMA audits and assessment of resident programs. It also encourages the IA teams to focus more on process and functional audits than they have in the past. This plan shows a substantial increase in OSMA capability by the addition of the responsibility and budgets for the Agency software IV&V services. The NESC, as a technical resource available to the SMA community, in coordination with the ITA, combined with IV&V and IA capabilities, provides an unprecedented increase in the independent assessment, audit, and review capability and will reinforce the SMA community's role in providing verification and assurance of compliance with technical requirements owned by the ITA, and in technical support for mishap investigations.

The ITA will own all technical requirements, including safety and reliability design and engineering standards and requirements. OSMA will continue to develop and improve generic safety, reliability, and quality (SRQ) process standards, including FMEA, risk, and hazards analysis processes; however, the ITA will specify and approve these analyses and their application in engineering technical products. OSMA's involvement with SRQ process standards will enable the Headquarters office and Center SMA organizations to better oversee compliance with safety, reliability, and quality requirements. In addition, OSMA, with the lessons learned in recent U.S. Navy (and other) benchmarking activities, will improve its functional audit capabilities, borrowing techniques used by the Naval Sea Systems Command in submarine certifications. NASA is also improving its trend analysis, problem tracking, and lessons learned systems (ref: F7.4-9, -10, and -11), all in a concerted effort to ensure the TA invokes the correct technical requirements. In order to improve OSMA insight and to reduce confusion cited in F7.4-13, NASA is formalizing its SMA Prelaunch Assessment Review (PAR) process for Shuttle and ISS, and the equivalent processes for expendable launch vehicles and experimental aerospace vehicle flight approvals, called Independent Mission Assurance Reviews (IMARs). Both of these processes will be standardized into a new NASA-wide review process called SMA Readiness Reviews (SMARRs).

In addressing the CAIB concern about the lack of mainstreaming and visibility of the system safety discipline (F7.4-4), OSMA has taken two actions, one long term and the other completed. First, as regards lack of mainstreaming of system safety engineering, the OSMA audit plan will include an assessment of the adequacy of system safety engineering by the audited project and/or line engineering organizations per the new NASA policy directives for Program management and ITA. As for the second concern about the lack of system safety visibility, for some years, the senior system safety expert in the Agency was also the OSMA Requirements Division Chief (now Deputy Chief, OSMA). To respond to the CAIB concern, OSMA has brought on a full-time experienced system safety manager who is the Agency's dedicated senior system safety assurance policy expert. In addition the Chief Engineer will select a Systems Safety Engineering Technical Warrant Holder who will be responsible for establishing systems safety engineering requirements.

The SMA Directorates supporting SSP are staffed with a combination of civil service and support contractors providing system safety, reliability, and quality expertise and services. Their role is predominantly assurance in nature, providing the Program with functional oversight of the compliance of the prime and sub-contractor engineering and operations with requirements. The civil service personnel assigned to work on Shuttle are functionally tied to their Center SMA organizations, and although some are collocated with their project or contractor element, their official supervisors are in the SMA organization.

The System Safety Review Panel (SSRP) process continues to evolve as the relationship between the ITA, SMA, and the SSP is defined and understood. This plan redefines the SSRP as the Engineering Risk Review Panels (ERRP). The

ERRP is designed to improve engagement by the engineering community into the safety process, including the development and maintenance of documentation such as hazard reports.

The organizational structure of the ERRP will consist of Level 2 (Program) and Level 3 (Project/Element) functionality. The ERRP's structure and processes continue to evolve in a phased approach. Until RTF, the ITA Shuttle System TWH will be represented at all ERRP levels through Engineering trusted agents who are assigned to support each ERRP. The trusted agents ensure that the engineering interests of the ITA are represented at all working levels of the ERRP and are reflected in the products resulting from these panels. After RTF, the Shuttle System TWH will reassess his/her role in all Shuttle Program panels and boards that deal with flight safety issues, including the ERRP.

The Level 2 Panel will ensure that the safety integration function remains at the Program level. It will have representation by all program elements as well as the Engineering Directorate, ITA, and SMA. The Lead ERRP Manager will also assure that Level 3 panels operate in accordance with safety program requirements. The Level 2 Panel exists to oversee and resolve integrated hazards, forwarding them to the System Integration Configuration Board (SICB), and finally to the ITA and the Program Manager for approval.

The Level 3 ERRPs will consist of a Johnson Space Center (JSC) Panel (Orbiter/extravehicular activity/government-furnished equipment/integration responsibility), a Marshall Space Flight (MSFC) Center Panel (External Tank/Reusable Solid Rocket Motor/Solid Rocket Booster/Space Shuttle Main Engine responsibility), and a Kennedy Space Center (KSC) Panel (ground servicing equipment/Ground Ops responsibility). As presently defined, the Level 3 Panels will be chaired by the independent SMA Directorates at each Space Operations Center, again with representation by trusted agents at these panels.

The Space Operations Mission Directorate Space Shuttle Certificate of Flight Readiness process is being updated to clearly show the new SMA, Integration, and ITA roles and responsibilities. Part of that will be a requirement for concurrence by the Chief Safety and Mission Assurance Officer on the flight readiness statement as a constraint to mission approval. Also, to clear up another ambiguity present in the system at the time of the *Columbia* accident, the JSC SMA Manager will not have a "third hat" as delegated NASA Headquarters OSMA representative on the Mission Management Team. An OSMA representative (the OSMA Shuttle Point of Contact (POC)) will fill that role in an advisory/functional oversight role.

Integration of the New ITA and SMA (R7.5-1/R7.5-2)

In a practical sense, the people that perform the responsibilities of SMA and the ITA need to be involved within a program or project beginning in the early stages and remain involved for the life of the program or project. R7.5-1 from the CAIB Report defined what activities at the program level must be clearly under formal ITA authority. At the same time, Chapter 7 discussion makes it clear that the SMA organization must be independent of the program and technically capable to provide proper check-and-balance with the program. Finally, the SMA organization must be able to perform its assurance functions in support of but independent of both program and engineering organizations.

In response to R7.5-1, NASA named the Chief Engineer to be the ITA. And that authority is delegated fully to responsible individuals who hold warrants under ITA authority for systems and engineering disciplines. Fundamentally, this concept brings a "balance of power" to program management such that the ITA sets technical requirements, the programs execute to that set of technical requirements, and SMA assures the requirements are satisfied. This means that the ITA owns the technical requirements and will be the waiver-granting authority for them.

The principal effect of the foregoing is the clear assignment of responsibility for execution of design and engineering, including the safety functions (FMEA, hazards analysis, reliability engineering, etc.) to Engineering with the ITA setting requirements and approving the resulting engineering products. In this context, SMA organizations have the responsibility for independently assuring that delivered products comply with requirements.

System Integration (R7.5-3)

The CAIB found several deficiencies in the organizational approach to Program system engineering integration for the Space Shuttle Program. Their recommendation R7.5-3 calls for a reorganization of the Space Shuttle Integration Office to "make it capable of integrating all elements of the Space Shuttle Program, including the Orbiter." The CAIB concluded, "…deficiencies in communication"…were a foundation for the *Columbia* accident. These deficiencies are byproducts of a cumbersome, bureaucratic, and highly complex Shuttle Program structure and the absence of authority in two key program areas that are responsible for integrating information across all programs and elements in the Shuttle program."

Integration Definition

NASA defines Integration as a system engineering function that combines the technical efforts of multiple system elements, functions, and disciplines to perform a higher-level system function in a manner that does not compromise the integrity of either the system or the individual elements. The Integration function assesses, defines, and verifies the required characteristics of the interactions that exist between multiple system elements, functions, and disciplines, as these interactions converge to perform a higher-level function.

Restructured Space Shuttle Systems Engineering and Integration Office

NASA has restructured its Shuttle Integration Office into a Space Shuttle Systems Engineering and Integration Office (SEIO) to include the systems engineering and integration of all elements of the Space Shuttle System. This new alignment is consistent with the new ITA where the Space Shuttle Systems TWH has responsibility not only for the Orbiter, but also for the integration of the technical requirements of the entire Space Shuttle system. The SEIO Manager now reports directly to the SSP Manager, thereby placing the SEIO at a level in the Space Shuttle organization that establishes the authority and accountability for integration of all Space Shuttle elements. The new SEIO charter clearly establishes that it is responsible for the systems engineering and integration of flight performance of all Space Shuttle elements. The number of civil service personnel performing analytical and element systems engineering and integration in the SEIO was doubled by acquiring new personnel from the JSC Engineering and Mission Operations Directorates and from outside of NASA. The role of the System Integration Plan (SIP) and the Master Verification Plans (MVPs) for all design changes with multi-element impact has been revitalized. The SEIO is now responsible for all SIPs and MVPs. These tools, along with the new role of the ITA as the owner of technical requirements, have energized SEIO to be a proactive function within the SSP for integration of design changes and verification. SIPs and MVPs have been developing for all major RTF design changes that impact multiple Shuttle elements.

Orbiter Project Office

The Space Shuttle Vehicle Engineering Office is now the Orbiter Project Office, and its charter is amended to clarify that SEIO is now responsible for integrating all flight elements. NASA reorganized and revitalized the Integration Control Board (ICB). The Orbiter Project Office is now a mandatory member of the ICB. The Space Shuttle Flight Software organization was moved from the Orbiter Project into the SEIO. This reflects the fact that the Shuttle Flight Software Office manages multiple flight element software sources besides the Orbiter.

Integration of Engineering at Centers

All SSP integration functions at MSFC, KSC, and JSC are now coordinated through the SEIO. Those offices receive technical direction from the SSP SEIO. The former MSFC Propulsion Systems Integration office is now called the Propulsion Systems Engineering and Integration (PSE&I) office. The PSE&I is increasing its contractor and civil servant technical strength and its authority within the Program. Agreements between the PSE&I Project Office and the appropriate MSFC Engineering organizations are being expanded to enhance anomaly resolution within the SSP.

Integrated Debris Environments/Certification

The SEIO is also responsible for generation of all natural and induced design environments analyses. Debris is now treated as an integrated induced environment that will result in element design requirements for generation limits and impact tolerance. All flight elements are being reevaluated as potential debris generators. Computations of debris trajectories under a wide variety of conditions will define the induced environment due to debris. The risk associated with the Orbiter Thermal Protection System will be reassessed for this debris environment, as will the systems of all flight elements.

ITA Interface

The SEIO Manager works closely with, supports the responsibilities of, and recognizes the authority over technical requirements of the Space Shuttle Systems TWH and the Discipline TWHs, as defined in NSTS 07700, Volume IV. SEIO will work closely with the TWHs to ensure: the adequacy of technical requirements employed for the Space Shuttle Program; the sufficiency, integrity, and consistency of the systems engineering approach; the robustness and thoroughness of integrated hazard analyses; and the acceptance of engineering analysis approaches and results by the Space Shuttle Systems TWH before technical options or proposed solutions are presented to the Space Shuttle Program Manager.

SMA Interface

The SEIO also works closely with the SMA organization to obtain independent verification that requirements have been met. SEIO will provide requirements, data, and analyses for use by SMA and will be the recipient of the independent verification results for the program. SEIO will factor these results into recommendations for program actions where appropriate. In addition, the conduct of hazard and risk analyses is coordinated with the SMA, who has the expertise for the government in conduct of these efforts.

Summary

The reorganized SEIO now addresses all elements of the Space Shuttle system including the Orbiter vehicle. The SEIO manager located at JSC has oversight and control of matrix Systems Engineering and Integration support from KSC and MSFC. SEIO works in compliance with ITA requirements and SMA. SEIO has revitalized its systems engineering and integration processes and has integrated the ITA. SEIO will incorporate the ITA in as an approval authority for variances to technical requirements, as documented in NSTS 07700, Volume IV. Additionally, SEIO will conduct integrated hazard analyses with the oversight of the Space Shuttle Systems TWH. The results of these analyses will be accepted or rejected by the Space Shuttle Systems TWH prior to use. The strengthened systems engineering and integration processes and organizational alignment fulfill the CAIB recommendations in supporting the Space Shuttle RTF actions.

Improving Engineering Integration Agency-wide

NASA has a broad range of programs, projects, and research activities with varying scope that are distributed within and between individual NASA Centers. NASA Headquarters, through the Office of the Chief Engineer, has established the policies that govern Program management, which include the policies for system integration functions as related to the project lifecycle. NASA will assess the effectiveness of integration functions for all of its programs and projects. Further, the policies that govern integration will be assessed and strengthened, as appropriate, to apply to all programs and projects.

POST STS-114 UPDATE

NASA leadership is in the process of revising the ITA to align it with the new management strategy for the Agency outlined in the "Strategic Management and Governance Handbook," NPD 1000.0, approved in August 2005. A key feature of NPD 1000.0 is the separation of the responsibility and management of programmatic from institutional capabilities such as engineering. Responsibility for engineering is given to Center Directors, reporting directly to the new NASA Associate Administrator, with program and project management responsibility given to Mission Directorate Associate Administrators. The version of ITA exercised in returning STS-114 safely to flight, and other Agency programs and projects during this time, is being updated to recognize these new responsibilities. The Office of the Chief Engineer has the action to create and implement an authority for technical decision making that reflects this separation of engineering and programmatic responsibilities. This authority will comply with CAIB recommendation R7.5-1 and provide, as its foundation, technical excellence in engineering. In defining the new authority, lessons learned during the implementation of ITA since January 2005 as well as the Stafford-Covey Return to Flight Task Group observations will be evaluated.

A definitive schedule for implementing the new technical authority has been established, with a transition planned for early March 2006. During the interim, current ITA responsibilities for programs and projects such as STS-121 will continue, until details for the new technical authority process are defined and fully transitioned.

SCHEDULE

Responsibility	Due Date	Activity/Deliverable
TA issues policy and warrants	(Completed)	Initial policy/warrants developed
SSP integrated with TA	(Completed)	TA in place and operational prior to RTF

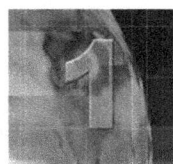

Columbia Accident Investigation Board
Recommendation 9.2-1

Prior to operating the Shuttle beyond 2010, develop and conduct a vehicle recertification at the material, component, subsystem, and system levels. Recertification requirements should be included in the Service Life Extension Program.

Note: NASA has closed this recommendation through the formal Program Requirements Control Board process. The following summary details NASA's response to the recommendation and any additional work NASA performed beyond the *Columbia* Accident Investigation Board recommendation.

BACKGROUND

In 2002, NASA initiated the Space Shuttle Service Life Extension Program (SLEP) to extend the vehicle's useful life. When SLEP was initiated, evaluation of the vehicle's mid-life recertification needs was a foundational activity. On January 14, 2004, the Vision for Space Exploration announced plans for the Space Shuttle to retire following completion of the International Space Station assembly, planned for 2010. Thus, recertification for operating the Space Shuttle beginning in 2010 is not necessary. The vision shortens the required service life of the Space Shuttle and, as a result, the scope of vehicle mid-life certification was changed substantially.

NASA IMPLEMENTATION

Despite the reduced time frame for the operation of the Shuttle, NASA continues to place a high priority on maintaining the safety and capability of the Orbiters. A key element of this is timely verification that hardware processing and operations are within qualification and certification limits. These activities will revalidate the operational environments (e.g., loads, vibration, acoustic, and thermal environments) used in the original certification. This action is addressed in SSP-13.

NASA has approved funding for work to identify and prioritize additional analyses, testing, or potential redesign of the Shuttle to meet recertification requirements. The identification of these requirements puts NASA on track for making appropriate choices for resource investments in the context of the Vision for Space Exploration.

In May 2003, the Space Flight Leadership Council approved the first SLEP package of work, which included funding for Orbiter mid-life certification and complementary activities on the Orbiter Fleet Leader Project, Orbiter Corrosion Control, and an expanded Probabilistic Risk Assessment for the Shuttle. In February 2004, SLEP Summit II revisited some of the critical issues for life extension and began a review of how to appropriately refocus available resources for the greatest benefit to NASA.

FINAL UPDATE

NASA continues to assess what is required for the remaining service life of the Space Shuttle. We will continue to invest in safety and sustainability.

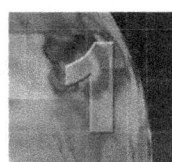
Columbia Accident Investigation Board
Recommendation 10.3-1

Develop an interim program of closeout photographs for all critical sub-systems that differ from engineering drawings. Digitize the closeout photograph system so that images are immediately available for on-orbit troubleshooting. [RTF]

Note: The Stafford-Covey Return to Flight Task Group held a plenary session on December 15, 2004, and NASA's progress toward answering this recommendation was reviewed. The Task Group agreed the actions taken were sufficient to fully close this recommendation.

BACKGROUND

Closeout photography is used, in part, to document differences between actual hardware configuration and the engineering drawing system. The *Columbia* Accident Investigation Board (CAIB) recognized the complexity of the Shuttle drawing system and the inherent potential for error and recommended to upgrade the system (ref. CAIB Recommendation 10.3-2).

Some knowledge of vehicle configuration can be gained by reviewing photographs maintained in the Kennedy Space Center (KSC) Quality Data Center film database or the digital Still Image Management System (SIMS) database. NASA now uses primarily digital photography. Photographs are taken for various reasons, such as to document major modifications, visual discrepancies in flight hardware or flight configuration, and vehicle areas that are closed for flight. NASA employees and support contractors can access SIMS. Prior to SIMS, images were difficult to locate, since they were typically retrieved by cross-referencing the work-authorizing document that specifies them.

NASA IMPLEMENTATION

NASA formed a Photo Closeout Team consisting of members from the engineering, quality, and technical communities to identify and implement necessary upgrades to the processes and equipment involved in vehicle closeout photography. KSC closeout photography includes the Orbiter, Space Shuttle Main Engine, Solid Rocket Boosters, and External Tank based on Element Project requirements. The Photo Closeout Team divided the CAIB action into two main elements: (1) increasing the quantity and quality of closeout photographs, and (2) improving the retrieval process through a user-friendly Web-based graphical interface system (figure 10.3-1-1).

Increasing the Quantity and Quality of Photographs

Led by the Photo Closeout Team, the Space Shuttle Program (SSP) completed an extensive review of existing closeout photo requirements. This multi-center, multielement, NASA and contractor team systematically identified the deficiencies of the current system and assembled and prioritized improvements for all Program elements. These priorities were distilled into a set of revised requirements that has been incorporated into Program documentation. Newly identified requirements included improved closeout photography of extravehicular activity tool contingency configurations and middeck and payload bay configurations. NASA has also added a formal photography work step for KSC-generated documentation and mandated that photography of all Material Review Board (MRB) reports be archived in the SIMS. These MRB problem reports provide the formal documentation of known subsystem and component discrepancies, such as differences from engineering drawings.

To meet the new requirements and ensure a comprehensive and accurate database of photos, NASA established a baseline for photo equipment and quality standards, initiated a training and certification program to ensure that all operators understand and can meet these requirements, and improved the SIMS. To verify the quality of the photos being taken and archived, NASA has developed an ongoing process that calls for SIMS administrators to continually audit the photos being submitted for archiving in the SIMS. Operators who fail to meet the photo requirements will be decertified pending further training. Additionally, to ensure the robustness of the archive, poor-quality photos will not be archived.

NASA determined that the minimum resolution for closeout photography should be 6.1 megapixels to provide the necessary clarity and detail. KSC has procured 36 Nikon 6.1 megapixel cameras and completed a test program in cooperation with Nikon to ensure that the cameras meet NASA's requirements.

Improving the Photograph Retrieval Process

To improve the accessibility of this rich database of Shuttle closeout images, NASA has enhanced SIMS by developing a Web-based graphical interface. Users will be able to easily view the desired Shuttle elements and systems and quickly drill down to specific components, as well as select photos from specific Orbiters and missions. SIMS will also include hardware reference drawings to help users identify hardware locations by zones. These enhancements will enable the Mission Evaluation Room (MER) and Mission Management Team to quickly and intuitively access relevant photos without lengthy searches, improving their ability to respond to contingencies.

To support these equipment and database improvements, NASA and United Space Alliance (USA) have developed a training program for all operators to ensure consistent photo quality and to provide formal certification for all camera operators. Additional training programs have also been established to train and certify Quality Control Inspectors and Systems Engineering personnel; to train Johnson Space Center (JSC) SIMS end users, such as staff in the MER; and to provide a general SIMS familiarization course. An independent Web-based SIMS familiarization training course is also in development.

STATUS

NASA has revised the Operation and Maintenance Requirements System (OMRS) to mandate that general closeout photography be performed at the time of the normal closeout inspection process and that digital photographs be archived in SIMS. Overlapping photographs will be taken to capture large areas. NSTS 07700 Volume IV and the KSC MRB Operating Procedure have also been updated to mandate that photography of visible MRB conditions be entered into the SIMS closeout photography database. This requirement ensures that all known critical subsystem configurations that differ from Engineering Drawings are documented and available in SIMS to aid in engineering evaluation and on-orbit troubleshooting.

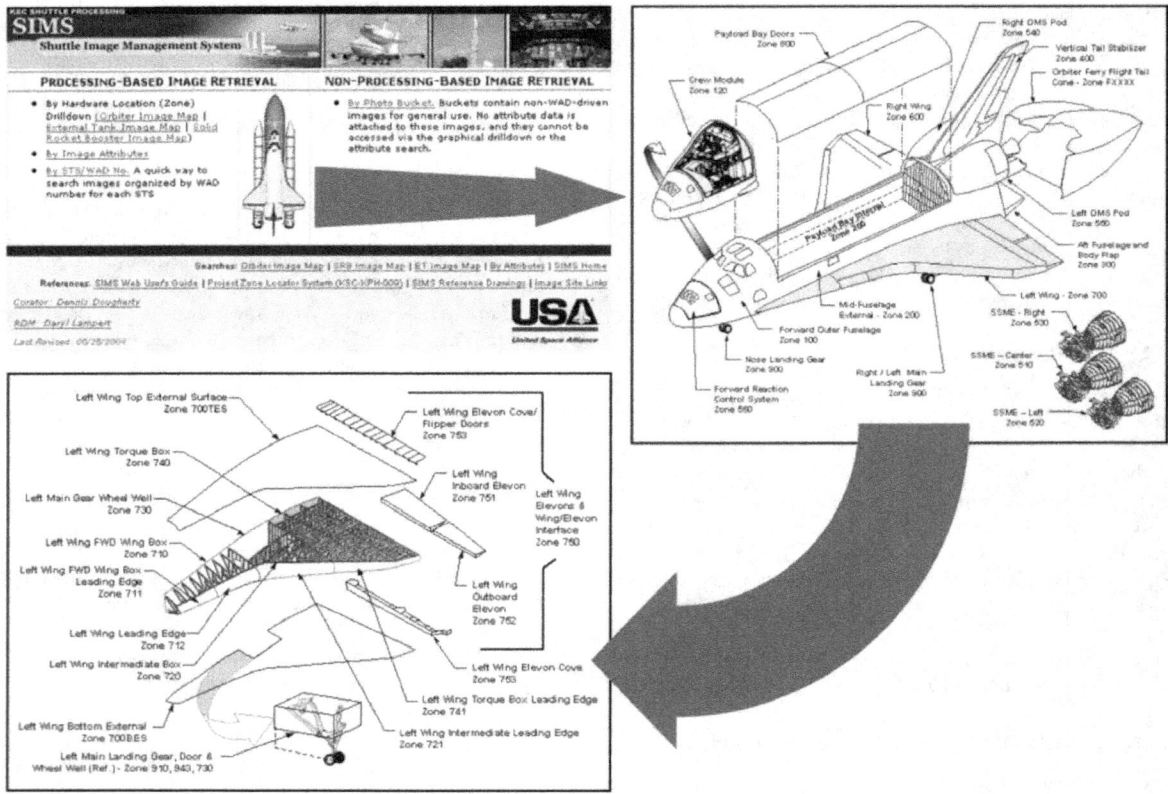

Figure 10.3-1-1. Enhanced SIMS graphic interface.

The revised Shuttle Program closeout photography requirements are documented in RCN KS16347R1 to OMRS File II, Volume I S00GEN.625 and S00GEN.620. Additionally, NASA Quality Planning Requirements Document (QPRD) SFOC-GO0007 Revision L and USA Operation Procedure USA 004644, "Inspection Points and Personnel Traceability Codes," were updated to be consistent with the revised OMRS and QPRD documents. The upgraded SIMS is operational and available for use by all SSP elements. On October 29, 2004, SIMS was successfully used during an inter-center Launch Countdown Simulation with the KSC Launch Team, JSC Flight Control Team, MER, Systems Engineering and Integration Office, and Huntsville Operations Support Center. As a part of the simulation scenario, the SIMS was accessed by participating organizations, and was used to retrieve and view photos to verify the configuration of an Orbital Maneuvering System Pod flight cap installed on the Orbiter.

Training for critical personnel is complete, and will be ongoing to ensure the broadest possible dissemination within the user community. Formal SIMS training has been provided to JSC MER and Marshall Space Flight Center (MSFC) personnel. Photographer training is complete and training classes are held regularly for any new or existing employees needing the certification. SIMS computer-based training (CBT) has been developed and released. Use of SIMS has been successfully demonstrated in a launch countdown simulation at KSC, which included participation from the KSC Launch Team, JSC Flight Control Team, MER, MSFC Huntsville Operations and Support Center (HOSC), and Systems Engineering & Integration (SE&I). Implementation of requirements into KSC operational procedures is continuing.

In July 2004, the Stafford-Covey Return to Flight Task Group reviewed NASA's progress and agreed to conditionally close this recommendation. The full intent of CAIB Recommendation 10.3-1 has been met and full closure of this recommendation was achieved in December 2004.

FINAL UPDATE

Incorporation of all R10.3-1 actions was successfully completed during the STS-114 flow prior to launch. No significant modifications were needed for STS-121 and subsequent missions. A significant improvement has been achieved in the quality and quantity of the archived images as compared to previous flows. The total SIMS database image count of 52,119 for STS-114 was over four times that of the 12,438 images taken during STS-107 processing. Almost all images currently taken during normal Shuttle processing are of much better resolution than those taken during previous flows. The improvements are due to the upgraded equipment purchased, as well as better personnel training. The USA SIMS administrator performs random daily audits of the SIMS images for quality content and correct attributes. A one-week, 100-percent audit of SIMS was performed in early 2005, and another in December 2005. Results of the audits are sent to the Quality Manager II for Vertical and Horizontal Processing. The findings are then shared with the photographers for further improvement and lessons learned.

User feedback from the Space Shuttle community has been very positive. Feedback from the JSC MER Manager confirmed that the SIMS system was actively used during the STS-114 and STS-121 missions and will be an integral tool for future missions. In addition to the Ground Operations organizations, the JSC Vehicle Integration Test Team Office and KSC NASA Space Shuttle Logistics Depot are two of the newest organizations that are using SIMS. KSC is also in the process of incorporating off-site landing and ground support groups into SIMS.

A demonstration of SIMS processes and technology was given to, and met with praise by, the Exploration Systems Chief Engineer and Constellation Project Managers for consideration in NASA exploration programs.

SCHEDULE

Responsibility	Due Date	Activity/Deliverable
KSC	Feb 04 (Completed)	Develop SIMS drilldown and graphical requirements
SSP	Apr 04 (Completed)	Projects transmit photo requirements to KSC Ground Operations
KSC	May 04 (Completed)	Complete graphical drilldown software implementation
KSC	Jun 04 (Completed)	Develop/complete SIMS training module
KSC	Jul 04 (Completed)	Provide training to MER. Demonstrate SIMS interface to JSC/MSFC
KSC	Aug 04 (Completed)	SIMS CBT course development and deployment. (SIMS familiarization course was provided as needed until CBT was completed)
KSC	Aug 04 (Completed)	Photographer training
SSP	Oct 04 (Completed)	S0044 Launch Countdown Simulation run set for 10/29 with full support from the KSC Launch Team, JSC Flight Control Team, MER, MSFC HOSC, and SE&I

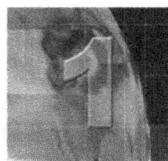

Columbia Accident Investigation Board
Recommendation 10.3-2

Provide adequate resources for a long-term program to upgrade the Shuttle engineering drawing system including
- Reviewing drawings for accuracy
- Converting all drawings to a computer-aided drafting system
- Incorporating engineering changes

Note: NASA has closed this recommendation through the formal Program Requirements Control Board (PRCB) process. The following summary details NASA's response to the recommendation and any additional work NASA performed beyond the *Columbia* Accident Investigation Board (CAIB) recommendation.

BACKGROUND

The CAIB noted deficiencies in NASA's documentation of the Space Shuttle's configuration and therefore recommended a two-step solution. The first was an interim program of closeout photographs for all critical subsystems that differ from engineering drawings (Recommendation 10.3-1). The second is outlined in Recommendation 10.3-2 (above).

NASA IMPLEMENTATION

The Space Shuttle Program (SSP) created a plan for converting Orbiter drawings to computer-aided design (CAD) models and incorporating outstanding engineering orders (EOs). Benefits of the plan include:

- Reducing the EO count to zero on all converted drawings.
- Verifying the accuracy of design data and eliminate dimensional inaccuracies.
- Reconciling many differences between as-designed and as-built configurations.
- Enabling the use of modern engineering and analysis tools.
- Improving safety.
- Recognizing some efficiency improvements.
- Positioning the Shuttle for an evolutionary path.

However, it will take at least three years and $150M to complete the effort.

STATUS

NASA considered the plan for converting all Orbiter drawings to CAD models and incorporating all outstanding EOs in June 2004. A cost benefit analysis did not support approval of this plan given the shortened life of the SSP. NASA did, however, approve a plan to incorporate some outstanding EOs based on frequency of use and complexity.

Because there is not enough time left in the Program to fully recognize the long-term plan, NASA has redoubled its effort to fully comply with CAIB Recommendation 10.3-1 in implementing an interim program of closeout photographs for all critical subsystems that differ from engineering drawings. This interim program was assessed and conditionally approved by the Stafford Covey Return to Flight Task Group in July 2004.

FINAL UPDATE

The SSP will continue to incorporate outstanding EOs into its drawings. Additionally, the SSP will continue to explore options to improve dissemination of its engineering data across the Program.

SCHEDULE

Responsibility	Due Date	Activity/Deliverable
SSP	May 04 (Completed)	Begin EO incorporation
SSP	Jun 04 (Completed)	Present drawing conversion concept to the PRCB

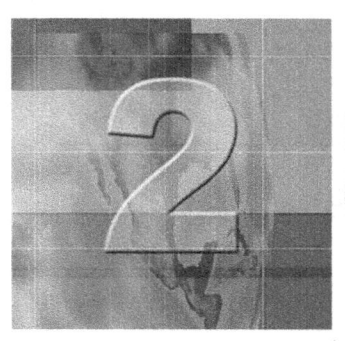

Raising the Bar – Other Corrective Actions

NASA undertook a fundamental reevaluation of its culture and processes. This process extended beyond immediate Return to Flight actions to institutional changes in the way NASA ensure safety and mission success. Much of the work needed for this effort was captured in the CAIB observations. Part 1 of this plan addressed the CAIB recommendations. Part 2 addresses other corrective actions, including internally generated actions, the observations contained in Chapter 10 of the CAIB Report, and CAIB Report, Volume II, Appendix D, Recommendations..

Space Shuttle Program Actions

In addition to the CAIB recommendations, NASA received and evaluated contributions from a variety of other sources, including from within the Space Shuttle Program. NASA systematically assessed all corrective actions and incorporated them into this Implementation Plan. This section contains self-defined actions and directives of the Space Shuttle Program that were addressed by NASA as well as the constraints to flight recommended by the Columbia Accident Investigation Board.

Space Shuttle Program Return to Flight Actions
Space Shuttle Program Action 1

NASA will commission an assessment, independent of the Space Shuttle Program (SSP), of the Quality Planning and Requirements Document (QPRD) to determine the effectiveness of government mandatory inspection point (GMIP) criteria in assuring verification of critical functions before each Shuttle mission. The assessment will determine the adequacy of existing GMIPs to meet the QPRD criteria. Over the long term, NASA will periodically review the effectiveness of the QPRD inspection criteria against ground processing and flight experience to verify that GMIPs are effectively assuring safe flight operations. This action also encompasses an independently led bottom-up review of the Kennedy Space Center Quality Planning Requirements Document (CAIB Observation 10.4-1).

Note: NASA has closed this action through the formal Program Requirements Control Board (PRCB) process. The following summary details NASA's response to the Space Shuttle Program (SSP) action and any additional work NASA performed beyond the SSP action.

BACKGROUND

The *Columbia* Accident Investigation Board (CAIB) Report highlighted the Kennedy Space Center (KSC) and Michoud Assembly Facility (MAF) Government Mandatory Inspection Point (GMIP) processes as an area of concern. GMIP inspection and verification requirements are driven by the KSC Ground Operations Quality Planning and Requirements Document (QPRD) and the Marshall Space Flight Center (MSFC) Mandatory Inspection Documents.

NASA IMPLEMENTATION

Assuring that NASA maintains appropriate oversight of critical work performed by contractors is key to our overall ability to ensure a safe flight operations environment. As a result, the Space Flight Leadership Council (SFLC) and the Associate Administrator for Safety and Mission Assurance, (SMA) with concurrence from theSMA Directors at KSC, Johnson Space Center (JSC), and MSFC, chartered an independent assessment of the Space Shuttle Program (SSP) GMIPs for KSC Orbiter Processing and MAF External Tank (ET) manufacturing. The SFLC also approved the establishment of an assessment team consisting of members from various NASA centers, the Federal Aviation Administration, the U.S. Army, and the U.S. Air Force. This Independent Assessment Team (IAT) assessed the KSC QPRD and the MAF Mandatory Inspection Document criteria, their associated quality assurance processes, and the organizations that perform them. The team issued a final report in January 2004, and the report recommendations have become formal SSP actions. The report was used as the primary basis for the SSP to evaluate similar GMIP activity at other Space Shuttle manufacturing and processing locations. The IAT report concluded that the NASA quality assurance programs in place today are relatively good, based on the ground rules that were in effect when the programs were formulated; however, these rules have changed since the programs' formulation. The IAT recommended that NASA reassess its quality assurance requirements, based on the modified ground rules established as a result of the *Columbia* accident. The modified ground rules for the Space Shuttle include an acknowledgement that the Shuttle is an aging, relatively high-risk development vehicle. As a result, the NASA SMA Quality Assurance Program must help ensure both safe hardware and an effective contractor quality program.

The IAT's findings echo the observations and recommendations of the CAIB. The team made the following recommendations:

- Strengthen the Agency-level policy and guidance to specify the key components of a comprehensive Quality Assurance Program that includes the appropriate application of GMIPs.
- Establish a formal process for periodically reviewing QPRD and GMIP requirements at KSC and the Mandatory Inspection Documents and GMIPs at MAF against updates to risk management documentation (hazard analyses, failure modes and effects analyses/critical item list) and other system changes.
- Continue to define and implement formal, flexible processes for changing the QPRD and adding, changing, or deleting GMIPs.
- Document and implement a comprehensive Quality Assurance Program at KSC in support of the SSP activities.
- Develop and implement a well-defined, systematically deployed Quality Assurance Program at MAF.

In parallel with the IAT's review, a new process to make changes to GMIP requirements was developed, approved, and baselined at KSC. This process ensures that anyone can submit a proposed GMIP change, and that the initiator who requests a change receives notification of the disposition of the request and the associated rationale. That effort was completed in September 2003. Since then, several change requests have been processed, and the lessons learned from those requests have been captured in a formal revision to the change process document, KDP-P-1822, Rev. A. This process uses a database for tracking the change proposal, the review team's recommendations, and the Change Board's decisions. The database automatically notifies the requester of the decision, and the process established a means to appeal decisions.

STATUS

In response to the CAIB Report, MSFC and KSC Shuttle Processing SMA initiated efforts to address the identified Quality Assurance Program shortfalls.

The following activities are completed or in progress at KSC:

- A formal process was implemented to revise GMIPs.
- A change review board comprised of the Shuttle Processing Chief Engineer, Shuttle SMA Chief, and, as applicable, contractor engineering and SMA representatives has been implemented to disposition proposed changes.
- A new process was developed to document and has implemented temporary GMIPs while permanent GMIP changes are pending, or as deemed necessary for one-time or infrequent activities. The new process also covers supplemental inspection points.
- A documented process and database was developed to trend GMIP accept/reject data to enhance first-time quality determination and identify paths for root cause correction.
- Surveillance has been increased through additional random inspections for hardware and compliance audits for processes. The data are used as part of the Certificate of Flight Readiness (COFR) package.
- Enhanced Quality Inspector training, based on benchmarking similar processes, has been identified and incorporated into the training plan that has been approved by the SSP.
- A QPRD Baseline Review that covered all systems was completed in July 2005.
- Metrics for trending and analysis of GMIP activity have been established and the data are used as part of the COFR process.

In response to the shortfalls identified at MAF, MSFC initiated the following:

- Applying CAIB observations and the IAT recommendations to all MSFC propulsion elements.
- Formalizing and documenting processes that have been in place for Quality Assurance Program planning and execution at each manufacturing location.
- Increasing the number of inspection points for ET assembly.
- Increasing the level and scope of vendor audits (process, system, and supplier audits).
- Improving training across the entire MSFC SMA community, with concentration on the staff stationed at manufacturer and vendor resident management offices.
- Further strengthening the overall Space Shuttle Quality Assurance Program by establishing a new management position and filling it on the Shuttle SMA Manager's staff with a specific focus on Quality.

FINAL UPDATE

The efforts to address the identified Quality Assurance Program shortfalls were delayed by two major hurricanes in 2005, Katrina and Rita. These two hurricanes caused significant disruption to the MAF facilities and personnel. Once MAF returned to normal operations in 2006, the work to respond to this recommendation was accomplished as outlined above.

SCHEDULE

Responsibility	Due Date	Activity/Deliverable
KSC Shuttle Processing	Sep 03 (Completed)	Develop and implement GMIP change process
Headquarters	Oct 03 (Completed)	Report out from IAT
Headquarters	Jan 04 (Completed)	Publish the IAT report
KSC Shuttle Processing	Apr 04 (Completed)	Develop and implement temporary GMIP process
SSP	Dec 04 (Completed)	Develop SSP Quality Assurance Policy for Civil Servants
MAF	Dec 04 (Completed)	Develop SMA Plan for Resident Management Office
KSC Shuttle Processing	Mar 04 (Completed)	Develop process for review of QPRD and kick off the baseline review
KSC Shuttle Processing	Jul 05 (Completed)	Completed baseline review of QPRD and populated rationale in the GMIP database
MAF	Apr 06 (Completed)	Established a closed-loop GMIP tracking system

Space Shuttle Program Return to Flight Actions
Space Shuttle Program Action 2

The Space Shuttle Program will evaluate relative risk to the public underlying the entry flight path. This study will encompass all landing opportunities from each inclination to each of the three primary landing sites.

Note: NASA has closed this action through the formal Program Requirements Control Board (PRCB) process. The following summary details NASA's response to the Space Shuttle Program (SSP) action and any additional work NASA performed beyond the SSP action.

BACKGROUND

The *Columbia* accident highlighted the need for NASA to better understand entry public risk. In its report, the *Columbia* Accident Investigation Board (CAIB) observed that NASA should take steps to mitigate the risk to the public from Orbiter entries. NASA now understands and has taken the appropriate steps to mitigate potential risks associated with entry public risk, a topic that is also covered in CAIB Observations 10.1-2 and 10.1-3.

NASA IMPLEMENTATION

NASA's concern for safety addresses public safety, the safety of our employees, and the safety of our critical national assets, such as the Space Shuttle. As a part of our work to improve safety in preparation for Return to Flight (RTF), we have undertaken a significant assessment of the potential risk posed to the public by Shuttle entry. All of the work to improve the safety of the Space Shuttle has reduced the risk to the public posed by any potential vehicle failures during ascent or entry. Beyond these technical improvements, operational changes are now in place to further reduce public risk. These operational changes include improved insight into the Orbiter's health prior to entry and new landing site selection flight rules that factor in public risk determinations as appropriate.

The entry risk to the public from impacting debris is a function of three fundamental factors: (1) the probability of vehicle loss of control (LOC) and subsequent breakup, (2) quantity and distribution of surviving debris, and (3) the population under the entry flight path. NASA has identified the phases of entry that present a greater probability of LOC based on elements such as increased load factors, aerodynamic pressures, and thermal conditions. The measures undertaken to improve crew safety and vehicle health will result in a lower probability of LOC, thereby improving the public safety during entry. Other factors, such as the effect of population sheltering, are also considered in the assessment.

NASA has completed a study of the public risks associated with entry to its three primary landing sites: Kennedy Space Center (KSC) in Florida; Edwards Air Force Base (EDW) in California; and White Sands Space Harbor/Northrop (NOR) in New Mexico. We have evaluated the full range of potential ground tracks for each site and conducted sensitivity studies to assess the entry risk to the public for each. NASA also has incorporated entry risk to the public, as well as crew risk considerations, into the entry flight rules that guide the flight control team's selection of landing opportunities.

STATUS

For NASA's risk assessment of the Space Shuttle landing tracks, more than 800 entry trajectories were analyzed covering the three primary landing sites from the Space Shuttle orbit inclination of 51.6° for International Space Station flights. The full range of entry crossrange[1] possibilities to each site was studied in increments of 25 nautical miles for all ascending (south to north) and descending (north to south) approaches. Figure SSP 2-1 displays the ground tracks analyzed for the 51.6° inclination orbit. The results indicate that some landing opportunities have an increased public risk compared to others.

The SSP recommended that the current landing site priorities be maintained, and that KSC remain our primary landing site. NASA will use operational methods and vehicle safety improvements implemented in preparation for RTF to manage the risk to the public posed by vehicle LOC during entry. NASA has developed flight rules to avoid certain opportunities to abate risk to the general public while satisfying other landing site selection priorities for weather, consumables, runway conditions, and entry constraints. Additionally, NASA has developed flight rules that give priority to lower risk opportunities in situations where the Space Shuttle's entry capabilities are compromised.

---

[1]Entry crossrange is defined as the distance between the landing site and the point of closest approach on the orbit ground track. This number is operationally useful to determine whether or not the landing site is within the Shuttle's entry flight capability for a particular orbit.

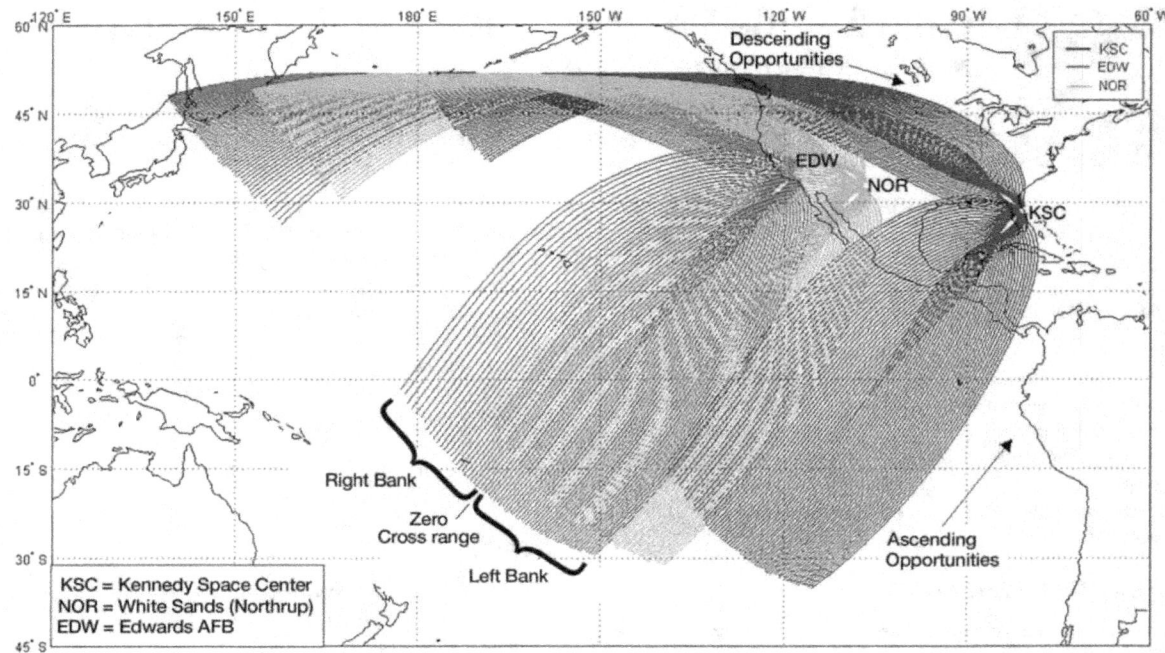

Figure SSP 2-1. Possible entry ground tracks from 51.6° orbit inclination.
Blue lines are landing at KSC, green at NOR, red at EDW.

NASA Headquarters (HQ) released a draft policy regarding public safety during all phases of space flight missions into its NASA Online Directives Information System. The policy is currently under review by all stakeholders. See O10.1-1, "Public Risk Policy," for details regarding this policy.

FINAL UPDATE

All RTF actions to evaluate relative risk to the public underlying the entry flight path are now complete.

SCHEDULE

Responsibility	Due Date	Activity/Deliverable
SSP	Jul 03 (Completed)	Preliminary results to RTF Planning Team and SSP PRCB
SSP	Sep 03 (Completed)	Update to RTF Planning Team and SSP PRCB
SSP	Jan 04 (Completed)	Update to RTF Planning Team and SSP PRCB
SSP	Jun 04 (Completed)	Update to SSP PRCB
SSP	Jun 04 (Completed)	Entry risk overview to NASA HQ
SSP	Dec 04 (Completed)	Report to SSP PRCB
NASA HQ	Feb 05 (Completed)	Report to HQ Ops Council

Space Shuttle Program Return to Flight Actions
Space Shuttle Program Action 3

NASA will evaluate the feasibility of providing contingency life support on board the International Space Station (ISS) to stranded Shuttle crewmembers until repair or rescue can be effected.

Note: NASA has closed this action through the formal Program Requirements Control Board (PRCB) process. The following summary details NASA's response to the Space Shuttle Program (SSP) action and any additional work NASA performed beyond the SSP action.

BACKGROUND

NASA has examined options for providing an emergency capability to sustain Shuttle crews on the International Space Station (ISS), should the Orbiter become unfit for entry. This Contingency Shuttle Crew Support (CSCS) capability could, in an emergency, sustain a Shuttle crew on board the ISS for a limited time to enable a repair to the Orbiter or allow the crew to be returned to Earth via a rescue mission. CSCS is not intended to mitigate known but unacceptable risks; rather, it is a contingency plan of last resort with limited capability to sustain the crew on the ISS. It must be noted that CSCS is not a certified and redundant ISS capability.

NASA IMPLEMENTATION

The fundamental rationale for Return to Flight was to control the liberation of critical debris from the External Tank (ET) during ascent. NASA resumed Shuttle missions only when we had sufficient confidence in the ET to fly. While CSCS offers a viable emergency capability for crew rescue, it will not be used to justify flying a Shuttle that is otherwise deemed unsafe.

CSCS provides an additional level of mitigation from residual risk following improvements to the ET. This was particularly desirable during the first few flights when we were validating the improvements made to the Shuttle system and specifically the ET thermal protection system. It is highly unlikely that the combination of failures necessary to lead NASA to invoke the CSCS capability will occur. It is secondary risk control and will be accomplished with zero fault tolerance in areas where ISS resources are taxed by an increased crew size. This approach is consistent with how NASA addresses other emergency measures, such as contingency launch aborts, to reduce residual risk to the crew.

STATUS

At the Space Flight Leadership Council on June 9, 2004, NASA approved the joint SSP/ISS proposal to pursue CSCS as a contingency capability for STS-114 and STS-121. In July 2006, the SSP extended preparations for CSCS for all remaining Shuttle flights and subsequently mandated a rapid launch on need capability for the Hubble Space Telescope (HST) Servicing Mission 4. CSCS capability will not be fault tolerant and is built on the presumption that, if necessary, all ISS consumables in addition to all Shuttle reserves will be depleted to support it. In the most extreme CSCS scenarios, it is possible that ISS will be decrewed following Shuttle crew rescue until consumables margins can be reestablished and a favorable safety review is completed. NASA ensures that the SSP has the capability to launch a rescue Shuttle mission within the time period that the ISS Program can reasonably predict that the combined Shuttle and ISS crew can be sustained on the ISS while allowing sufficient time to decrew the ISS following Shuttle departure, if decrew is necessary. At the launch minus 6 months, launch minus 3 months, and launch minus 1 month milestones, both programs report CSCS status to the Joint Program Requirements Control Board (JPRCB). ISS assesses its latest consumable levels, and SSP assesses its latest launch schedules. Any disconnects are resolved through slipping launch schedules or manifesting of additional consumables. This time period of supportable CSCS duration, which is referred to as the ISS "engineering estimate," represents a point between worst- and best-case operational scenarios for the ISS based on engineering judgment and operational experience. Additionally, each program element is required to report at the Flight Readiness Review its ability to support a rescue flight and identify any issues that might impact the support.

For planning purposes, NASA assumes that the failures preventing the entry of the stranded Orbiter can be resolved before launching the rescue Shuttle. In an actual CSCS situation, it may not be possible to protect the rescue Shuttle from the hazards that resulted in the damage that precipitated the need for a rescue, and a difficult risk-risk trade analysis would be performed at the Agency level or above before proceeding to launch.

The Mission Management Team (MMT) conducted several CSCS simulations prior to the launch of STS-114, and another simulation in January 2006, in preparation for the STS-121 Return to Flight mission.

Contingency Capability for CSCS

CSCS is a contingency capability that would be employed only under the direst emergency situations. In NASA's formal risk management system, CSCS does not improve an otherwise "unacceptable" risk into the "accepted" category. The implementation of risk mitigation efforts such as CSCS have been accomplished to the greatest degree practicable, but these efforts are not primary controls to the SSP Integrated Hazards of "Degraded Functioning of Orbiter Thermal Protection System" and "Damage to the Windows Caused by the Natural or Induced Debris Environment." Accordingly, CSCS verification standards are based on risk management decisions by an informed Program management.

The use of CSCS as a contingency capability is analogous to some of our other abort modes. The ability to perform emergency deorbits provides some protection against cabin leaks and multiple system failures. Contingency ascent aborts offer the ability to abort launches to contingency landing sites as protection against two or three Space Shuttle Main Engine failures. In both of these examples, as in many others, the capability is not certified for all, or even most, scenarios. Nevertheless, they do offer mitigation against residual risk and uncertainty. Another analogy can be drawn between CSCS and the ejection seats that were installed in the Orbiter for the first four flights of the Shuttle Program. They offered some crew escape capability during the first part of ascent and the last part of descent and landing, but they by no means represented comprehensive protection. However, they were appropriate and valuable additional risk mitigation options during conduct of the initial test flights that validated the performance of the Shuttle system.

CSCS Requirements

The SSP and ISS Programs have been working to define CSCS requirements using our established JPRCB process. CSCS capability is not premised on the use of any International Partner resources other than those that are an integral part of joint ISS operations, such as common environmental health and monitoring systems. The additional capabilities that could be brought to bear by the International Partners to support CSCS could provide added performance margin.

To support the CSCS capability, NASA has evaluated the capability to launch on need to provide crew rescue. Using this capability, NASA has a second Shuttle ready for launch on short notice during the mission. The ability to launch a rescue mission within the ISS Program engineering estimate is held as a constraint to launch. The SSP, working with Safety and Mission Assurance and the ISS Program, has developed detailed criteria for the constraint. These criteria are reported to the JPRCB and documented in an SSP/ISS Program Memorandum of Agreement (MOA). Based on this MOA, both the SSP and the ISS Program are taking the necessary measures to satisfy their respective responsibilities.

NASA's designated rescue missions are subject to the same development requirements as any other Shuttle mission; however, they will be processed on an accelerated schedule. These assessments assume a work acceleration to three shifts per day, seven days a week, but no deletion of requirements or alteration of protocols.

Stranded Orbiter Undocking, Separation, and Disposal

The Mission Operations Directorate has developed procedures for undocking a stranded Orbiter from the ISS, separating to a safe distance, then conducting a deorbit burn for disposal into an uninhabited oceanic area. These procedures have been worked in detail at the ISS Safe Haven Joint Operations Panel (JOP), and have been simulated in a joint integrated simulation involving flight controllers and flight crews from both the ISS Program and the SSP. The requirements and procedures for safely conducting a disposal of a stranded Orbiter are well understood and documented.

Current plans call for the Orbiter crew to conduct a rewiring in-flight maintenance procedure on the day prior to disposal that would "hot wire" the docking system hook motors to an unpowered main electrical bus. Before abandoning the Orbiter and closing the hatches, the crew would set up the cockpit switches to enable all necessary attitude control, orbital maneuvering, and ground uplink control systems. On the day of disposal, after the hatches are closed, Mission Control would uplink a ground command to repower the bus, immediately driving the hooks to the open position. The rewiring procedure is well understood and within the SSP's experience base of successful on-orbit maintenance work.

The Orbiter will separate vertically upward and away from the ISS. Orbital mechanics effects will increase the relative opening rate and ensure a safe separation. The Mission Control Center will continue to control the attitude of the Orbiter within safe parameters. Once the Orbiter is farther than 1000 ft from the ISS, the attitude control motors will be used to increase the separation rate and to set up for the disposal burn for steep entry into Earth's atmosphere. The primary targeted impact zone would be near the western (beginning) end of an extremely long range of remote ocean. Planning a steep entry reduces the debris footprint; targeting the western end protects against eastward footprint migration due to underburn. This disposal plan has been developed with the benefit of lessons learned from the deorbit, ballistic entry, and ocean disposal of the Compton Gamma Ray Observatory in June 2000 and the Russian *Mir* Space Station in 2001.

FINAL UPDATE

Although CSCS was not invoked for the STS-114 or STS-121 missions, all plans and capabilities were in place to support a launch-on-need rescue mission if it had been required. Planning for CSCS and a launch-on-need Shuttle rescue flight will be provided for the remaining Space Shuttle flights. Special preparations have been developed for the HST rescue since the Space Station will not be available for CSCS; a rescue Shuttle will be ready to launch 7 days after the HST launch.

SCHEDULE

Responsibility	Due Date	Activity/Deliverable
ISS Program	Aug 03 (Completed)	Status International Partners at Multilateral Mission Control Boards
ISS Program	Nov 03 (Completed)	Assess ISS systems capabilities and spares plan and provide recommendations to ISS and SSP
ISS Program	Jun 04 (Completed)	Develop CSCS Integrated Logistics Plan
ISS Program and SSP	Jun 04 (Completed)	Develop waste management and water balance plans
ISS Program and SSP	Jun 04 (Completed)	Develop ISS Prelaunch Assessment Criteria
ISS Program	Jun 04 (Completed)	Develop food management plan
SSP/ISS Program	Jun 04 (Completed)	Develop crew health and exercise protocols
ISS Program	Jun 04 (Completed)	Assess and report ISS ability to support CSCS
SSP/ISS Program	Dec 04 (Completed)	Safe Haven JOP report to JPRCB on requirements to implement CSCS
MMT	May 05 (Completed)	CSCS MMT Simulation

Space Shuttle Program Return to Flight Actions
Space Shuttle Program Action 4

NASA will validate that the controls are appropriate and implemented properly for "accepted risk" hazards and any other hazards, regardless of classification, that warrant review due to working group observations or fault tree analysis.

Note: NASA has closed this action through the formal Program Requirements Control Board (PRCB) process. The following summary details NASA's response to the Space Shuttle Program (SSP) action and any additional work NASA performed beyond the SSP action.

BACKGROUND

Hazard analysis is the determination of potential sources of danger that could cause loss of life, personnel capability, system, or result in injury to the public. Hazard analysis is accomplished through (1) performing analyses, (2) establishing controls, and (3) establishing a maintenance program to implement the controls. Controls and verifications for the controls are identified for each hazard cause.

Accepted risk hazards are those hazards that, based on analysis, have a critical or catastrophic consequence and the controls of which are such that the likelihood of occurrence is considered higher than improbable and might occur during the life of the Program. Examples include critical single failure points, limited controls or controls that are subject to human error or interpretation, system designs or operations that do not meet industry or Government standards, complex fluid system leaks, inadequate safety detection and suppression devices, and uncontrollable random events that could occur even with established precautions and controls in place.

All hazards, regardless of classification, were reviewed considering working group observations or fault-tree analysis that called into question the classification of the risk or the efficacy of the mitigation controls. This work was reviewed at Program Safety Panels and Control Boards.

NASA IMPLEMENTATION

Each SSP project performed the following assessment for each accepted risk hazard report and any additional hazard reports indicated by the STS-107 accident investigation findings:

1. Verify proper use of hazard reduction precedence sequence per NSTS 22254, Methodology for Conduct of Space Shuttle Program Hazard Analyses.
2. Review the basis and assumptions used in setting the controls for each hazard, and determine whether they are still valid.
3. Verify each reference to Launch Commit Criteria, Flight Rules, Operation and Maintenance Requirements Specification Document, crew procedures, and work authorization documents as a proper control for the hazard cause.
4. Verify proper application of severity and likelihood per NSTS 22254, Methodology for Conduct of Space Shuttle Program Hazard Analyses, for each hazard cause.
5. Verify proper implementation of hazard controls by confirming existence and proper use of the control in current SSP documentation.
6. Identify any additional feasible controls that can be implemented that were not originally identified and verified.
7. Assure that all causes have been identified and controls documented.

The System Safety Review Panel (SSRP) and subsequent Safety Engineering Review Panels (SERPs) served as the forum to review the projects' assessment of the validity and applicability of controls. The SSRP and SERPs assessed the existence and effectiveness of controls documented in the hazard reports. In accordance with SSP requirements, the SSRP and SERPs reviewed, processed, and dispositioned updates to baselined hazard reports.

Although the scope of the Return to Flight (RTF) action encompassed only the accepted risk hazards, the STS-107 accident brought into question the implementation and effectiveness of controls in general. As such, existing controlled hazards were also suspect. The evaluation of all hazards, including the controlled hazards, was deemed beneficial and included in the post-RTF plan.

In summary, the purpose of this review was to confirm that the likelihood and severity of each accepted risk hazard are thoroughly and correctly understood and that mitigation controls are properly implemented.

STATUS

Each project and element reviewed its accepted risk hazard reports per the PRCB approved schedules. Their results were presented to the PRCB and accepted by the Program. NASA has identified and updated all the necessary risk documentation and reestablished the baseline for future operations of the Space Shuttle.

NASA completed an extensive rewrite of the External Tank and integration hazards for the Shuttle. As a result of this rigorous hazard documentation process, risk was more fully understood and mitigated before RTF.

FINAL UPDATE

NASA remains committed to continuous, thorough reviews and updates of all hazards for the remaining life of the Space Shuttle Program (SSP).

SCHEDULE

Responsibility	Due Date	Activity/Deliverable
SSRP	Oct 03 (Completed)	SSRP review element hazards and critical items list review processes Kennedy Space Center – Sep 9, 11 Reusable Solid Rocket Motor – Sep 24, 25 Integration – Oct Solid Rocket Booster – Sep 8 Space Shuttle Main Engine – Oct 7, 8
SSP	(Completed)	Identify and review "Accepted Risk" hazard report causes and process impacts
SSP	(Completed)	Analyze implementation data
SSP	(Completed)	Validate and verify controls and verification methods
SSP	(Completed)	Develop, coordinate, and present results and recommendation

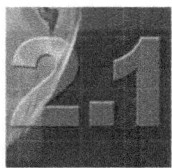

Space Shuttle Program Return to Flight Actions
Space Shuttle Program Action 5

NASA will determine critical debris sources, transport mechanisms, and resulting impact areas. Based on the results of this assessment, we will recommend changes or redesigns that would reduce the debris risk. NASA will also review all Program baseline debris requirements to ensure appropriateness and consistency.

Note: NASA has closed this action through the formal Program Requirements Control Board (PRCB) process. The following summary details NASA's response to the Space Shuttle Program (SSP) action and any additional work NASA performed beyond the SSP action.

BACKGROUND

A review of critical debris from all vehicle elements and the Launch Pad was necessary to prevent the recurrence of a failure similar to that which occurred on STS-107. As a result, NASA improved the end-to-end process for predicting debris release, transport, impacts, and the resulting damage as well as characterization of risk level. Engineering support for this approach crossed multiple organizations within NASA and support contractors. JSC and MSFC engineering provided key algorithm development for liberation mechanisms and debris transport. The External Tank (ET) Project and element engineers developed engineering material properties data and test validation data to support release mechanism predictions. "Project Iceball," lead by KSC, provided basic ice characterization basic data. The Orbiter Debris Impact Assessment Team developed the impact and damage tolerance algorithms for Reinforced Carbon-Carbon wing leading edge impacts as well as tile impact and window impact data. Propulsion element engineering along with MSFC Propulsion Systems Engineering and Integation (PSE&I) engineers developed empirical data for impact sensitivity to debris on propulsion element. KSC element engineering covered launch and landing debris issues along with coordination on launch pad imagery and debris inspection criteria for day of launch. All the NASA research centers, NASA White Sands Test Facility, and Stennis Space Center participated in providing tools and/or validation data for debris assessment. The Aerospace Corporation played a key role for systems engineering for debris and end-to-end Monte Carlo analysis along with Boeing-Houston. Finally, the United Space Alliance-Houston Program Integration had the role for data products and configuration management.

NASA IMPLEMENTATION

NASA analyzed credible debris sources from a wide range of release locations to predict the impact location and conditions. It developed critical debris source zones to provide maximum allowable debris sizes for various locations on the vehicle. A list of credible ascent debris sources was compiled for each SSP hardware element—Solid Rocket Booster, Reusable Solid Rocket Motor, Space Shuttle Main Engine, External Tank (ET), Orbiter, and the pad area around the vehicle at launch. Potential debris sources were identified by their location, size, shape, material properties, and, if applicable, likely time of debris release. The initial set of analyses were completed in 2004, and culminated in a comprehensive Space Shuttle Systems Integration Control Board review in December 2004. At this review, all of the Space Shuttle Projects presented their list of all potential debris sources for which design controls are in place to ensure that debris will not be liberated. Those debris sources that fell into the "expected" category—not controlled through design changes—were the focus of the transport analyses performed. The debris transport analyses were used to predict impact location and conditions, such as velocities and relative impact angles. They also supported the SSP's risk assessment by helping to classify the expected debris sources as remote, improbable, or probable.

NASA executed and analyzed over one billion debris transport cases. These included debris type, location, size, and release conditions (freestream Mach number, initial velocity of debris piece, etc.). This work has provided NASA with unprecedented insight into the ascent debris environment to which the Space Shuttle is subjected, and will aid in the design of future space vehicles.

In some cases, debris sources that could cause significant damage have been redesigned. Examples of vehicle redesigns driven by debris risk include the ET Bipod Closeout to eliminate the debris source, ET Intertank (IT) Thrust Panel venting to minimize mass size and rate of release, new process for Liquid Hydrogen IT Flange installation, installation of the Forward Bellows Heater to prevent ice, installation of the Bellows Drip Lip to reduce ice formation, Protuberance Airload (PAL) Ramp Removal to eliminate the debris source, Bipod Heater Wire redesign to prevent debris liberation and LH2 Ice Frost Ramp redesign to correct a known design deficiency. Some critical impact locations on the Orbiter have been redesigned or debris protection is being added (see Recommendation 3.3.2, Orbiter Hardening).

STATUS

All hardware project and element teams have identified known and suspected debris sources originating from the flight hardware. The Flight Crew Operations Directorate requested that NASA review the unexpected debris sources to ensure that they are accurately assessed and validated. This review was completed in May 2005. The debris source tables for all of the Space Shuttle elements and the pad environment have been formally reviewed and approved by the Program.

The debris transport tools that existed prior to STS-107 have been completely rewritten, and the results have been peer-reviewed. NASA has completed the transport analysis for all planned debris cases; the resulting data were provided to the SSP elements for evaluation and briefed to the wider NASA community in two Design Verification Reviews (DVRs) for Debris in Spring 2005. This data has allowed NASA to develop a comprehensive damage tolerance assessment for the Orbiter Thermal Protection System, and to refine the debris limits for ET foam and ice as well as other debris sources. Results of the Debris Transport Analysis and debris risk assessment helped the SSP to respond to the *Columbia* Accident Investigation Board recommendations, such as those on ET modifications (R3.2-1), Orbiter hardening modification (R3.3-2), and ascent and on-orbit imagery requirements (R3.4-1 and R3.4-3).

NASA has also been able to validate and refine some of the debris transport model assumptions through testing. These tests included wind tunnel tests and F-15 flight tests to assess foam flight characteristics. NASA also conducted tests on ice, focusing on ice behavior in a flow field. In May 2005, NASA completed its testing on ice debris to determine its flight characteristics. NASA grew ice of various densities and performing vibration/acoustic testing at the NASA Marshall Space Flight Center to understand how ice on the feedline brackets and bellows might liberate during ascent. NASA's results indicate that, contrary to early assumptions, "soft" ice may pose a greater risk than "hard" ice because of its tendency to release earlier in the flight and to break up into multiple pieces. Hard ice, on the other hand, tends to stay attached longer, breaking off only after the Shuttle has reached a stage in ascent where liberated ice does not pose as large a threat to the Orbiter. Overall, the tests performed have validated the model assumptions, and indicated that, in some cases, the model may be too conservative.

NASA also completed a supersonic wind tunnel test at the NASA Ames Research Center. This test validated the debris transport flow fields in the critical Mach number range. The results showed excellent agreement between wind tunnel results and analytically derived flow field predictions.

The first Debris Design Verification Review in June 2005, accomplished two major objectives to support the rationale for RTF. First, NASA developed an end-to-end estimate of the Orbiter's capability to withstand damage relative to the ascent environment. This was done by using the ET Project's best estimate of the foam and/or ice debris that may be liberated and the worst-case assumptions about that debris' potential for transport to the Orbiter. Second, for those cases in which the initial assessment indicated that the Orbiter could not withstand the potential impact, NASA performed a probabilistic risk assessment to determine the likelihood of a critical debris impact. NASA assessed four foam transport cases and two ice transport cases. These cases represented the worst potential impacts of a general category, and were used to bound similar, but less severe, transport cases from the same areas. This analysis enabled NASA to quantify the risk posed by the debris and aid in the determination of the ascent debris risk remaining after return to flight.

POST STS-114 UPDATE

The Protuberance Airload (PAL) ramp loss on STS-114 was due to a delamination failure mechanism that had not been accounted for in the debris risk assessment. As a result, a comprehensive review of all debris sources and applicable failure modes was undertaken. In addition, extensive foam dis section data and analysis was obtained for multiple areas on the external tank. Three additional foam failure modes were identified as well as the development of risk assessment masses for 30 different foam debris sources, which were based on ET dissection data, thermal vacuum testing and the empirical divot/no-divot curves. Other post STS-114 developments included development of a new ice-on-foam damage map, modeling of other debris sources (such as gap fillers, putty repairs, MLI blankets and ceramic inserts) and debris model comparisons with imagery, radar, and post flight inspection data. The post STS-114 debris work resulted in significant model improvements that accounted for the additional foam failure modes and debris sources, reduced the conservatism, and improved the accuracy. For example, the number of foam cases analyzed was increased from 4 to 30 and the number of foam failure modes analyzed was increased from 4 to 7. The second Debris Design Verification Review in May 2006 provided a more comprehensive characterization of the debris environment and risk by accounting for the additional foam failure modes and foam debris sources, as well as impacts from other non-foam/ice debris sources.

FINAL UPDATE

The Debris Integration Group (DIG) was established in August 2006 and chartered to assess the debris risk to the Space Shuttle vehicle associated with the expected debris sources during lift-off, ascent and re-entry. The DIG activities emphasize analysis of debris generation and release, debris transport, impact tolerance and the validation components of the debris assessment process consistent with Shuttle Program requirements. The DIG is responsible for characterizing the debris environment and assessing the debris risk to the Space Shuttle vehicle. This activity includes implementation, review, and evaluation of analysis, trade studies, test planning and execution, data analysis, post-flight assessment, sensitivity studies, and the interface relationships necessary to determine the debris risk. The results are reflected in an Integrated Debris Hazard Report.

The lengthy process to determine, assess and characterize debris risks has been an enormous undertaking by NASA. There have been significant improvements since STS-114 that include removal of PAL ramps, redesign of Bipod Closeout, and additional launch pad and Orbiter debris mitigations. There is a better understanding and characterization of all the foam failure modes. Additional foam and ice testing along with flight data have led to model refinements to remove conservatism and improve accuracy, as well as provide validation data to increase confidence in the analysis tools. Although reduced, limitations still exist in all areas: debris release, aerodynamic transport, and impact tolerance. As a result, NASA will continue to mitigate high risk debris sources, refine its analyses with further data from each Space Shuttle flight to reduce uncertainties, and better characterize the debris environment and risks.

SCHEDULE

Responsibility	Due Date	Activity/Deliverable
SSP	Jul 03 (Completed)	Elements provide debris history/sources
SSP	Nov 03 (Completed)	Begin Return to Flight Debris Transport analyses
SSP	Dec 04 (Completed)	Complete second set of Debris Transport analyses
SSP	Mar 05 (Completed)	Complete final round of Debris Transport analyses
SSP	Mar/Apr 05 (Completed)	Summary report/recommendation to PRCB
SSP	May 05 (Completed)	Complete flight characteristic testing on ice debris
SSP	May 05 (Completed)	Validate unexpected debris sources

Space Shuttle Program Return to Flight Actions
Space Shuttle Program Action 6

All waivers, deviations, and exceptions to Space Shuttle Program (SSP) requirements documentation will be reviewed for validity and acceptability before return to flight.

Note: NASA has closed this action through the formal Program Requirements Control Board (PRCB) process. The following summary details NASA's response to the Space Shuttle Program (SSP) action and any additional work NASA performed beyond the SSP action.

BACKGROUND

Requirements are the fundamental mechanism by which the SSP directs the production of hardware, software, and training for ground and flight personnel to meet performance needs. The rationale for waivers, deviations, and exceptions to these requirements must include compelling proof that the associated risks are mitigated through design, redundancy, processing precautions, and operational safeguards. The Program manager, with concurrence by the Independent Technical Authority (ITA), has approval authority for waivers, deviations, and exceptions.

NASA IMPLEMENTATION

Because waivers and deviations to SSP requirements and exceptions to the Operations and Maintenance Requirements and Specifications contain the potential for introducing unintended risk, the Program directed all elements to review these exemptions to Program requirements and to recommend whether the exemptions should be retained. Each project and element was alert for items that required mitigation before Return to Flight.

The following instructions were provided to each project and element:

1. Any item that has demonstrated periodic, recurrent, or increasingly severe deviation from the original design intention must be technically evaluated and justified. If there is clear engineering rationale for multiple waivers for a Program requirement, it could mean that a revision to the requirement is needed. The potential expansion of documented requirements should be identified for Program consideration.

2. The review should include the engineering basis for each waiver, deviation, or exception to ensure that the technical rationale for acceptance is complete, thorough, and well considered.

3. Each waiver, deviation, or exception should have a complete engineering review to ensure that incremental risk increase has not crept into the process over the Shuttle lifetime and that the level of risk is appropriate.

The projects and elements were encouraged to retire out-of-date waivers, deviations, and exceptions.

In addition to reviewing all SSP waivers, deviations, and exceptions, each element reviewed all NASA Accident Investigation Team working group observations and findings and Critical Item List (CIL) waivers associated with ascent debris.

Updating the waivers and deviations allows the SSP Management to understand which waivers represent real safety issues and which were merely administrative. This in turn clarifies the level of risk associated with each waiver.

STATUS

Each project and element presented a plan and schedule for completion to the daily Program Requirements Control Board (PRCB) on June 25, 2003. Each project and element identified and reviewed the CIL associated with ascent debris generation. The SSP completed the review of the waivers, deviations, and exceptions at the daily PRCB.

FINAL UPDATE

Approximately 50% of the original set of 650 waivers were primarily related to electromagnetic interference (EMI). Because of this large number, the SSP directed the Electromagnetic Environmental Effects Panel to review all future EMI hardware for risk and to bring forward for approval those hardware items that have an increase in risk. In addition, the SSP has updated its process for assessing and documenting hardware noncompliances to ensure any threats to the vehicle or its operations are properly addressed. The majority of the SSP EMI waivers were rendered obsolete and are being retired.

SCHEDULE

Responsibility	Due Date	Activity/Deliverable
SSP	Jun 05 (Completed)	Review of all waivers, deviations, and exceptions

Space Shuttle Program Return to Flight Actions
Space Shuttle Program Action 7

The Space Shuttle Program (SSP) should consider NASA Accident Investigation Team (NAIT) working group findings, observations, and recommendations.

Note: NASA has closed this action through the formal Program Requirements Control Board (PRCB) process. The following summary details NASA's response to the Space Shuttle Program (SSP) action and any additional work NASA performed beyond the SSP action.

BACKGROUND

As part of their support of the *Columbia* Accident Investigation Board (CAIB), each NASA Accident Investigation Team (NAIT) technical working group compiled assessments and critiques of Program functions. These assessments offer a valuable internal review and will be considered by the Space Shuttle Program (SSP) for conversion into directives for corrective actions.

NASA IMPLEMENTATION

All NAIT technical working groups have an action to present their findings, observations, and recommendations to the Space Shuttle PRCB. Each project and element will disposition recommendations within its project to determine which should be Return to Flight actions. Actions that require SSP or Agency implementation will be forwarded to the PRCB for disposition.

The following NAIT working groups have reported their findings and recommendations to the SSP at the PRCB: the Space Shuttle Main Engine Project Office, the Reusable Solid Rocket Motor Project Office, the Mishap Investigation Team, the External Tank Project, the Solid Rocket Booster Project Office, and Space Shuttle Systems Integration. The Orbiter Project Office has reported the findings and recommendations of the following working groups to the PRCB: *Columbia* Early Sighting Assessment Team, Certification of Flight Readiness Process Team, Unexplained Anomaly Closure Team, Previous Debris Assessment Team, Hardware Forensics Team, Materials Processes and Failure Analysis Team, Starfire Team, Integrated Entry Environment Team, Image Analysis Team, Palmdale Orbiter Maintenance Down Period Team, Space/Atmospheric Scientist Panel, KSC Processing Team, *Columbia* Accident Investigation Fault Tree Team, *Columbia* Reconstruction Team, and Hazard Controls Analysis Team.

All NAIT working group findings and recommendations were evaluated by the affected SSP projects. Inconsistencies between working group recommendations and the projects' disposition were arbitrated by the Systems Engineering and Integration Office (SE&IO), with new actions assigned as warranted. Review of all working group recommendations and final project dispositions was completed in May 2004.

Project and PRCB recommendations currently being implemented include revision of the SSP Contingency Action Plan, modifications to the External Tank, and evaluation of hardware qualification and certification concerns. Numerous changes to Orbiter engineering, vehicle maintenance and inspection processes, and analytical models are also being made as a result of the recommendations of the various accident investigation working groups. In addition, extensive changes are being made to the integrated effort to gather, review, and disposition prelaunch, ascent, on-orbit, and entry imagery of the vehicle, and to evaluate and repair any potential vehicle damage observed. All of this work complements and builds upon the extensive recommendations, findings, and observations contained in the CAIB Report.

FINAL UPDATE

Following PRCB approval of working group recommendations, the responsible project office tracks associated actions and develops implementation schedules with the goal of implementing approved recommendation prior to Return to Flight. The responsible SSP projects status closure of the working group recommendations as part of the Design Certification Review activity in support of Return to Flight.

SCHEDULE

Responsibility	Due Date	Activity/Deliverable
SSP SE&IO	May 04 (Completed)	Review working group recommendations and SSP Project dispositions

Space Shuttle Program Return to Flight Actions
Space Shuttle Program Action 8

NASA will identify certification of flight readiness (CoFR) process changes, including program milestone reviews, flight readiness review (FRR), and prelaunch Mission Management Team (MMT) processes to improve the system.

Note: NASA has closed this action through the formal Program Requirements Control Board (PRCB) process. The following summary details NASA's response to the Space Shuttle Program (SSP) action and any additional work NASA performed beyond the SSP action.

BACKGROUND

The certification of flight readiness (CoFR) is the fundamental process for ensuring compliance with Program requirements and assessing readiness for proceeding to launch. The CoFR process includes multiple reviews at increasing management levels that culminate with the Flight Readiness Review (FRR), chaired by the Associate Administrator for Space Flight, approximately two weeks before launch. After successful completion of the FRR, all responsible parties, both Government and contractor, sign a CoFR.

NASA IMPLEMENTATION

To ensure a thorough review of the CoFR process, the Shuttle PRCB has assigned an action to each organization to review NSTS 08117, Certification of Flight Readiness, to ensure that its internal documentation complies and responsibilities are properly described. This action was assigned to each Space Shuttle Program (SSP) supporting organization that endorses or concurs on the CoFR and to each organization that prepares or presents material in the CoFR review process.

Each organization reviewed the CoFR process in place during STS-112, STS-113, and STS-107 to identify any weaknesses or deficiencies in its organizational plan.

FINAL UPDATE

NASA has revised NSTS 08117, Certification of Flight Readiness, including providing updates to applicable documents lists as well as the roles and responsibilities within project and Program elements, and has increased the rigor of previous mission data review during the project-level reviews. The revised document was approved by the PRCB in January 2004 and released in February 2004.

SCHEDULE

Responsibility	Due Date	Activity/Deliverable
SSP Element reviews	Aug 03 (Completed)	Report results of CoFR reviews to PRCB
SSP	Feb 04 (Completed)	Revise NSTS 08117, Certification of Flight Readiness

Space Shuttle Program Return to Flight Actions
Space Shuttle Program Action 9

NASA will verify the validity and acceptability of failure mode and effects analyses (FMEAs) and critical items lists (CILs) that warrant review based on fault tree analysis or working group observations.

Note: NASA has closed this action through the formal Program Requirements Control Board (PRCB) process. The following summary details NASA's response to the Space Shuttle Program (SSP) action and any additional work NASA performed beyond the SSP action.

BACKGROUND

The purpose of failure mode and effects analyses (FMEAs) and critical items lists (CILs) is to identify potential failure modes of hardware and systems and their causes, and to assess their worst-case effect on flight. A subset of the hardware analyzed in the FMEA becomes classified as critical, based on the risks and identified undesirable effects and the corresponding criticality classification assigned. These critical items, along with supporting acceptance rationale, are documented in a CIL that accepts the design.

The analysis process involves the following phases:

1. Perform the design analysis.

2. For critical items, assess the feasibility of design options to eliminate or further reduce the risk. Consideration is given to enhancing hardware specifications, qualification requirements, manufacturing, and inspection and test planning.

3. Formulate operating and maintenance procedures, launch commit criteria, and flight rules to eliminate or minimize the likelihood of occurrence and the effect associated with each failure mode. Formally document the various controls identified for each failure mode in the retention rationale of the associated CIL, and provide assurance that controls are effectively implemented for all flights.

NASA IMPLEMENTATION

In preparation for Return to Flight (RTF), NASA developed a plan to selectively evaluate the effectiveness of the SSP FMEA/CIL process and assess the validity of the documented controls associated with the SSP CIL. Each project and element identified those FMEAs/CILs that warranted revalidation based on their respective criticality and overall contribution to design element risk. In addition, STS-107 investigation findings and working group observations affecting FMEA/ CIL documentation and risk mitigation controls were assessed, properly documented, and submitted for SSP approval.

This plan concentrated on revalidation efforts on FMEA/CILs that were called into question by investigation results or that contribute the most significant risks for that Program element. Revalidation efforts included:

1. Reviewing existing STS-107 investigation fault trees and working group observations to identify areas inconsistent with or not addressed in existing FMEA/CIL risk documentation.

    a. Verifying the validity of the associated design information and assessing the acceptability of the retention rationale to ensure that the associated risks are being effectively mitigated consistent with SSP requirements.

    b. Establishing or modifying SSP controls as required.

    c. Developing and revising FMEA/CIL risk documentation accordingly.

    d. Submitting revised documentation to the SSP for approval as required.

2. Assessing most significant SSP element risk contributors.

    a. Identifying a statistically significant sample of the most critical CILs from each element project, including those CILs in which ascent debris generation is a consequence of the failure mode experienced.

    b. Verifying that criticality assignments are accurate and consistent with current use and environment.

    c. Validating the SSP controls associated with each item to ensure that the level of risk initially accepted by the SSP has not changed.

        1. Establishing or modifying Program controls as required.

        2. Developing and revising FMEA/CIL risk documentation accordingly.

3. Submitting revised documentation to the SSP for approval as required.

d. Determining if the scope of the initial review should be expanded based on initial results and findings. Reassessing requirements for performance of FMEAs on systems previously exempted from SSP requirements, such as the Thermal Protection System, select pressure and thermal seals, and certain primary structures.

The System Safety Review Panel (SSRP) and Safety Engineering Review Panel (SERPs) served as the forums to review the project assessment of the validity and applicability of the CIL retention rationale. The SSRP and SERPs reviewed any updates to baselined CILs.

STATUS

Completed.

SCHEDULE

Responsibility	Due Date	Activity/Deliverable
SSP	Apr 05 (Completed)	Projects status reports to PRCB
SSP	Apr 05 (Completed)	Completion of review

Space Shuttle Program Return to Flight Actions
Space Shuttle Program Action 10

NASA will review Program, project, and element contingency action plans and update them based on *Columbia* mishap lessons learned.

Note: NASA has closed this action through the formal Program Requirements Control Board (PRCB) process. The following summary details NASA's response to the Space Shuttle Program (SSP) action and any additional work NASA performed beyond the SSP action.

BACKGROUND

The SSP Program Requirements Control Board has directed all of its projects and elements to review their internal Contingency Action Plans (CAPs) for ways to improve their emergency response processes.

NASA IMPLEMENTATION

The SSP has updated and approved the Program-level CAP to reflect the lessons learned from the *Columbia* accident. SSP projects and elements are updating their subordinate plans as required to reflect changes to the Program CAP. The Program document has been distributed to all NASA Centers that support human space flight, and orientation training has been conducted across the SSP. A simulation to exercise a realistic contingency situation of the CAP was successfully completed in January 2005.

FINAL UPDATE

In implementing changes to the CAP, the SSP incorporated many of the specific lessons learned from the *Columbia* experience while striving to maintain a generic plan that would be useful in a wide range of potential contingency situations. The resulting document is optimized to serve as a rigorous first-response checklist, then to give a menu of possible longer-term response outlines from which to choose based upon the severity of the contingency, its location, and the involvement and responsibilities of other federal, state, and local agencies and foreign governments. Structured responses to Space Shuttle launch contingencies such as transoceanic aborts and East Coast abort landings have been retained in the appropriate appendices.

Space Shuttle Program Return to Flight Actions
Space Shuttle Program Action 11

Based on corrosion recently found internal to body flap actuators, NASA will inspect the fleet leader vehicle actuators to determine the condition of similar body flap and rudder speed brake actuators.

Note: NASA has closed this action through the formal Program Requirements Control Board (PRCB) process. The following summary details NASA's response to the Space Shuttle Program (SSP) action and any additional work NASA performed beyond the SSP action.

BACKGROUND

Internal corrosion was found in Orbiter Vehicle (OV)-104 body flap (BF) actuators in Fall 2002, and subsequently in the OV-103 BF actuators. In addition, corrosion pits were discovered on critical working surfaces of two BF actuators (e.g., planetary gears and housing ring gears), and general surface corrosion was found inside other BF actuators.

Since the rudder speed brake (RSB) actuator design and materials are similar to BF actuators, similar internal corrosion in RSB actuators could adversely affect performance of Criticality 1/1 hardware. Any existing corrosion will continue to degrade the actuators. The loss of RSB functionality due to "freezing up" of the bearing or jamming caused by broken gear teeth would cause Orbiter loss of control during entry. The operational life of the installed RSB actuators is outside of Orbiter and industry experience. The Space Shuttle Program (SSP) and the Space Flight Leadership Council (SFLC) approved removal of all RSB actuators to investigate corrosion, wear, and hardware configuration.

NASA IMPLEMENTATION

The SSP directed the removal and refurbishment of all four OV-103 RSB actuators. The SSP spares inventory included four RSB actuators. All spare RSB actuators were returned to the vendor for acceptance test procedure (ATP) revalidation. All passed ATP and were returned to logistics. The removed (original) OV-103 RSB actuators were disassembled, and one of the actuators, actuator 4, was found to have the planetary gear set installed in reverse. Analysis showed that this condition presented negative margins of safety for the most severe load cases. In addition to the reversed planetary gears and corrosion, fretting and wear were documented on some of the gears from OV-103 RSB actuators. Surface pits resulting from the fretting have led to microcracks in some of the gears.

Figure SSP 11-1. OV-103 RSB actuator.

As a result of the reversed planetary gear set discovery, the spare actuators, installed in OV-103, were X-rayed, and actuator 2 was also found to have the planetary gear set installed in reverse. The RSB actuators were removed from OV-103 and shipped to the vendor, where they were disassembled and inspected. Following repair of spare actuator 2, the spare actuators were reinstalled on OV-103.

RSB actuators from OV-104 and OV-105 were shipped to the vendor for disassembly and inspection. For OV-104, the actuators were assembled from existing OV-105 parts and new parts, all within specification.

FINAL UPDATE

All actuators for OV-104 were installed before its first post-*Columbia* flight, STS-115. A new ship-set of actuators was procured for OV-105 and installed before its first post-*Columbia* flight, STS-115.

SCHEDULE

Responsibility	Due Date	Activity/Deliverable
SSP	Jul 03 (Completed)	Initial plan reported to SFLC
SSP	Aug 03 (Completed)	ATP Spare RSB actuators at vendor and returned to Logistics
SSP	Sep 03 (Completed)	OV-103 RSB actuators removed and replaced with spares
SSP	Mar 04 (Completed)	RSB findings and analysis completed

Space Shuttle Program Return to Flight Actions
Space Shuttle Program Action 12

NASA will review flight radar coverage capabilities and requirements for critical flight phases.

Note: NASA has closed this action through the formal Program Requirements Control Board (PRCB) process. The following summary details NASA's response to the Space Shuttle Program (SSP) action and any additional work NASA performed beyond the SSP action.

BACKGROUND

In addition to Shuttle vehicle ascent imaging by photo and visual means, NASA uses radar systems of the Air Force Eastern Range for metric tracking of the Space Shuttle launches. There are several C-Band radars and a Multiple Object Tracking Radar (MOTR) used to monitor the ascent trajectory. Although not specifically designed to track debris, these radars have some limited ability to resolve debris separating from the ascending vehicle. All of these assets, however, are systemically and geometrically limited in their ability to resolve the very faint radar targets represented by debris.

During the STS-107 launch, the MOTR, which is specifically intended for the purpose of tracking several objects simultaneously, was unavailable.

NASA IMPLEMENTATION

The Space Shuttle Systems Engineering and Integration Office commissioned the Ascent Debris Radar Working Group (ADRWG) to characterize the debris environment during a Space Shuttle launch and to identify/define the return signals seen by the radars. Once the capabilities and limitations of the existing radars for debris tracking were understood, this team researched proposed upgrades to the location, characteristics, and post-processing techniques needed to provide improved radar imaging of Shuttle debris.

The specific technical goal of the ADRWG was to improve the radars' ability to resolve, identify, and track potential debris sources. Another goal was to decrease the postlaunch data processing time such that a preliminary radar assessment is available more rapidly, and to more easily correlate the timing of the ascent radar data to optical tracking systems. The successful implementation of a radar debris tracking system has an advantage over optical systems as it is not constrained by ambient lighting or cloud interference. It further has the potential to maintain insight into the debris shedding environment beyond the effective range of optical tracking systems. This was demonstrated when the larger Mid-Course Radar (MCR) was brought online during STS-121 and resolved a continuation of the second-stage debris shower first seen by the smaller radar used during STS-114. The initial experience showed a diminishing of second-stage particles around T+280 seconds, but STS-121 demonstrated that this phenomena continues for the entire duration of the visible ascent.

STATUS

The ADRWG was initiated in August 2003. After a review of existing debris documentation and consultation with radar experts within and outside of NASA, a planning presentation outlining the approach and process to be used was provided to the Space Shuttle Program (SSP) office in September 2003. A number of workshops were held at NASA centers and at Wright-Patterson Air Force Base to characterize the debris sources and how they appeared on radar, and to analyze the potential debris threat to the Shuttle represented by the radar data.

The ADRWG constructed a composite list of potential debris sources. This list was coordinated with all of the Shuttle elements and is one basis for analysis of radar identification capabilities such as radar cross section (RCS) signatures. A series of critical radar system attributes was compiled, and a number of existing radar systems were evaluated against these criteria. Data analysis included comparisons of radar data with known RCS signatures and ballistic trajectories.

On January 13, 2004, the ADRWG provided its initial findings and draft recommendations to the SSP. The team found that the existing range radars were not well suited to perform the Shuttle debris assessment task because of their siting and systemic configuration. Only a properly sited and configured radar system can be expected to provide the insight needed to assess the debris threat during a Shuttle launch. A candidate architecture, using several elements of the Navy Mobile Instrumentation System, formed the basis of the radar system for Return to Flight.

Radar field testing included a series of six Booster Separation Motor firings to characterize how the plume contributed to the existing radar data. These tests were completed at the U.S. Navy's China Lake facility in February 2004. A comprehensive set of RCS measurements of candidate Shuttle debris material has been completed at Wright-Patterson Air Force Base and was correlated to dynamic field results at the Naval Air Station at Patuxent River in June 2004.

The final SSP presentation, including field results, prior mission analysis, and final recommendations, was completed in April 2004, with the associated implementation and funding plans approved in July 2004.

To provide adequate threat assessment, a ground-based radar system must include both wideband capabilities to provide the precise position of debris as well as Doppler capabilities for differential motion discrimination. Also necessary are near-real-time data reduction and display in remote facilities, ballistic coefficient traceability, and the highest calibration to meet Range Certification Standard STD 804-01. To meet these requirements, NASA, in cooperation with the U.S. Navy and the U.S. Air Force, has developed a radar system that involved relocation of the U.S. Navy midcourse radar from Puerto Rico to Cape Canaveral. This radar provides wideband, coherent C-band radar coverage and is supplemented with two newly procured continuous wave Doppler X-band ship-based radars (figure SSP 12-1).

FINAL UPDATE

A Memorandum of Understanding between NASA and the U.S. Navy, regarding the use and sharing of radar assets, was approved in May 2005. The system has supported the STS-114 and STS-121 flights, as well as a number of expendable launch vehicles in both a test and operational mode.

To expedite the operational availability of the relocated MCR, the permanent facility construction was arranged in a series of stages so that the radar was not waiting on the entire facility to be complete to support shuttle missions. The radar became operational in January 2006, along with completion of the first two construction phases. The final phase of site construction is expected to be completed during the fall of 2007.

Flight support for all future Shuttle missions is anticipated through the end of the program.

SCHEDULE

Responsibility	Due Date	Activity/Deliverable
ADRWG	Nov 03 (Completed)	Complete Radar Study
ADRWG	Nov 03 (Completed)	Finalize finding and recommendations
ADRWG	Apr 04 (Completed)	Provide final list of debris sources
SSP	Apr 04 (Completed)	Baseline requirements and initiate implementation – Present to SSP Program Requirements Control Board

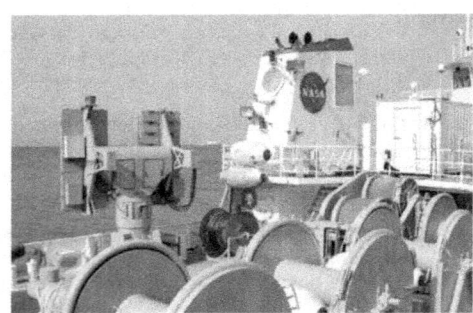 

Figure SSP 12-1. X-band radar on solid rocket booster recovery ship (left), MCR C-band radar (right).

Space Shuttle Program Return to Flight Actions
Space Shuttle Program Action 13

NASA will verify that hardware processing and operations are within the hardware qualification and certification limits.

Note: NASA has closed this action through the formal Program Requirements Control Board (PRCB) process. The following summary details NASA's response to the Space Shuttle Program (SSP) action and any additional work NASA performed beyond the SSP action.

BACKGROUND

An Orbiter Project Office investigation into several Orbiter hardware failures identified certification environments that were not anticipated or defined during original qualifications. Some examples of these include drag chute door pin failure, main propulsion system flow liner cracks, and environmental control and life support system secondary O_2/N_2 flex hose bellows failure.

Because of these findings by the Orbiter Project Office, all projects and elements assessed all Space Shuttle hardware operations according to requirements for certification/qualifications. If a finding was determined to be a constraint to flight, the project or element immediately reported the finding to the PRCB for disposition.

NASA IMPLEMENTATION

On December 17, 2002, prior to the *Columbia* accident, the SSP Council levied an action to all SSP projects and elements to review their hardware qualification and verification requirements and to verify that processing and operating conditions are consistent with the original hardware certification (memorandum MA-02-086). At the SSP Council meeting April 10-11, 2003, each Program project and element identified that its plan for validating that hardware operating and processing conditions, along with environments or combined environments, is consistent with the original certification (memorandum MA-03-024). The PRCB has reissued this action as a Return to Flight action.

Revalidating the qualification and verification of SSP hardware was critical to identifying areas of unrecognized risk in the Space Shuttle system prior to Return to Flight. Through this process, we have ensured that the Space Shuttle hardware should continue to function within its design specifications for the remainder of the Space Shuttle's service life.

STATUS

Interim status reports from the SSP project and element organizations were presented to the SSP PRCB throughout the year 2004. As a result of this proactive review, NASA identified several areas for additional scrutiny, such as the Solid Rocket Booster Separation Motor debris generation and Orbiter nosewheel steering failure modes. This attitude of critical review, even of systems that have consistently functioned within normal specifications, has significantly improved the safety and reliability of the Shuttles and reduced the risk of future problems.

The SSP projects and elements have completed their assessment of the qualification and verification of all Criticality 1 Space Shuttle hardware. The assessments showed no constraints to the hardware certification limits.

FINAL UPDATE

Certification assessments for certain lower criticality Orbiter hardware will continue through 2007.

SCHEDULE

Responsibility	Due Date	Activity/Deliverable
All SSP project and element organizations	Mar 05 (Completed)	Present certification assessment results to SSP PRCB for Return to Flight commitments
All SSP project and element organizations	Dec 06 (Completed)	Present certification assessment results to SSP PRCB for any remaining post-Return to Flight commitments

2.1 Space Shuttle Program Return to Flight Actions
Space Shuttle Program Action 14

Determine critical Orbiter impact locations and TPS damage size criteria that will require on-orbit inspection and repair. Determine minimum criteria for which repairs are necessary and maximum criteria for which repair is possible.

Note: NASA has closed this action through the formal Program Requirements Control Board (PRCB) process. The following summary details NASA's response to the Space Shuttle Program (SSP) action and any additional work NASA performed beyond the SSP action.

This Space Shuttle Program Action is addressed by *Columbia* Accident Investigation Board Recommendations 3.3-2 and 6.4-1 of this Implementation Plan.

Space Shuttle Program Return to Flight Actions
Space Shuttle Program Action 15

NASA will identify and implement improvements in problem tracking, in-flight anomaly (IFA) disposition, and anomaly resolution process changes.

Note: NASA has closed this action through the formal Program Requirements Control Board (PRCB) process. The following summary details NASA's response to the Space Shuttle Program (SSP) action and any additional work NASA performed beyond the SSP action.

BACKGROUND

Bipod ramp foam was released during the launch of STS-112 in October 2002. After the mission, the Space Shuttle Program (SSP) considered this anomaly and directed the External Tank Project to conduct the testing and analysis necessary to understand the cause of bipod foam release and present options to the SSP for resolution. The Program did not hold completion of these activities as a constraint to subsequent Shuttle launches because the interim risk was not judged significant. The *Columbia* accident investigation results clearly disclose the errors in that engineering judgment.

NASA IMPLEMENTATION

NASA conducted a full review of its anomaly resolution processes with the goal of ensuring appropriate disposition of precursor events in the future. To remove subjectivity from the Problem Reporting and Corrective Action (PRACA) reporting process, NSTS 08126, PRACA System Requirements, was revised to require all PRACA reportable problems during the defined mission period to be baselined as in-flight anomalies (IFAs). In addition, the Systems Engineering and Integration Office was tasked with reviewing all IFAs for potential integrated effects and initiating an integrated PRACA when applicable. The new process also requires that all anomalies that result in an increase in risk, per the hazard report process, require Space Shuttle Program Requirements Control Board review and hazard report approval. This process ensures that Program Management and the newly established positions of the Safety Technical Authority and the Engineering Technical Authority have opportunity to review and approve issues with an increased risk.

In support of the Return to Flight activity, the SSP, supported by all projects and elements, identified and implemented improvements to the problem tracking, IFA disposition, and anomaly resolution processes. A team reviewed SSP and other documentation and processes, as well as audited performance for the past three Shuttle missions. The team concluded that, while *clarification* of the PRACA System Requirements is needed, the *implementation* of those requirements appears to be the area that has the largest opportunity for improvement. The team identified issues with PRACA implementation that indicate misinterpretations of definitions, resulting in misidentification of problems, and noncompliance with tracking and reporting requirements.

The corrective actions include:

1. The JSC Safety and Mission Assurance (S&MA) Support Contractor developed a training class and trained all SSP S&MA elements and support organizations on PRACA requirements and processes.

2. Updated NSTS 08126 to clarify the IFA definition, delete "Program" IFA terminology, and add payload IFAs and Mission Operations Directorate (MOD) anomalies to the scope of the document.

3. Updated the PRACA nonconformance system (Web PCASS) to include flight software, payload IFAs, and MOD anomalies.

FINAL UPDATE

NASA and its contractors are providing ongoing training to ensure that all SSP elements and support organizations understand the PRACA system and are trained in entering data into PRACA.

SCHEDULE

Responsibility	Due Date	Activity/Deliverable
JSC	Aug 04 (Completed)	Approve CR to update NSTS 08126, PRACA Systems Requirements

CAIB Observations

The observations contained in Chapter 10 of the CAIB Report expanded upon the CAIB recommendations, touching on critical areas of public safety, crew escape, Orbiter aging and maintenance, quality assurance, test equipment, and the need for a robust training program for NASA managers. NASA examined these observations and implemented corrective measures. As a result, NASA institutionalized improvements to our culture and programs that will ensure we can meet the challenges of continuing to fly the Shuttle safely through 2010.

Columbia Accident Investigation Board
Observation 10.1-1

NASA should develop and implement a public risk acceptability policy for launch and re-entry of space vehicles and unmanned aircraft.

Note: NASA has closed this observation through the formal Program Requirements Control Board process. The following summary details NASA's response to the observation and any additional work NASA performed beyond the *Columbia* Accident Investigation Board observation.

BACKGROUND

Space flight is not a risk-free endeavor. All space flight missions, particularly those going to orbit or deeper into space or returning to Earth from space, pose some level of risk to uninvolved people. No matter how small, there is always some potential for failure during flight. If a failure occurs, there will be a possibility of injuring the general public. Overall, NASA's safety approach ensures that any risk to the public associated with space flight is identified and controlled.

People knowingly and unknowingly accept risk throughout their daily lives. Common sources of risk include driving in an automobile, participating in sports, and potential exposure to hazards in the home and the workplace. Our goal is to ensure that a space flight does not add significantly to the public's overall risk of injury. However, a decision to accept greater public risk may be appropriate if the benefits of the mission are great. Such a decision is based on a comprehensive assessment of the risks and a clear understanding of the benefits associated with taking those risks.

As the government agency directing and controlling space flight operations, NASA is legally responsible for public safety during all phases of operation. Throughout its history, NASA has met this responsibility. No NASA space flight has ever caused an injury to any member of the general public.

Historically, NASA has had a general risk management policy designed to protect the public as well as NASA personnel and property, codified in NASA Policy Directive (NPD) 8700.1A. This policy calls for NASA to implement structured risk management processes using qualitative and quantitative risk-assessment techniques to make decisions regarding safety and the likelihood of mission success. The policy requires program managers to implement risk management policies, guidelines, and standards within their programs. Although this Agency-level risk policy does not specifically address range flight operations, individual NASA safety organizations, such as those at Wallops Flight Facility and Dryden Flight Research Center, have well-established public and workforce risk management requirements and processes at the local level. Also, NASA has always worked closely with the safety organizations at the U.S. Air Force's Eastern and Western Ranges to satisfy public risk requirements during Space Shuttle and other NASA space flight operations.

The *Columbia* Accident Investigation Board (CAIB) suggested that NASA should "develop and implement a public risk acceptability policy." Although the CAIB did not make this a Return to Flight recommendation, NASA completed the development and implementation of this policy as part of its efforts to "raise the bar" prior to Return to Flight. The NASA Administrator signed the new policy, NASA Procedural Requirement (NPR) 8715.5 "Range Safety Program," on July 8, 2005.

NASA IMPLEMENTATION
Policy Overview

NASA has developed a public risk policy that incorporates the Agency's approach for identifying and managing the risk to the general public that is associated with space flight operations, such as launch and entry of space flight vehicles and the operation of crewless aircraft. This new Agency-level policy is documented in Chapter 3 of NPR 8715.5, Range Safety Program. NASA implemented this policy prior to the Space Shuttle Return to Flight and will apply it to all future NASA space flight missions.

Development of any Agency policy required significant coordination among the NASA Centers and programs that will be responsible for its implementation. The NASA Headquarters Office of Safety and Mission Assurance established a risk policy working group with members throughout the Agency and chartered the group to perform the initial development and coordination of the new public risk policy. The working group coordinated with the interagency range safety community and consulted with experts in applying public and workforce risk assessment to the operation of experimental and developmental vehicles. The CAIB's lead investigator for the issue of public risk participated in many of the working group's activities. This inclusive approach helped to ensure that NASA's new policy fully responds to the related CAIB findings and observations.

The NASA public risk policy incorporates a widely accepted risk management approach that has been used successfully at United States launch sites for addressing the risk to the public associated with space flight operations. The policy includes requirements for risk assessment, risk mitigation, and acceptance/disposition of risk to the public and workforce. The policy incorporates performance standards for assessing risk and contains acceptable risk criteria. Finally, the policy requires review and approval by NASA Senior Management for any proposed operations where the risk to the public or workforce might exceed the public risk criteria. Such approval may be on a case-by-case or programmatic basis.

Public risk policies in general incorporate established risk criteria that a majority of the affected operations are expected to satisfy. Such criteria define a standard level of risk that the approval authority, in this case the NASA Administrator, accepts for normal day-to-day operations. The establishment of public risk criteria at the Agency level helps to facilitate the acceptance of risk in operational environments where it would be impractical for upper management to be involved in making every risk acceptance decision on a case-by-case or programmatic basis.

There are primarily two types of risk criteria that the public risk policy addresses. The first type of risk is referred to as "individual risk." The second type of risk is referred to as "collective risk." The NASA public risk policy incorporates criteria for both types of risk. NASA's public risk criteria are consistent with those used throughout the government, the commercial range community, and with other industries whose activities are potentially harmful to the general public.

The measurement for individual risk represents the probability that an individual at a specific location could experience a serious injury for a single event, such as the launch or entry of a Space Shuttle, if a large number of events could be carried out under identical circumstances.

For example, the public individual risk criterion used throughout the space flight operations community and in the new NASA policy is less than or equal to one in a million. In other words, if an individual were to attend one million identical launches, that person would be expected to experience a serious injury once or less during those million launches (i.e., a relatively low risk). The individual risk criterion is typically enforced by establishing a "keep-out" zone for each launch or entry such that if all individuals remain outside the keep-out zone their individual risk will satisfy the criterion. All NASA launches including Space Shuttle launches have always, and will continue to employ keep-out zones in the vicinity of each launch site where the risk approaches the one-in-a-million threshold; additionally, NASA now employs keep-out zones at each landing site. Enforcement of keep-out zones for launches and landings will ensure that the one-in-a-million individual risk criterion is satisfied for all of the public, including visitors to a NASA launch or landing site.

The measurement for collective risk is the average number of serious injuries expected within a defined population for a single event, such as a Space Shuttle launch or entry, if a large number of events could be carried out under identical circumstances. Although the individual risk to members of an exposed population may be very low for a single event, as the number of people within the exposed population increases, the collective risk will increase. The collective risk can be controlled to a reasonable level by controlling the exposed population.

For example, if a group of 100,000 people attends a launch and all of the people are located at the border of the keep-out zone such that each person has an individual risk equal to one in a million, the collective risk for the group would equate to one in a million multiplied by 100,000 or an average of one serious injury within the group in 10 launches. This is an exaggerated example, but it serves to demonstrate how collective risk will continue to increase as the number of people that have any significant individual risk continues to increase. Placing a collective public risk limit on a space flight provides the impetus for the Agency to consider the number of people exposed to a given hazardous condition and place limits on the exposed population.

The criteria for individual and collective risk are established at levels considered acceptable for a majority of the expected operations. Within our space flight community, public risk is assessed to ensure that the risk is understood and is within acceptable limits for day-to-day operations. As with all risk policies, NASA's public risk policy allows for the acceptance of any risk that exceeds the established criteria through a variance or other appropriate process after all reasonable risk-reduction strategies have been employed. NASA Senior Management will make such decisions when warranted on a case-by-case or programmatic basis, with a thorough understanding of any additional risk and the benefits to be derived from taking the additional risk. Within NASA, the ultimate authority for accepting any risk above the established criteria lies with the NASA Administrator, who may delegate related authority. Authority with regard to the public risk policy is delegated to the Independent Technical Authority and the Center Director or Headquarters-designated manager responsible for the vehicle program with concurrence by the official responsible for the range, launch site, or landing site. Note that NASA's acceptance of the public risk associated with Space Shuttle flights is on a programmatic basis through establishment of the risk policy, which includes a Space Shuttle-specific provision and requires Senior Management approval of the Shuttle Range Safety Risk Management Plan. NASA does not foresee the need to process any variance to the new risk policy for any future Space Shuttle flight.

Space Shuttle Launches

NASA has continued to coordinate fully with the Air Force range safety community to determine the risk to the public associated with each Space Shuttle launch from the Kennedy Space Center (KSC). NASA and the Air Force have worked closely to improve the input data used in the risk assessments to ensure that results are based on the best possible estimate of nominal and off-nominal vehicle behavior. NASA has updated personnel categories and ensured workforce and visitor locations on KSC are accurately modeled. For each Space Shuttle launch, the Air Force has continued to use its risk analysis tools to provide a best estimate of the risks to the general public, visitors to the launch site, and the workforce. The Air Force, in coordination with NASA, has continued to update these models and to ensure the latest technologies and input data are employed.

All Space Shuttle launches are expected to satisfy the public risk criteria contained in NASA's new policy. Space Shuttle launches have always satisfied Air Force public risk criteria for individual risk and the Air Force collective risk criteria for the general public outside of KSC. Those criteria are reflected in NASA's new policy. NASA has not previously applied a collective risk criterion to people on KSC during Space Shuttle launches. Application of a collective public risk criterion to people on KSC represents the primary change affecting launch that was in place for Space Shuttle Return to Flight.

The new NASA policy incorporates an annual public collective risk criterion of one serious injury in a thousand flight years, which is a historical basis for the per-launch public risk criteria used by the federal ranges. All Space Shuttle launches satisfy this annual criterion. NASA expects to average five Space Shuttle launches per year to complete the International Space Station. One-in-a-thousand years divided by an average of five launches per year yields a per-launch risk criterion of 200 casualties in a million launches. This 200-in-a-million risk can be allocated between on-site and off-site collective public risk. The policy limits collective risk to the public outside KSC to 30 in a million per launch, which remains consistent with the Air Force public launch risk criterion enforced by the Eastern Range. This means that NASA applies a limit of 170 in a million to the public on site at KSC during a Space Shuttle launch. A NASA KSC management review board evaluates the risk assessment results provided by the Air Force for each Space Shuttle launch and determines the appropriate risk mitigation options needed to ensure that the risk criteria are satisfied. This includes identifying where people may be located on KSC during a launch and how many will be allowed at each location.

NASA's implementation of the public risk policy has ensured that any risk associated with attending a Space Shuttle launch at KSC is kept at a reasonable level. Individual risk to the vast majority of the public, those who are not on KSC, is significantly lower than the one-in-a-million individual risk criterion. Satisfying the collective risk criterion has resulted in limitations on the numbers of visitors allowed to attend a Space Shuttle launch at KSC and where they can be located. Through proper establishment of viewing sites and close controls on the numbers of people at each site, KSC has continued to accommodate a reasonable number of visitors for each Space Shuttle launch, consistent with NASA's mission to inspire the next generation of explorers.

Space Shuttle Orbiter Entries

Assessment of public risk associated with Orbiter entries is a new requirement for the Space Shuttle Program (SSP) after the *Columbia* accident. Unlike Space Shuttle launch, for which the Air Force's risk assessment tools and models were previously well established, the SSP has had to develop the tools and models needed to assess entry public risk. Encouraged by the CAIB Report, this was a significant effort for NASA civil servant and contractor personnel, resulting in a new capability prior to Return to Flight.

Because the trajectories, failure modes, and hazard characteristics are very different for entry as compared to launch, new and innovative approaches to risk modeling were developed. For example, vehicle breakup during a launch failure is typically modeled as instantaneous (i.e., as in an explosion). The *Columbia* accident demonstrated that a high-altitude structural failure of the Orbiter results in a progressive breakup over a relatively long period of time as pieces separate from the vehicle and then even break into smaller pieces as they fall. Personnel at the NASA Johnson Space Center developed new modeling techniques capable of accounting for progressive vehicle breakup. The *Columbia* accident represents just one type failure that can occur during an entry. There are other failure modes, such as potential loss of control late in flight at a relatively low altitude. Such a failure would have vehicle breakup characteristics that are very different from a high-altitude failure. NASA developed risk assessment models that account for the different failure modes and other contributors to public risk associated with Space Shuttle Orbiter entries. NASA has performed the public risk assessment for Space Shuttle Orbiter entries as part of the risk management process, and has continued to update the entry risk models and ensure the latest technologies and input data are employed.

All future NASA entries, including Space Shuttle Orbiter entries, will satisfy the one-in-a-million individual public risk criterion contained in the new NASA policy. The Space Shuttle entry risk assessments demonstrated that a person would have to be standing in an area close to the approach end of the runway during an Orbiter landing for that person's individual risk to exceed the criterion. With establishment of appropriate keep-out zones, NASA has ensured that the individual risk criterion is satisfied during each Space Shuttle entry operation.

With regard to the public collective risk criteria associated with entry operations, the new NASA policy takes a two-part approach. The first part of the entry risk policy applies specifically to Space Shuttle. This provision recognizes Space Shuttle's established design and operational constraints, which were developed more than 25 years ago without a specific requirement for managing public entry collective risk. Under this provision, KSC has continued as the Space Shuttle's primary landing site, with Edwards Air Force Base (EAFB) and White Sands Missile Range (WSMR) as backups. The SSP has implemented new flight rules that address the need for public risk abatement in the selection of the landing site for each mission.

The second part of NASA's new entry public collective risk policy contains risk criteria that will apply to future space flight vehicles. These risk criteria were developed in consultation with the national range community and are intended to serve the Nation's space program into the future as new vehicles are developed and entry operations become more common.

NASA has assessed the public collective risk associated with all possible Space Shuttle entry trajectories into the three landing sites from the International Space Station orbit inclination of 51.6 degrees, as well as from the Hubble Space Telescope orbit inclination of 28.5 degrees. On average, entry opportunities into KSC are half the public risk level of entries into EAFB. On average, entry opportunities into WSMR are one-seventh the public risk level of EAFB and one-third the public risk level of KSC. Although entries to WSMR represent a lower average public collective risk, WSMR does not have the infrastructure needed to safely and efficiently support regular Space Shuttle landings. WSMR and EAFB are best used as backups in conjunction with the SSP's use of flight rules designed to balance all safety concerns in the selection of a landing site.

The risk to the general public during entry was significantly reduced for Space Shuttle Return to Flight as compared to the past. Most of the improvements developed for Return to Flight either directly or indirectly serve to improve public safety during entry. For example, improvements to the External Tank and thermal protection systems and efforts to reduce critical debris all serve to ensure a successful and, therefore, safe flight for both the crew and the public. Also, we now inspect and assess the operational status of safety-critical thermal protection systems while on orbit. The flight rules for entry now account for the Orbiter systems' operational status and balance crew and public safety concerns when selecting among the available entry opportunities and landing sites. NASA is confident that this balanced approach is the wisest. The bottom line is that the Orbiter will normally land at KSC; but if the Orbiter experiences system failures or is considered compromised, the flight rules provide guidelines for further abating risk to the public while carefully considering the overall risk to the crew. In cases that pose a threat to the public, WSMR becomes the preferred landing site.

The criterion for entry collective risk represents the only portion of NASA's new policy that contains a Space Shuttle-specific provision. In addition to this provision, all other aspects of NASA's public risk policy apply to the Space Shuttle starting with the Return to Flight mission.

FINAL UPDATE

The NASA Administrator signed the new policy, NPR 8715.5 "Range Safety Program," on July 8, 2005. The new NASA policy requires that each program document its safety risk management process in a written Range Safety Risk Management Plan (RSRMP) approved by the responsible NASA officials. The SSP completed two RSRMPs. The launch RSRMP was signed by the Agency Range Safety Manager and the KSC Director in October 2006. The entry RSRMP was signed by the Agency Range Safety Program Manager and the Shuttle Range Safety Manager in June 2006. The SSP will update its RSRMP every 2 years throughout the life of the program as required by the new NASA policy.

SCHEDULE

Schedule Track to process NPR 8715.XX, NASA Range Safety Program

Action	March 2005 NODIS Review Cycle
Published Deadline for Submission to NASA Online Directives System (NODIS)	Mar 2005 (Completed)
Comments Due	Apr 2005 (Completed)
Signature Package Prepared	May 2005 (Completed)
Final Signature Expected	Jul 2005 (Completed)

Columbia Accident Investigation Board
Observations 10.1-2 and 10.1-3

O10.1-2   NASA should develop and implement a plan to mitigate the risk that Shuttle flights pose to the general public.

O10.1-3   NASA should study the debris recovered from *Columbia* to facilitate realistic estimates of the risk to the public during Orbiter re-entry.

Note: NASA has closed these observations through the formal Program Requirements Control Board (PRCB) process. The following summary details NASA's response to the observation and any additional work NASA performed beyond the *Columbia* Accident Investigation Board observations.

BACKGROUND

The *Columbia* accident raised important questions about public safety. The recovery and investigation effort found debris from the Orbiter scattered over a ground impact footprint approximately 275 miles long and 30 miles wide. Although there were no injuries to the public due to falling debris, the accident demonstrates that Orbiter breakup during entry may pose a risk to the general public.

NASA IMPLEMENTATION

The Space Shuttle Program (SSP) issued a PRCB directive to the Johnson Space Center Mission Operations Directorate to develop and implement a plan to mitigate the risk to the general public. NASA completed studies of the individual and collective risks to persons associated with entry to the three primary Shuttle landing sites, and developed plans and policies to mitigate the public risk, thus addressing Observation 10.1-2. The conclusions of these analyses resulted in some ground tracks being removed from consideration as normal, preplanned, end-of-mission landing opportunities when everything else involved in the landing decision is equal. (For a complete discussion of this topic and Observation 10.1-2, see the related actions in Space Shuttle Program Action 2 and Observation 10.1-1.)

FINAL UPDATE

Additionally, a multi-agency effort continues between NASA, the Federal Aviation Administration (FAA), and the U.S. Air Force to study the debris recovered from *Columbia*. This study addresses Observation 10.1-3. The multi-agency team has defined requirements for data collection and performed a measurement-taking trial run to refine those requirements. Initial data collection for this study was conducted in 2006, and studies for additional data collection are ongoing. Any required refinements to the currently defined public risk assessments and mitigation plans will be provided when these studies are completed.

SCHEDULE

Responsibility	Due Date	Activity/Deliverable
SSP	May 04 (Completed)	Finalize Responsibilities and Requirements for Data Collection
SSP/FAA	Sep 04 (Completed)	Signed Memorandum of Agreement between NASA and the FAA

Columbia Accident Investigation Board
Observation 10.2-1

Future crewed-vehicle requirements should incorporate the knowledge gained from the *Challenger* and *Columbia* accidents in assessing the feasibility of vehicles that could ensure crew survival even if the vehicle is destroyed.

Note: NASA has closed this observation through the formal Program Requirements Control Board (PRCB) process. The following summary details NASA's response to the observation and any additional work NASA performed beyond the *Columbia* Accident Investigation Board observation.

NASA IMPLEMENTATION

In July 2003, NASA published the Human-Rating Requirements and Guidelines for Space Flight Systems policy, NASA Procedural Requirement (NPR) 8705.2. This document includes a requirement for flight crew survival through a combination of abort and crew escape capabilities. The requirements in NPR 8705.2 are evolving to include NASA lessons learned from the Space Shuttle Program, including the lessons learned from the *Challenger* and *Columbia* accidents, Space Station operations, and other human space flight programs. NPR 8705.2 will be the reference document for the development of the planned Crew Exploration Vehicle (CEV).

On July 21, 2004, the Space Shuttle Upgrades PRCB approved the formation of the Space Craft Survival Integrated Investigation Team (SCSIIT). This multidisciplinary team, comprised of JSC Flight Crew Operations, JSC Mission Operations Directorate, JSC Engineering, Safety and Mission Assurance, the Space Shuttle Program (SSP), and Space and Life Sciences Directorate, was tasked to perform a comprehensive analysis of the two Shuttle accidents for crew survival implications. The team's focus is to combine data from both accidents (including debris, video, and Orbiter experiment data) with crew module models and analyses. After completion of the investigation and analysis, the SCSIIT will issue a formal report documenting lessons learned for enhancing crew survival in the Space Shuttle and for future human space flight vehicles, such as the CEV. Funding for fiscal year 2005 (FY05) and FY06 has been committed for this team's activities.

FINAL UPDATE

In conjunction with Space Shuttle Program activities, the Space and Life Sciences Directorate sponsored a contract with the University Space Research Association and the Biodynamics Research Corporation to perform an assessment of biodynamics from *Columbia* evidence. Their assessment was completed in October 2006 and the results will be incorporated into the SCSIIT Report, which is anticipated to be issued in August 2007.

Future crewed-vehicle spacecraft will use the products of the SSP and Space and Life Sciences Directorate to aid in the developments of crew safety and survival requirements.

Columbia Accident Investigation Board
Observation 10.4-1

Perform an independently led, bottom-up review of the Kennedy Space Center Quality Planning Requirements Document to address the entire quality assurance program and its administration. This review should include development of a responsive system to add or delete government mandatory inspections.

This Observation is addressed in Section 2.1, Space Shuttle Program Action 1.

Columbia Accident Investigation Board
Observation 10.4-2

Kennedy Space Center's Quality Assurance programs should be consolidated under one Mission Assurance office, which reports to the Center Director.

Note: NASA has closed this observation through the formal Program Requirements Control Board process. The following summary details NASA's response to the observation and any additional work NASA performed beyond the *Columbia* Accident Investigation Board (CAIB) observation.

BACKGROUND

Prior to the *Challenger* accident, Quality Assurance functions were distributed among the programs at Kennedy Space Center (KSC). In response to the findings of the Rogers Commission Report, KSC consolidated its Safety and Mission Assurance (SMA) functions into a single organizational entity. In May 2000, KSC once again dispersed the SMA function into each program and appropriate operational directorate. This was done to provide direct SMA support to each of the directorates, to ensure that the programs had the resources to be held accountable for safety and to enhance acceptance of the SMA role. Although this improved the relationships between SMA and the programs, the dependence of SMA personnel on program support limited their ability to effectively perform their role.

NASA IMPLEMENTATION

In close coordination with the effort led by the Associate Administrator for Safety and Mission Assurance (AA for SMA) in responding to CAIB Recommendation 7.5-2, KSC has established a center-level team to assess the KSC SMA organizational structure. This team was chartered in October 2003 to determine plans for implementing a consolidated SMA organization. The team developed several different candidate organizational structures. To maintain the benefits of the existing organization, which had SMA functions distributed to the appropriate programs and operational directorates, and to limit disruption to ongoing processes, the KSC Center Director chose a consolidated structure organized internally by program (see figure 10.4-2-1).

On January 13, 2004, KSC formed a Return to Flight Reorganization Team, which included an SMA Reorganization Team. The first task of this team was to perform a bottom-up review of the entire SMA organization. This bottom-up review revealed the need for additional SMA resources to fully perform the required functions. The proportion of SMA personnel to the total center population was deliberately decreased from a period shortly before the creation of the Space Flight Operations Contract (SFOC) based on the tasks transitioned to the contractor workforce; however, the bottom-up review demonstrated the need for expansion of the oversight/insight function and the associated collection of SMA data independent of the contractor-derived SMA data. As a result, additional SMA positions (Full-Time Equivalents (FTEs)) are being provided. These additional FTEs will reduce the amount of overtime currently required of the SMA professionals. They will also bring the percentage of SMA personnel to the entire KSC population back to the level that existed prior to the SFOC (see figure 10.4-2-2, chart 1). The additional positions will also decrease the dependence on the contractor for SMA data.

The bottom-up review also revealed unnecessary duplication of independent assessment resources. It was determined that if the entire KSC SMA workforce became centralized and once again independent of the programs, there would be no need for a large independent assessment organization.

When developing the single consolidated SMA organization at KSC, the SMA Reorganization Team identified the need for an Integration Division. Depicted as SA-G in figure 10.4-2-1, this Division will be responsible for ensuring consistency across the programs and for developing and implementing technical training for the SMA disciplines. The Integration Division will include discipline experts in Safety Engineering, Quality Engineering, Quality Assurance, Software Assurance, Reliability, Human Factors, and Risk Management, and it will be responsible for policy creation and review and procurement assurance.

The SMA Reorganization Team also evaluated the work required by the planned Independent Technical Authority (ITA) to incorporate its requirements into the centralized SMA organization. To fulfill these requirements, KSC

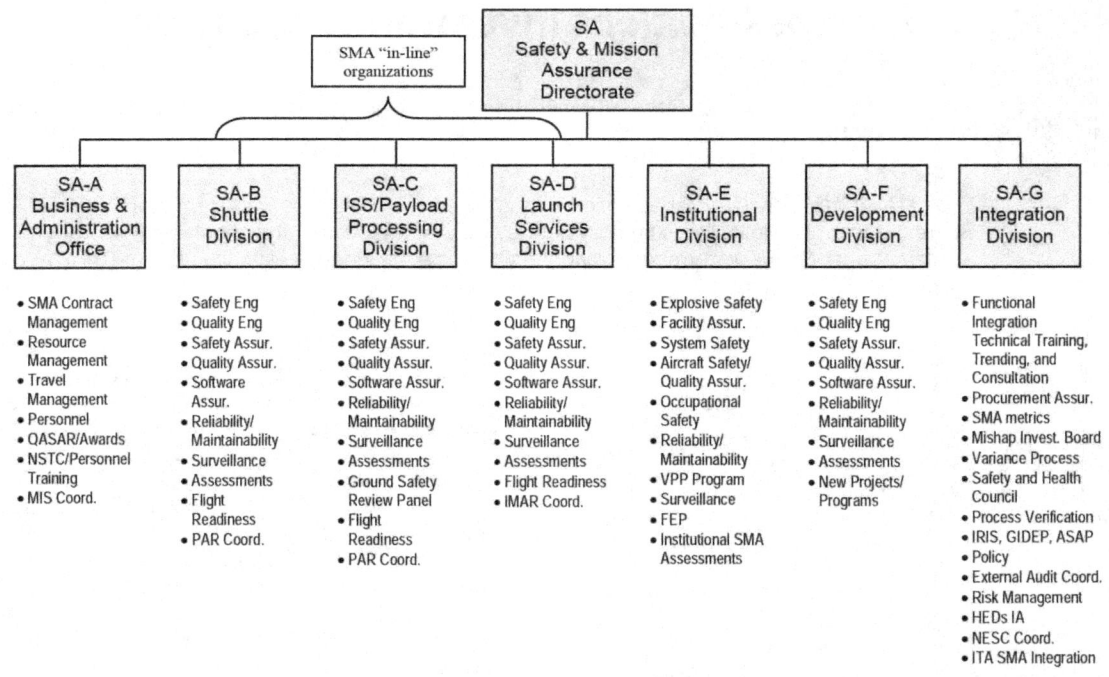

Figure 10.4-2-1. Consolidated SMA organization.

provided three FTEs for SMA/ITA within the total additional FTEs received.

In addition to the managerial independence established by consolidation, the SMA Reorganization Team worked with the KSC financial organization and NASA Headquarters to create a new "directed service pool" funding process. The directed service pool gives the SMA Directorate the authority to determine, in consultation with the programs, the level of support it will provide to each program.

FINAL UPDATE

The SMA Reorganization Team developed an avenue to use the Johnson Space Center SMA contract to provide for immediate resource needs. KSC now has several independent contracts that it uses to provide for SMA contractor support.

Finally, KSC implemented several ongoing initiatives to address the culture within SMA and throughout the center. Specifically, Behavioral Science Technologies Inc. identified the need for the KSC SMA organization to work on improving its organizational culture. Therefore, the KSC SMA Directorate has worked with 4-D Systems to continue its efforts to improve the culture within and outside of the organization.

NASA KSC's SMA Directorate remains a centralized, independent organization that made several organizational adjustments since the closure of this observation. We expect to maintain this centralized structure and will continually make adjustments to enhance our support to the needs of the programs at KSC.

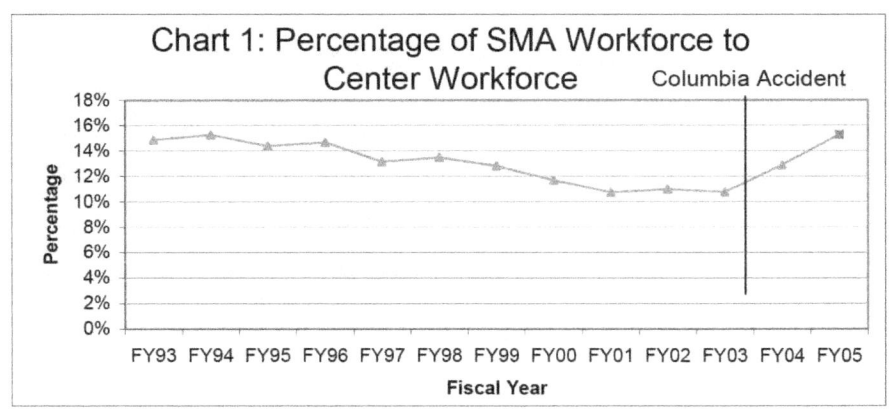

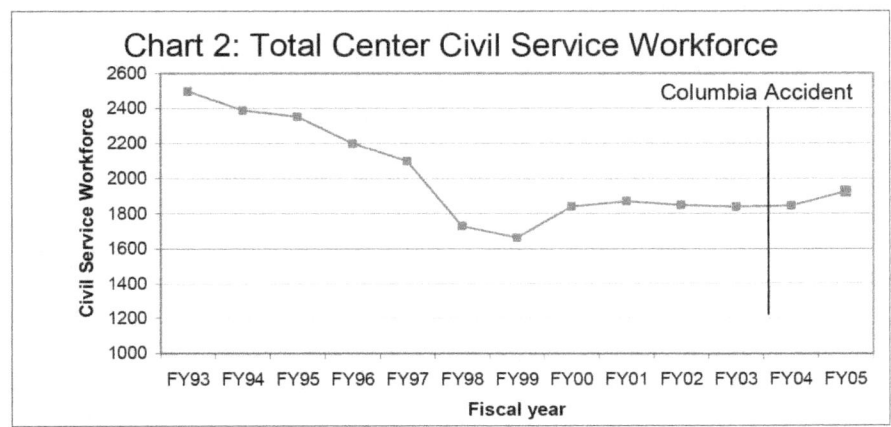

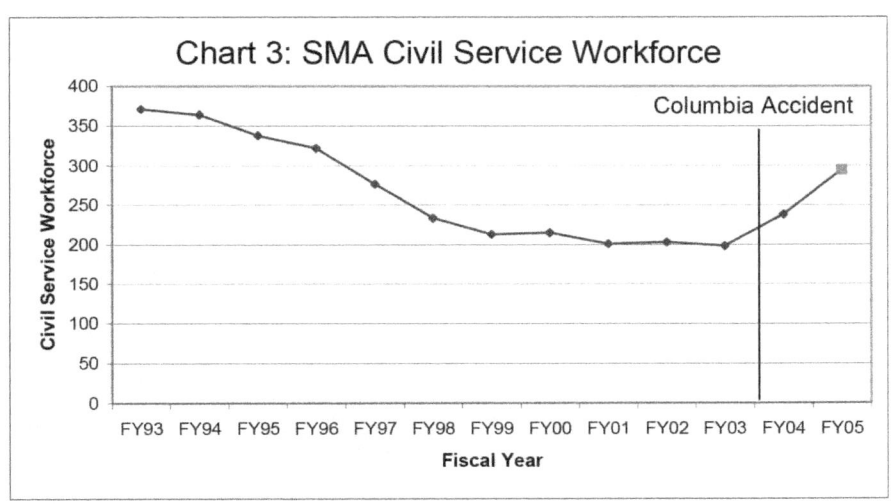

Figure 10.4-2-2. SMA workforce.

SCHEDULE

Responsibility	Due Date	Activity/Deliverable
KSC	Completed	Recommendations to KSC Center Director
KSC	Apr 04 (Completed)	Reorganization definition complete
KSC	May 04 (Completed)	Implementation complete

Columbia Accident Investigation Board
Observation 10.4-3

Kennedy Space Center quality assurance management must work with NASA and perhaps the Department of Defense to develop training programs for its personnel.

Note: NASA has closed this observation through the formal Program Requirements Control Board process. The following summary details NASA's response to the observation and any additional work NASA performed beyond the Columbia Accident Investigation Board observation.

BACKGROUND

The *Columbia* Accident Investigation Board reported most of the training for quality engineers, process analysts, and quality assurance specialists was on-the-job training rather than formal training. In general, Kennedy Space Center (KSC) training is extensive for the specific hardware tasks (e.g., crimping, wire bonding, etc.), and includes approximately 160 hours of formal, on-the-job, and safety/area access training for each quality assurance specialist. However, there were deficiencies in basic quality assurance philosophy and skills.

NASA IMPLEMENTATION

NASA's KSC has worked with the Department of Defense (DoD) and Defense Contract Management Agency (DCMA) to benchmark their training programs and to determine how NASA can develop a comparable training program for quality engineers, process analysts, and quality assurance specialists. Of interest is training on Fundamentals of Quality Assurance. A team completed the DCMA Quality Assurance skills course. The KSC Safety and Mission Assurance (SMA) Directorate has documented the training requirements for all SMA positions and the improved training was implemented.

STATUS

NASA continues to monitor and improve our Quality Assurance programs.

FINAL UPDATE

The SMA Directorate Development Plan was approved and published in October 2004. Fundamentals of Quality Assurance training is scheduled to be completed by the end of July 2007.

SCHEDULE

Responsibility	Due Date	Activity/Deliverable
KSC	Apr 04 (Completed)	Benchmark DoD and DCMA training programs
KSC	Oct 04 (Completed)	Develop and document improved training requirements

Columbia Accident Investigation Board
Observation 10.4-4

Kennedy Space Center should examine which areas of International Organization for Standardization 9000/9001 truly apply to a 20-year-old research and development system like the Space Shuttle.

Note: NASA considers this observation closed. The following summary details NASA's response to the observation and any additional work NASA performed beyond the *Columbia* Accident Investigation Board observation.

BACKGROUND

The *Columbia* Accident Investigation Board Report high-lighted Kennedy Space Center's (KSC's) reliance on the International Organization for Standardization (ISO) 9000/9001 certification. The report stated, "While ISO 9000/9001 expresses strong principles, they are more applicable to manufacturing and repetitive-procedure industries, such as running a major airline, than to a research-and-development, flight test environment like that of the Space Shuttle. Indeed, many perceive International Standardization as emphasizing process over product." ISO 9000/9001 is currently a contract requirement for United Space Alliance (USA).

NASA IMPLEMENTATION

NASA assembled a team of Agency and industry experts to examine the ISO 9000/9001 standard and its applicability to the Space Shuttle Program. Specifically, this examination addressed the following: 1) ISO 9000/9001 applicability to USA KSC operations; 2) how NASA should use USA's ISO 9000/9001 applicable elements in evaluating USA performance; 3) how NASA currently uses USA's ISO certification in evaluating its performance; and 4) how NASA will use the ISO certification in the future and the resultant changes.

FINAL UPDATE

The ISO 9000/9001 review team established a methodology to determine the applicability of the ISO standard to USA operations at KSC. The ISO 9000/9001 standard establishes high-level requirements that comprise a quality management system, but it does not dictate the specifics of implementation for those requirements. Consequently, NASA believes the ISO 9000/9001 certification applies to USA KSC operations.

NASA should not and does not rely on third-party ISO 9000/9001 certification for assurance that flight preparations are appropriately and effectively performed. NASA should and does rely on surveillance of contractor activities and compliance with contractual and USA requirements to assure flight processes and products are accurate, effective, and timely. NASA surveillance requirements to evaluate contractor performance in processing flight/ground hardware are documented at the Agency, Program, and Center levels for both the Certificate of Flight Readiness and Award Fee.

NASA does encourage ISO 9000/9001 certification. Certification reflects management commitment to developing and sustaining a robust quality management system. Third-party audits also provide objective looks at the contractor and opportunities for improvement. Although ISO 9000/9001 certification will not be used as a direct measure of performance, the products of a robust quality system will be used, such as measures/metrics and process improvement initiatives.

After the initial response to this observation, discussions with NASA Quality Assurance Specialists (QASs) revealed that, at times, they were not initially provided the requested information to perform inspections. The rationale was that printing the information would invalidate it as it may not be the latest revision. This is not USA policy, nor is it a requirement of the ISO 9000/9001 standard. NASA QASs have been trained and have demonstrated the integrity to stop work if they need to obtain additional information to properly perform the assigned task.

SCHEDULE

Responsibility	Due Date	Activity/Deliverable
KSC	Jan 05 (Completed)	Identify applicability to USA KSC Operations
KSC	Jan 05 (Completed)	Proper usage of standard in evaluating contractor performance
KSC	Mar 05 (Completed)	Current usage of standard in evaluating contractor performance
KSC	Mar 05 (Completed)	Future usage of standard and changes to surveillance or evaluation of contractor
KSC	Jun 05 (Completed)	Presentation of Review

Columbia Accident Investigation Board
Observation 10.5-1

Quality and Engineering review of work documents for STS-114 should be accomplished using statistical sampling to ensure that a representative sample is evaluated and adequate feedback is communicated to resolve documentation problems.

Note: NASA has closed this *Columbia* Accident Investigation Board (CAIB) Observation through the formal Program Requirements Control Board process. The following summary details NASA's response to the CAIB Observation and any additional work NASA performed beyond the CAIB Observation.

BACKGROUND

The Kennedy Space Center (KSC) Processing Review Team conducted a review of the ground processing activities and work documents from all systems for STS-107 and STS-109, and from some systems for Orbiter Major Modification. This review examined approximately 3.9 million work steps and identified 9672 processing and documentation discrepancies resulting in a work step accuracy rate of 99.75%. While this is comparable to our performance in recent years, our goal is to further reduce processing discrepancies; therefore, we initiated a review of STS-114 documentation.

NASA IMPLEMENTATION

NASA has performed a review and systemic analysis of STS-114 work documents from the time of Orbiter Processing Facility roll-in through system integration test of the flight elements in the Vehicle Assembly Building. Pareto analysis of the discrepancies revealed areas where root cause analysis is required.

STATUS

The STS-114 Processing Review Team systemic analysis revealed six Corrective Action recommendations consistent with the technical observations noted in the STS-107/109 review. Teams were formed to determine the root cause and long-term corrective actions. These recommendations were assigned Corrective Action Requests that are used to track the implementation and effectiveness of the corrective actions. In addition to the remedial actions from the previous review, there were nine new system-specific remedial recommendations. These remedial actions primarily addressed documentation errors, and have been implemented. Quality and Engineering will continue to statistically sample and analyze work documents for all future flows.

FINAL UPDATE

The root cause analysis results and Corrective Actions were presented to and approved by the Space Shuttle Program in February 2004.

SCHEDULE

Responsibility	Due Date	Activity/Deliverable
KSC	Feb 04 (Completed)	Program Requirements Control Board

Columbia Accident Investigation Board
Observation 10.5-2

NASA should implement United Space Alliance's suggestions for process improvement, which recommend including a statistical sampling of all future paperwork to identify recurring problems and implement corrective actions.

Note: NASA has closed this observation through the formal Program Requirements Control Board process. The following summary details NASA's response to the observation and any additional work NASA performed beyond the *Columbia* Accident Investigation Board observation.

BACKGROUND

The Kennedy Space Center (KSC) Processing Review Team conducted a review of the ground processing activities and work documents from all systems for STS-107 and STS-109 and from some systems for the Orbiter Major Modifications. This review examined approximately 3.9 million work steps and identified 9672 processing and documentation discrepancies resulting in a work step accuracy rate of 99.75%. These results were validated with the review of STS-114 work documents (ref. Observation 10.5-1). Pareto analysis of the discrepancies revealed areas where corrective action is required and where NASA Shuttle Processing surveillance needs augmentation.

NASA IMPLEMENTATION

NASA has refocused the KSC Shuttle Processing Engineering and Safety and Mission Assurance (SMA) surveillance efforts and enhanced the communication of surveillance results between the two organizations. KSC Shuttle Processing Engineering has increased surveillance of processing tasks and of the design process for government-supplied equipment and ground systems. This has included expanding the list of contractor products requiring NASA engineering approval. SMA surveillance has expanded to include sampling of closed paper and hardware surveillance (ref. Observation 10.5-3).

NASA has improved communication between the Engineering Office and SMA through the activation of a Web-based log and the use of a new Quality Planning and Requirements Document change process for government inspection requirements.

Engineering and SMA organizations have revised their surveillance plans. Required changes to the Ground Operations Operating Procedures have been identified.

FINAL UPDATE

NASA has implemented periodic reviews of surveillance plans and adjusted the tasks as necessary to target problem areas identified by data trends and audits. Working in partnership with United Space Alliance, NASA has completed five statistical processing audits, reviewed the audits results, and implemented corrective actions.

SCHEDULE

Responsibility	Due Date	Activity/Deliverable
KSC	Nov 03 (Completed)	Surveillance task identification

Columbia Accident Investigation Board
Observation 10.5-3

NASA needs an oversight process to statistically sample the work performed and documented by Alliance technicians to ensure process control, compliance, and consistency.

Note: NASA has closed this recommendation through the formal Program Requirements Control Board process. The following summary details NASA's response to the recommendation and any additional work NASA performed beyond the *Columbia* Accident Investigation Board (CAIB) recommendation.

BACKGROUND

The CAIB noted the need for a statistically valid sampling program to evaluate contractor operations. NASA Safety and Mission Assurance identified two distinct processing activities within the observation: (1) work performed and (2) work documented.

NASA IMPLEMENTATION

NASA assessed the implementation, required resources, and potential benefits of developing a new statistical sampling program for work performed and documented by United Space Alliance (USA) technicians. This assessment included an independent analysis of a process sampling program for work performed in Shuttle ground operations implemented by USA in 1988.

NASA also engaged a Faculty Fellow in 2004 to evaluate ongoing process sampling efforts. The Faculty Fellow study indicated the need for close collaboration between NASA and USA to ensure that duplications were minimized, that the process sampling effort focused on areas of importance, and that there was appropriate NASA management of the activity. As a result, the USA process sampling effort was converted to a program jointly owned by NASA and USA.

NASA has also developed and implemented an independent statistical sampling program for closed Work Authorization Documents (WADs). The WAD data have been used for support of the Certification of Flight Readiness as an indicator of contractor performance. An initial closed WAD sample schedule has been delivered for measuring WAD accuracy of completeness in execution. The schedule ensures that closed WADs are sampled by NASA for adherence to the requirements. The plan incorporated unplanned and planned WADs. Problem Reports and Discrepancy Reports were sampled and results communicated to USA for appropriate corrective action.

STATUS

NASA will continue improving its ability to assure the quality of USA work.

FINAL UPDATE

Based on satisfactory reports of a stable process at USA, NASA has discontinued the joint sampling program. However, NASA has implemented additional sampling activities for review of in-process WADs.

SCHEDULE

Responsibility	Due Date	Activity/Deliverable
KSC	Nov 03 (Completed)	Provide resource estimate
KSC	Nov 03 (Completed)	Implement in-process sampling program
KSC	Nov 03 (Completed)	Implement Closed WAD sampling program – vehicle problem reports only
KSC	Mar 04 (Completed)	Define/develop in-process metrics
KSC	Apr 04 (Completed)	Closed WAD sampling program – addition of Space Shuttle Main Engine and Ground Support Equipment problem reports
KSC	May 04 (Completed)	Define/develop closed WAD sampling standard metrics
KSC	Oct 04 (Completed)	Develop closed WAD sampling plan and schedule

Columbia Accident Investigation Board
Observation 10.6-1

The Space Shuttle Program Office must make every effort to achieve greater stability, consistency, and predictability in Orbiter Major Modification planning, scheduling, and work standards (particularly in the number of modifications). Endless changes create unnecessary turmoil and can adversely impact quality and safety.

Note: NASA has closed this observation through the formal Program Requirements Control Board process. The following summary details NASA's response to the observation and any additional work NASA performed beyond the *Columbia* Accident Investigation Board observation.

BACKGROUND

NASA agrees that greater stability in Orbiter Maintenance Down Period (OMDP) processes will reduce risk.

NASA IMPLEMENTATION

The OV-105 OMDP is complete. In planning for this OMDP, NASA emphasized stability in the work plan by following the practice of approving most or all of the known modifications at the onset of the OMDP/Orbiter Major Modification period.

FINAL UPDATE

The Space Shuttle Program (SSP) will continue to assess and periodically review the status of all required modifications. NASA will continue to integrate lessons learned from each modification and will emphasize factors that could destabilize plans and schedules. NASA conducts delta Launch Site Flow Reviews for each Orbiter on an ongoing basis.

SCHEDULE

Responsibility	Due Date	Activity/Deliverable
SSP	Oct 03 (Completed)	OV-105 OMDP Modification Site Flow Review

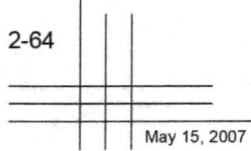

Columbia Accident Investigation Board
Observation 10.6-2

NASA and United Space Alliance managers must understand workforce and infrastructure requirements, match them against capabilities, and take actions to avoid exceeding thresholds.

Note: NASA has closed this observation through the formal Program Requirements Control Board (PRCB) process. The following summary details NASA's response to the observation and any additional work NASA performed beyond the *Columbia* Accident Investigation Board observation.

BACKGROUND

The transfer of Orbiter Maintenance Down Periods (OMDPs) from Palmdale to Kennedy Space Center placed additional demands on the existing infrastructure, ground support equipment, and personnel. NASA made significant efforts to anticipate these demands, to transfer the needed equipment from Palmdale, and to hire additional personnel required to accomplish the OMDP-related tasks independent of normal Orbiter flow processing. Because of the fluctuating demands on the Orbiters supporting the flight manifest, some workers with unique critical skills were frequently shared among the Orbiters in OMDP and the Orbiters being processed for flight. Additional inspection and modification requirements, and unanticipated rework for structural corrosion and Thermal Protection Systems, created previously unanticipated demands on limited critical skill sets.

NASA IMPLEMENTATION

Lessons learned from the third Orbiter Vehicle (OV)-103 OMDP have been incorporated into the current OV-105 OMDP. These lessons have allowed NASA and United Space Alliance managers to better integrate infrastructure, equipment, and personnel from a more complete set of work tasks. Unlike the piecemeal approach used during OV-103's OMDP, the requirements for OV-105's OMDP were approved at the beginning, with the exception of two modifications. The PRCB approved 72 modifications at the Modification Site Requirements Review in early July 2003, and reviewed the overall modification plan again in mid-October 2003 at the Modification Site Flow Review. The Space Shuttle Program (SSP) follows the practice of approving most or all of the known modifications for incorporation at the beginning of an OV's processing flow, typically at the Launch Site Requirements Review.

Many "out of family" discrepancies identified as the result of scheduled structural and wiring inspections require design center coordination and disposition. The incorporation of new Orbiter modifications also requires close coordination for design issue resolution. Timely design response can reduce the degree of rescheduling and critical skill rebalancing required. During the OV-103 OMDP, design center engineers were available on the floor in the Orbiter Processing Facility where the work was being accomplished to efficiently and effectively disposition discrepancies when identified. The additional emphasis on "on floor" design response, which helped to reduce rescheduling and resource rebalancing during OV-103's third OMDP, was also used for OV-105's OMDP.

FINAL UPDATE

The OV-105 OMDP is the last planned OMDP due to the SSP retirement in 2010.

SCHEDULE

Responsibility	Due Date	Activity/Deliverable
SSP	Oct 03 (Completed)	Modification Site Flow Review
SSP	Dec 03 (Completed)	Complete OV-103 Lessons Learned

Columbia Accident Investigation Board
Observation 10.6-3

NASA should continue to work with the U.S. Air Force, particularly in areas of program management that deal with aging systems, service life extension, planning and scheduling, workforce management, training, and quality assurance.

Note: NASA has closed this observation through the formal Program Requirements Control Board process. The following summary details NASA's response to the observation and any additional work NASA performed beyond the *Columbia* Accident Investigation Board observation.

BACKGROUND

In June 2003, NASA requested that the U.S. Air Force conduct an assessment of the Orbiter Maintenance Down Period/Orbiter Major Modification (OMDP/OMM) being performed at Kennedy Space Center. The U.S. Air Force team provided similarities, compared best practices, identified differences between NASA and the U.S. Air Force practices, identified potential deficiencies, and provided recommendations and areas for potential improvements. NASA has used this information to improve our practices and processes in evaluating the Orbiter fleet, and to formulate our approach for continued benchmarking.

NASA also initiated a number of aging vehicle assessment activities as part of the integrated Space Shuttle Service Life Extension Program (SLEP) activities. Each of the Space Shuttle element organizations continues to pursue appropriate vehicle assessments to ensure that Space Shuttle Program (SSP) operations remain safe and viable throughout the Shuttle's operational life.

NASA IMPLEMENTATION

Personnel from Wright-Patterson Air Force Base have provided direct support to SLEP and have contributed to management decisions on needed investments through membership on SLEP panels. NASA continues to work with the U.S. Air Force in its development of aging vehicle assessment plans and benefit from its knowledge of operating and maintaining long-life aircraft systems. Assessments for the Space Shuttle Orbiters, for example, will include expanded fleet leader hardware programs and corrosion control programs.

In addition to working with the U.S. Air Force on these assessments, NASA is actively drawing upon resources external to the SSP that have valuable experience in managing the operations of aging aircraft and defense systems. NASA is identifying contacts across government agencies and within the aerospace and defense industries to bring relevant expertise from outside the SSP to assist the team. For instance, NASA Orbiter Project Office is participating in NASA's new Aging Aircraft and Durability Program, is a member of the NASA Engineering and Safety Center Materials and Nondestructive Evaluation Super Problem Resolution Teams, and is actively involved in the annual joint NASA/Federal Aviation Administration (FAA)/Department of Defense (DoD) Aging Aircraft conference. The Orbiter Project has already augmented its aging Orbiter assessment team with systems experts from Boeing Integrated Defense Systems.

In 1999, NASA began a partnership with the U.S. Air Force Research Laboratory, Materials and Manufacturing Directorate, at Wright-Patterson Air Force Base to characterize and investigate wire anomalies. The Joint NASA/FAA/DoD Conference on Aging Aircraft focused on studies and technology to identify and characterize these aging systems. NASA will continue this partnership with constant communication, research collaboration, and technical interchange.

FINAL UPDATE

NASA continues to assess vehicle systems for aging effects and will update inspection and maintenance requirements accordingly. Lessons learned from past Orbiter maintenance periods as well as knowledge gained in cooperation with the U.S. Air Force will be applied in the remaining OMDPs/OMMs.

Columbia Accident Investigation Board
Observation 10.6-4

The Space Shuttle Program Office must determine how it will effectively meet the challenges of inspecting and maintaining an aging Orbiter fleet before lengthening Orbiter Major Maintenance intervals.

Note: NASA has closed this recommendation through the formal Program Requirements Control Board process. The following summary details NASA's response to the recommendation and any additional work NASA performed beyond the *Columbia* Accident Investigation Board recommendation.

BACKGROUND

An aging Orbiter fleet presents inspection and maintenance challenges that must be incorporated in the planning of the Orbiter Maintenance Down Periods (OMDPs). Prior to the *Columbia* accident, the Space Shuttle Program had begun an activity to lengthen the interval between OMDPs from the current requirement of every 3 years or 8 flights to a maximum of 6 years or 12 flights. This activity consisted of two major areas of assessment, structural inspection, and systems maintenance.

The Structures Problem Resolution Team (PRT) was assigned the action to examine all structural inspection requirements for effects to extending the OMDP interval. The Structures PRT examined every requirement dealing with structural inspections in the Orbiter Maintenance Requirements and Specifications Document and compared findings from previous OMDP and in-flow inspections to determine whether new inspection intervals were warranted. The findings from this effort resulted in updated intervals for structures inspections. Structural inspections can support an OMDP interval of 6 years or 12 flights. Part of this new set of inspections is the inclusion of numerous interval inspections that would be conducted between OMDPs. Adverse findings from the sampling inspections could lead to a call for an early OMDP.

In similar fashion, the systems maintenance requirements were to be assessed for interval lengthening by the various responsible PRTs. These assessments were put on hold at the time of the *Columbia* accident until NASA could determine whether more consideration should be given to extending OMDP intervals.

NASA IMPLEMENTATION

Orbiter aging vehicle assessments, originally initiated as part of the Shuttle Service Life Extension Program, ensure that inspection and maintenance requirements are evaluated for any needed requirements updates to address aging vehicle concerns. An explicit review of all hardware inspection and systems maintenance requirements was conducted during the Orbiter life certification assessment to determine if aging hardware considerations or certification issues warranted the addition of new inspection/maintenance requirements or modification to existing requirements. Subsequent to completion of the life certification assessment, inspection requirement adequacy has continued to be evaluated through ongoing aging vehicle assessment activities, including the Orbiter fleet leader program and corrosion control program.

FINAL UPDATE

Following STS-114, all system maintenance requirements were reviewed. This review resulted in the extension of the OMDP interval from 3 years/eight flights to 5.5 years/eight flights. NASA has put into place an ongoing assessment process to ensure that Space Shuttle operations remain safe and viable throughout the Shuttle's service life.

Columbia Accident Investigation Board
Observation 10.7-1

Additional and recurring evaluation of corrosion damage should include non-destructive analysis of the potential impacts on structural integrity.

Note: NASA has closed this observation through the formal Program Requirements Control Board process. The following summary details NASA's response to the observation and any additional work NASA performed beyond the *Columbia* Accident Investigation Board (CAIB) observation.

BACKGROUND

The SSP has initiated an action to assess the CAIB observations related to corrosion damage in the Shuttle Orbiters. This action has been assigned to the Orbiter Project Office.

NASA IMPLEMENTATION

The Orbiter element is in full compliance with this observation. Before the disposition of any observed corrosion on Orbiter hardware, a full action plan is coordinated by the responsible subsystem engineering discipline. To resolve specific corrosion issues, evaluation and/or analysis is performed by the appropriate subsystem, stress, and materials engineers. Investigations into hardware conditions and exposure environments are performed to determine root cause of any corrosion, and nondestructive analysis is used to assist in characterization of the depth and breadth of existing corrosion. Destructive analysis is pursued where appropriate.

In all cases, Space Shuttle requirements mandate that positive safety margins must be retained by Orbiter hardware. To do this, where necessary, affected components may be replaced or supplementary load paths/doublers applied. Any course of action (e.g., leave as-is, application of corrosion preventative compounds, re-work, replace, etc.) must be agreed upon by the appropriate technical communities. Cross-disciplinary reviews of significant corrosion-related issues take place on a regular basis. As new or repeat corrosion issues are discovered, the governing Operations and Maintenance Requirements and Specifications Document is reviewed and modified as appropriate. Future inspection schedules are adjusted accordingly to maintain conservative time intervals.

To support Orbiter corrosion issues and concerns, the Orbiter Corrosion Control Review Board (CCRB) provides an independent technical review of ongoing corrosion issues. The CCRB has representation from both NASA and NASA contractors in materials and processes engineering, subsystem engineering, and safety and mission assurance.

For "minor" corrosion issues, the Orbiter CCRB may be consulted for a recommendation at the discretion of the subsystem engineer. If the corrosion in question cannot be repaired by the Orbiter Standard Repair Procedure (V-ST-0029) or if reapplication of per print corrosion protective finishes cannot be accomplished or is inadequate, a review by the CCRB is required.

On a case-by-case basis, the engineering review team/CCRB may identify other similar hardware, materials, and locations on the flight vehicles as suspect; this determination results in targeted inspections. In areas where nondestructive analysis is not currently feasible (e.g., under the Thermal Protection System, between faying surface joints, etc.), "sampling" inspections are carried out to quantify the scope and magnitude of the corrosion issue. Analysis is completed to determine whether the corrosion is local or systemic.

FINAL UPDATE

Additional funding for augmentation of Orbiter corrosion control activities was authorized in May 2004 and extended through the end of fiscal year 2006. Since that time, the expanded efforts have been covered within scope as part of the Space Flight Operations Contract extension. This authorization implemented proactive corrosion control measures to ensure continued safety and sustainability of Orbiter hardware throughout the planned SSP service life, including identification of improvements to nondestructive evaluation techniques.

Columbia Accident Investigation Board
Observation 10.7-2

Long-term corrosion detection should be a funding priority.

Note: NASA has closed this observation through the formal Program Requirements Control Board process. The following summary details NASA's response to the observation and any additional work NASA performed beyond the *Columbia* Accident Investigation Board observation.

BACKGROUND

Both Orbiter engineering and management concur that ongoing corrosion of the Space Shuttle fleet should be addressed as a safety issue. As the Orbiters continue to age, NASA must direct the appropriate level of resources to sustain the expanding scope of corrosion and its impact to Orbiter hardware.

NASA IMPLEMENTATION

Following the *Columbia* accident, the Orbiter Corrosion Control Review Board has been strengthened significantly. Additional funding for augmentation of Orbiter corrosion control activities was authorized in May 2004 and extended through the end of fiscal year 2006. Since that time, the expanded efforts have been covered within scope as part of the Space Flight Operations Contract extension. This authorization implemented proactive corrosion control measures to ensure safety and sustainability of Orbiter hardware throughout the planned Space Shuttle Program (SSP) service life. Specific activities addressing corrosion prevention and detection include: developing methods to reduce hardware exposure to corrosion causes; identifying and evaluating the environment of corrosion prone areas and environmental control mitigation options; identifying improved nondestructive evaluation (NDE) techniques; and implementing an industry benchmark team for reducing corrosion and improving NDE methods.

FINAL UPDATE

Under the leadership of the Orbiter Corrosion Control Review Board, NASA, United Space Alliance, and Boeing have developed and are implementing the expanded scope of an effective, long-term corrosion control program. This expanded program will attempt to inspect for, detect, evaluate, trend, and predict corrosion on Orbiter hardware throughout the remaining life of the SSP.

SCHEDULE

Responsibility	Due Date	Activity/Deliverable
Orbiter Project Office	Completed	Direct appropriate long-term funding (sustained)
Orbiter Project Office	Jun 04 (Completed)	Develop an advanced Orbiter Corrosion Control Program to detect, trend, analyze, and predict future corrosion issues

Columbia Accident Investigation Board
Observation 10.7-3

Develop non-destructive evaluation inspections to find hidden corrosion.

Note: NASA has closed this observation through the formal Program Requirements Control Board (PRCB) process. The following summary details NASA's response to the observation and any additional work NASA performed beyond the *Columbia* Accident Investigation Board (CAIB) observation.

BACKGROUND

An integral part of an effective corrosion control program is the continual development and use of nondestructive evaluation (NDE) tools. The development of tools that explore hidden corrosion is a complex problem.

NASA IMPLEMENTATION

NASA continues to investigate a wide range of advanced NDE techniques and has several activities ongoing to use NDE to find hidden corrosion. These activities include:

- Chartered by the NASA, the NDE Working Group (NNWG) has representatives from each of the NASA field centers and affiliated contractors. This group meets periodically to address NASA's short- and long-term NDE needs. Since its charter, the NNWG has executed efforts to develop NDE techniques directly in support of this subject, such as corrosion under tile. Orbiter engineering will partner with the NNWG on NDE development work as specific achievable needs are identified.

- An Orbiter NDE working group was established to address both immediate and long-term Orbiter needs. This technical team has become an important resource in support of ongoing Orbiter problem resolutions, including addressing the need for advanced NDE tools and techniques required to address hidden corrosion.

- Additional funding for augmentation of Orbiter corrosion control activities was authorized in May 2004 and extended through the end of fiscal year 2006. This effort included a review of the state-of-the-art NDE techniques for the detection of corrosion. The Orbiter Corrosion Control Review Board (CCRB) will continue to use the results of this review and will partner with the NNWG and Orbiter NNWG to investigate advanced NDE techniques for the detection of hidden corrosion.

- Johnson Space Center and Marshall Space Flight Center have developed a compilation of hidden corrosion test standards. These standards will be used for future evaluation of potential NDE techniques.

In areas where nondestructive analysis is not currently feasible (e.g., under the Thermal Protection System (TPS), between faying surface joints, etc.), "sampling" inspections are carried out to quantify the scope and magnitude of the particular corrosion issue. Analysis is subsequently completed to determine whether the corrosion is local or systemic.

As an example, the CAIB Report referenced corrosion discovered prior to STS-107 on the *Columbia* vehicle in the lower forward fuselage skin panel and stringer areas (inner surfaces). Subsequently, inspections of the TPS bond line (outer surfaces) identified isolated incidents of localized surface corrosion. This raised concerns regarding a potential threat to the TPS bond-line. As a result, a complete history of previous TPS corrosion inspections, bond-line corrosion indications, bond surface preparation processes and controls, and TPS bond operation materials and processes was reviewed. The review was coordinated jointly between the Materials and Processes, TPS, and Structures engineering organizations with a contributing independent assessment by the CCRB. This activity resulted in a reversal of previous engineering direction; as a result, damaged Koropon primer is now required to be repaired/reconditioned before tiles are bonded, and NASA authorized development of an extensive multi-year sampling program intended to characterize the magnitude and scope of corrosion occurring under tile.

In May 2004, the Space Shuttle Program (SSP) authorized $3.3M of additional funding for augmentation of Orbiter corrosion control activities via PRCB directive S061984R1. This authorization implemented proactive corrosion control measures to ensure continued safety and sustainability of Orbiter hardware throughout the planned SSP service life, including identification and development of improvements to NDE techniques. Since the end of fiscal year 2006, the expanded Orbiter corrosion control efforts have been covered under the Space Flight Operations Contract extension.

FINAL UPDATE

As a part of this expanded program, the current and future Orbiter project needs for NDE are being evaluated for further development. The review of all activities will be completed and evaluated against long-term project needs through the life of the SSP.

SCHEDULE

Responsibility	Due Date	Activity/Deliverable
Orbiter Project Office	Jun 04 (Completed)	Develop an advanced Orbiter Corrosion Control Program, chartered to detect, trend, analyze, and predict future corrosion issues. Development of NDE techniques for corrosion detection shall be included in the Program.

Columbia Accident Investigation Board
Observation 10.7-4

Inspection requirements for corrosion due to environmental exposure should first establish corrosion rates for Orbiter-specific environments, materials, and structural configurations. Consider applying Air Force corrosion prevention programs to the Orbiter.

Note: NASA has closed this observation through the formal Program Requirements Control Board process. The following summary details NASA's response to the observation and any additional work NASA performed beyond the *Columbia* Accident Investigation Board observation.

BACKGROUND

Historically, inspection intervals for Orbiter corrosion have not been driven by mathematical corrosion rate assessments. In practice, predicting corrosion rates is only effective when the driving mechanism is limited to general surface corrosion in a known environment over a known period of time. To date, general surface corrosion is not an Orbiter problem. Common Orbiter corrosion problems include pitting, crevice, galvanic, and intergranular corrosion attack. These mechanisms are extremely sporadic and inconsistent, and present tremendous difficulty in effectively predicting corrosion rates. Environments are complex, including time histories with intermittent exposure to the extreme temperatures and vacuum of space. Also, with a limited data set (three vehicles), it is difficult to develop and use a database with a reasonable standard deviation. Any calculated results would carry great uncertainty.

NASA IMPLEMENTATION

NASA agrees with the importance of understanding when and where corrosion occurs as a first step towards mitigating it. Given the difficulty in establishing trenchant mathematical models of corrosion rates for the multiple Orbiter environments, the NASA/contractor team (through the Orbiter Corrosion Control Review Board (CCRB)) has and will continue to assess mechanisms, magnitudes, and rates of corrosion occurrence. This will be used to prioritize high corrosion occurrence areas. The CCRB will target inspections toward low-traffic and/or hard-to-access areas that are not consistently inspected. Furthermore, the CCRB will address predicting the rates of long-term degradation of Orbiter corrosion protection systems (i.e., corrosion preventative compounds, paints, sealants, adhesives, etc.).

Beyond the original Orbiter design life of 10 years, corrosion inspection intervals have been driven by environment, exposure cycles, time, materials, and configuration without the use of specific corrosion rate predictions. Although not fool-proof, these inspection intervals have generally been extremely conservative. In the few cases where this has not been conservative enough, the scope of concern has been expanded accordingly and the inspection interval requirements have been changed. Moreover, when corrosion is identified, the standard procedure is to immediately repair it. If the corrosion is widespread in an area or a configuration, specific fixes are incorporated (e.g., between faying surfaces/dissimilar metals, etc.) or refurbishments are implemented (e.g., strip and reapplication of primers, etc.). In the few cases where this is not possible, such as when the rework cannot be completed without major structural disassembly, engineering assessments are completed to characterize the active corrosion rate specific to the area of concern, and inspection intervals are assigned accordingly, until the corrosion can be corrected. Relative to the general aviation industry, NASA's approach to corrosion repair is extremely aggressive.

In the past, NASA has worked closely with the U.S. Air Force to review corrosion prevention programs for potential application to the Orbiter Program. Several successes from Air Force programs have already been implemented, such as the use of water wash-downs and corrosion preventative compounds. In the future, the Orbiter CCRB will continue to partner with NASA research centers, industry, and the Department of Defense (DoD) to further develop and optimize the Orbiter corrosion control program. To maintain exposure to the current state-of-the-art in this area, the CCRB will participate annually in the NASA/Federal Aviation Agency/DoD Aging Aircraft Conference.

FINAL UPDATE

Following the *Columbia* accident, the Orbiter CCRB has been strengthened significantly. Additional funding for augmentation of Orbiter corrosion control activities was authorized in May 2004 and NASA, United Space Alliance, and Boeing have implemented an expanded corrosion control program. This authorization implemented proactive corrosion control measures to ensure continued safety and sustainability of Orbiter hardware throughout the planned Shuttle Program service life. This activity included a review of the current state of the art in corrosion control tools and techniques, followed by consideration for implementation into the future Orbiter corrosion control program. Authorized funding extended through the end of fiscal year 2006 to expand Orbiter corrosion control. Since that time, the expanded efforts have been covered within scope as part of the Space Flight Operations Contract extension.

SCHEDULE

Responsibility	Due Date	Activity/Deliverable
Orbiter Project Office	Completed	Direct appropriate funding to develop a sustained Orbiter CCRB.
Orbiter Project Office	Jun 04 (Completed)	Develop an advanced Orbiter Corrosion Control Program, chartered to detect, trend, analyze, and predict future corrosion issues.

Columbia Accident Investigation Board
Observation 10.8-1

Teflon (material) and Molybdenum Disulfide (lubricant) should not be used in the carrier panel bolt assembly.

Note: NASA has closed this observation through the formal Program Requirements Control Board process. The following summary details NASA's response to the observation and any additional work NASA performed beyond the Columbia Accident Investigation Board observation.

BACKGROUND

Concerns regarding the use of these materials were initiated due to the brittle fracture mode observed on some A-286 stainless steel leading edge subsystem carrier panel bolts. Specifically, it was argued that lubricant materials consisting of Teflon and/or Molybdenum Disulfide should not be used due to their potential to contribute to a stress corrosion cracking fracture mechanism at elevated temperatures. Traces of perfluorinated polyether grease and Molybdenum Disulfide (lubricants) were found on the carrier panel bolt shank and sleeve. However, no Teflon was found during the failure analysis of carrier panel fasteners.

A-286 fasteners in the presence of an electrolyte must also be exposed to elevated temperatures for stress corrosion cracking to be of concern. However, fastener installations are protected from temperature extremes (the maximum temperatures seen, by design, are less than 300°F).

NASA IMPLEMENTATION

NASA conducted interviews with ground technicians at Kennedy Space Center (KSC); these interviews indicated that the use of Braycote grease as a lubricant may have become an accepted practice due to the difficult installation of this assembly. Braycote grease contains perfluorinated polyether oil, Teflon, and Molybdenum Disulfide materials. According to design drawings and assembly procedures, the use of lubricants should not have been allowed in these fastener installations.

As a result of these findings, NASA directed United Space Alliance (USA) to institute appropriate corrections to their fastener installation training and certification program. USA shall emphasize to its technicians to follow exactly the installation instructions for all Orbiter fastener installations. Any deviation from specific instructions will require disposition from engineering before implementation. USA will further emphasize that lubricants cannot and should not be used in any fastener installation, unless specifically authorized.

In addition, NASA has implemented an engineering review of all discrepancy repairs made on Orbiter hardware at KSC. An engineering review will occur to provide the appropriate checks and balances if a lubricant is required to address a specific fastener installation problem.

FINAL UPDATE

NASA and USA have implemented corrective actions to ensure that lubricant will not be used in fastener applications unless explicitly approved by engineering.

SCHEDULE

Responsibility	Due Date	Activity/Deliverable
KSC/USA Ground Operations	Mar 04 (Completed)	Update fastener training and certification program for USA technicians; require deviations from instructions to be approved before implementation

Columbia Accident Investigation Board
Observation 10.8-2

Galvanic coupling between aluminum and steel alloys must be mitigated.

Note: NASA has closed this observation through the formal Program Requirements Control Board (PRCB) process. The following summary details NASA's response to the observation and any additional work NASA performed beyond the *Columbia* Accident Investigation Board observation.

BACKGROUND

Galvanic coupling between dissimilar metals is a well-recognized Orbiter concern. As galvanic couples between aluminum and steel alloys cannot be completely eliminated, the Space Shuttle Program (SSP) must implement appropriate corrosion protection schemes.

The SSP Orbiter element requirements are in full compliance with this observation. Currently, according to the Boeing Orbiter Materials Control Plan, "Metals shall be considered compatible if they are in the same grouping as specified in Military-Standard (MIL-STD)-889 or the difference in solution potential is ≤ 0.25 Volts." Otherwise, mitigation for galvanic corrosion is required. Per NASA requirement Marshall Space Flight Center-Specification (MSFC-SPEC)-250, "…when dissimilar metals are involved… the fasteners shall be coated with primer or approved sealing compounds and installed while still wet or for removable or adjustable fasteners, install with corrosion preventative compound." Where there are exceptions, such as fastener installations that are functionally removable, we depend on scheduled inspections of the fastener hole.

NASA IMPLEMENTATION

Since Orbiter galvanic couples are generally treated with corrosion mitigation schemes, the time-dependent degradation of approved sealing compounds must be addressed. Inspections performed in 2004 raised concern in areas where significant galvanic couples exist, even in the presence of sealing materials. This concern has led to the consideration of design changes. Design modifications that have been implemented include electrical ground paths in the Orbiter nose cap and on the metallic fittings of the External Tank doors. NASA has taken action to be more proactive in addressing this vehicle-wide concern.

FINAL STATUS

The SSP Aging Vehicle Assessment Committee has approved a proposal to expand the scope and authority of the Orbiter Corrosion Control Review Board. This activity included a review of the time-dependent degradation of approved corrosion preventative and sealing compounds. NASA has developed an advanced Orbiter Corrosion Control Program, including implementation of an aging materials evaluation as applied to galvanic couple seal materials on Orbiter hardware.

SCHEDULE

Responsibility	Due Date	Activity/Deliverable
SSP	Apr 04 (Completed)	Present to the SSP PRCB for direction and funding.
Kennedy Space Center	Jun 04 (Completed)	Develop an advanced Orbiter Corrosion Control Program.

Columbia Accident Investigation Board
Observation 10.8-3

The use of Room Temperature Vulcanizing 560 and Koropon should be reviewed.

Note: NASA has closed this observation through the formal Program Requirements Control Board process. The following summary details NASA's response to the observation and any additional work NASA performed beyond the *Columbia* Accident Investigation Board observation.

BACKGROUND

Concerns regarding the use of Room Temperature Vulcanizing (RTV) 560 and Koropon materials were initiated due to the brittle fracture mode observed on some A-286 stainless steel leading edge subsystem carrier panel bolts. Specifically, it was argued that trace amounts of contaminants in these materials could, at elevated temperatures, contribute to a stress corrosion cracking (SCC) of the bolts. It was also proposed that these contaminants might accelerate corrosion, particularly in tight crevices.

SCC of A-286 material is only credible at high temperatures. This is not a concern as all fastener installations are protected from such temperature extremes (the maximum temperatures seen, by design, are less than 300°F).

NASA IMPLEMENTATION

NASA completed materials analyses on multiple A-286 bolts that exhibited a brittle-like fracture mode. Failure analysis included fractography, metallography, and chemical analysis. Furthermore, a research program was executed to duplicate and compare the bolt failures experienced on *Columbia*. This proved conclusively that the brittle-looking fracture surfaces were produced during bolt failure at temperatures approaching 2000°F and above. This failure mode is not a concern with the A-286 stainless steel leading edge subsystem carrier panel bolts, as all fastener installations are protected from such temperature extremes.

In addition to failure analysis, both RTV 560 and Koropon were assessed for the presence of trace contaminants. Inductively Coupled Plasma analyses were completed on samples of both materials. The amount and type of trace contaminants were analyzed and determined to be insignificant.

RTV 560 and Koropon were selected for widespread use in the Space Shuttle Program (SSP) because they prevent corrosion. All corrosion testing and failure analysis performed during the life of the SSP have shown no deleterious effects from either product. Several non-Shuttle aerospace companies have used Koropon extensively as an anticorrosion primer and sealant. To date, problems with its use in the military and industry have not been identified.

Both of these materials may eventually fail in their ability to protect from corrosion attack, but do not fail by chemically breaking down to assist corrosion mechanisms. Thus, NASA concluded that trace contaminants in Koropon and RTV 560 do not contribute to accelerated corrosion or SCC mechanisms.

In addition to answering this specific observation, NASA is assessing the long-term performance of all nonmetallic materials used on the Orbiter through a vehicle-wide aging materials evaluation. This effort is ongoing and will continue in support of the Orbiter for the remainder of its service life.

FINAL UPDATE

NASA considers that these materials have been reviewed and present no risk for supporting accelerated corrosion and/or SCC mechanisms. Appropriate long-term additional studies have been initiated.

SCHEDULE

Responsibility	Due Date	Activity/Deliverable
SSP	Mar 04 (Completed)	Review use of RTV 560 and Koropon

Columbia Accident Investigation Board
Observation 10.8-4

Assuring the continued presence of compressive stresses in A-286 bolts should be part of their acceptance and qualification procedures.

Note: NASA has closed this observation through the formal Program Requirements Control Board process. The following summary details NASA's response to the observations and any additional work NASA performed beyond the *Columbia* Accident Investigation Board observation.

BACKGROUND

Initial concerns regarding the use of these A-286 stainless steel fastener materials were initiated due to the brittle fracture mode observed on some leading edge subsystem carrier panel bolts. The concern about residual compressive stresses, and to some extent the concerns about Koropon, Room Temperature Vulcanizing 560, Teflon, and Molybdenum Disulfide, emanated from a conjecture that the brittle fracture of some of the bolts could have been caused by stress corrosion cracking (SCC).

For SCC to occur, each of the following conditions must exist:

- Material of concern must be susceptible to SCC
- Presence of an active electrolyte
- Presence of a sustained tensile stress

Additionally, SCC of A-286 fasteners is a concern only under exposure to high temperatures. All fastener installations are protected from such temperature extremes.

NASA IMPLEMENTATION

To address the concern that sustained tensile stress might have contributed to SCC, NASA completed materials analyses on multiple A-286 bolts that exhibited a brittle-like fracture mode (i.e., minimal ductility, flat fracture). The failure analysis included fractography, metallography, and chemical analysis. Furthermore, a research program was executed to duplicate and compare the bolt failures experienced on *Columbia*. This proved conclusively that the brittle-looking fracture surfaces were produced during bolt failure at temperatures approaching 2000°F and above. The observed intergranular fracture mechanism is consistent with grain boundary embrittlement at elevated temperatures, along with potential effects from liquid metal embrittlement from vaporized aluminum. The effects of high temperature exposures on A-286 stainless steel materials are not consistent with the SCC concerns.

In addition to this effort, NASA completed residual stress analyses on several A-286 bolts via neutron diffraction at the National Research Council of Canada. In general, residual stresses were determined to be negligible or compressive in the axial bolt direction. The bolts used on the Space Shuttle have a sufficient compressive stress layer, which is governed by appropriate process controls at the manufacturer.

NASA reviewed the manufacturing and material specifications for the A-286 bolts. This review confirmed that only qualified vendors are contracted, manufacturing process controls are sufficient, and Certificates of Compliance are maintained for material traceability. Furthermore, NASA executes material lot testing on all fasteners procured for use in the Shuttle Program to ensure appropriate quality control.

FINAL UPDATE

NASA has analyzed the requirements and process for A-286 bolts and found that current processes and controls are adequate.

Columbia Accident Investigation Board
Observation 10.9-1

NASA should consider a redesign of the (Hold-Down Post Cable) system, such as adding a cross-strapping cable, or conduct advanced testing for intermittent failure.

Note: NASA has closed this observation through the formal Program Requirements Control Board process. The following summary details NASA's response to the observation and any additional work NASA performed beyond the *Columbia* Accident Investigation Board (CAIB) observation.

This response also addresses Recommendation D.a-10, Hold-Down Post (HDP) Cable Anomaly.

BACKGROUND

Each of the two Solid Rocket Boosters (SRBs) is attached to the Mobile Launch Platform by four hold-down bolts. These bolts are secured by a 5-in. diameter restraint nut. Each restraint nut contains two pyrotechnic initiators designed to sever the nuts when the SRBs ignite, releasing the Space Shuttle stack to lift off the launch platform.

Release is normally accomplished by simultaneously firing two redundant pyrotechnic charges called NASA standard initiators (NSIs) on each of eight SRB. Two independent ground-based pyrotechnic initiation control (PIC) systems, A and B, are used to receive the command and distribute the firing signals to each HDP. On STS-112, the system A Fire 1 command was not received by the ground-based PIC system; however, the redundant system B functioned properly and fired all system B NSIs, separated the frangible nuts, and enabled the release of the four hold-down bolts. As a result, the Shuttle safely separated from the launch platform.

NASA was unable to conclusively isolate the anomaly in any of the failed components. The most probable cause was determined to be an intermittent connection failure at the launch platform-to-Orbiter interface at the tail service mast (TSM). The dynamic vibration environment could have caused this connection failure after main engine start. Several contributing factors were identified, including groundside connector corrosion at the TSM T-0 umbilical, weak connection spring force, potential nonlocked Orbiter connector savers, lack of proper inspections, and a blind (non-visually verified) mate between the ground cable and the Orbiter connector saver.

The STS-112 investigation resulted in the replacement of all T-0 ground cables after every flight, a redesign of the T-0 interface to the PIC rack cable, and replacement of all Orbiter T-0 connector savers. Also, the pyrotechnic connectors are prescreened with pin retention tests, and the connector saver mate process is verified using videoscopes. The CAIB determined that the prelaunch testing procedures for this system may not be adequate to identify intermittent failure. Therefore, the CAIB suggested that NASA consider a redesign of the system or implement advanced testing for intermittent failures.

NASA IMPLEMENTATION

Five options for redesign of this system were presented to the Orbiter Project Configuration Control Board on August 20, 2003. The recommended redesign configuration provides redundancy directly at the T-0 umbilical, which was determined to be a primary contributing cause of the STS-112 anomaly. The selected option resulted in the least impact to hardware (fewer connectors, less wiring, less weight added), was capable of implementation in a reasonably short time period, and required only limited modifications to existing ground support equipment. Orbiter and groundside implementations were not affected since the interface is at the same T-0 pins.

Kennedy Space Center (KSC) has implemented a number of processing changes to greatly reduce the possibility of another intermittent condition at the TSM. The ground cables from the Orbiter interface to the TSM bulkhead plate are now replaced after each use, instead of reuse based upon inspection as was previously allowed. The ground connector springs that maintain the mating force against the Orbiter T-0 umbilical are all removed and tested to verify the spring constants meet specification between flights. The Orbiter T-0 connector savers are inspected before each flight and are now secured with safety wire before the launch platform cables are connected. New ground cables are thoroughly inspected before mate to the Orbiter. In addition, the connection process was enhanced to provide a bore scope optical verification of proper mate.

Prior to the STS-114 return to flight (RTF) mission, the Space Shuttle Program (SSP) implemented several design changes and enhancements to further reduce the risk of a similar event. Redundant command paths have been added for each Arm, Fire 1, Fire 2, and return circuits from the Orbiter through separate connectors on the Orbiter/TSM umbilical. The ground support equipment cables were modified to extend the signals to the ground PIC rack solid-state switches. This modification adds copper path redundancy through the most dynamic and susceptible environment in the PIC system. Additionally, all electrical cables have been redesigned and replaced from the Orbiter T-0 umbilical through the TSMs and all respective distribution points. The new cables are constructed with more robust insulation and are better suited for the launch environment. This new cable design also eliminates the old style standard polyimide ("Kapton") wire insulation that can be damaged by handling and degrades with age.

SSP technical experts investigated laser-initiated ordnance devices and have concluded that there would be no functional improvement in the ground PIC system operation. Additionally, use of laser-initiated ordnance would have changed only the firing command path from the ground PIC rack to each of the ordnance devices. This would not change or have had any impact on master command path failures experienced during the STS-112 launch since those would still be electrical copper paths.

NASA has been engaged since late 2000 with the Department of Defense, NASA, Federal Aviation Agency, and industry aging aircraft wiring community to develop, test, and implement fault-detection methods and equipment to find emerging wire anomalies and intermittent failures before electrical function is prevented. Several tools have been developed and tested for that purpose, but no tool is available with a conclusive ability to guarantee total wire function in environments with such dynamic conditions prior to use.

STATUS

A cross-strapping cable was not recommended as one of the redesign options because of concerns for introducing a single-point failure that could inhibit both hold-down post pyrotechnic systems. The recommended redesign, plus the previously identified processing and verification modifications, are considered sufficient to mitigate the risks identified during the STS-112 anomaly investigation. Actions are in place to provide verification of connector mating and system integrity. Better mating methods are being utilized (Go/No-Go gauge and borescope inspections).

Proposed hardware modifications and development activity status:

- The TSM cable redesign is complete and was designated an RTF mandatory modification by the Shuttle Processing Project.
- The Orbiter Project has implemented the T-0 redundancy modification in the Orbiter T-0 separation system and T-0 connectors. KSC has modified groundside circuits.
- The SSP has evaluated laser-initiated pyrotechnic firing for the Shuttle Program and concluded there would be no functional improvement with the modification of the current system to a laser-initiated pyrotechnic firing system.
- NASA is currently supporting two separate strategies to determine wiring integrity. In addition, NASA is engaged with the Department of Defense and the Federal Aviation Agency to encourage further studies and projects.

A NASA Headquarters-sponsored Independent Assessment (IA) team was formed to review this anomaly and generically review the T-0 umbilical electrical/data interfaces. While this independent review was not considered a constraint to implementing the redesign, it provided an opportunity to ensure that the original investigation was thorough and provided additional recommendations or improvements that might be implemented. This activity resulted in electrical connector design and operational procedure enhancements (modifications to Orbiter and ground pyrotechnic circuitry as well as T-0 connector installation procedures). Additional design verification was performed that included testing and analysis of the carrier plates and connectors subjected to the vibration of the connectors at frequencies that match the launch environment. The analysis included Finite Element Model (FEM) analysis of TSM and orbiter umbilical carrier plates to confirm that no permanent deformation will occur for the design environment, and that the plates are still within the elastic limit of design (this testing will be repeated with the next use of MLP-2).

Additionally, SSP Systems Engineering and Integration led a Program-wide team to address the findings identified by the IA team, which reviewed the anomaly and also assessed potential common cause failures across the other separation interfaces (both flight and launch interfaces). The recommendations of this team resulted in the implementation of revised assembly and inspection procedures, revised ground cable manufacturing tolerances, and new controls on ground cable assembly and mating.

Additionally, SSP Systems Engineering and Integration has led a Program-wide team to address the findings identified by the IA team that reviewed the anomaly and also assess potential common cause failures across the other separation interfaces (both flight and launch interfaces). The recommendations of this team resulted in the implementation of revised assembly and inspection procedures, revised ground cable manufacturing tolerances, and new controls on ground cable assembly and mating.

SCHEDULE

FINAL UDPATE

The NASA team will continue to engage in development of emerging wire fault detection and fault location tools with the government and industry wiring community. NASA will advocate funding for tool development and will consider implementing new methods that prove effective.

Responsibility	Due Date	Activity/Deliverable
SSP, KSC, USA	Oct 03 (Completed)	Present to SSP Integration Control Board
SSP, KSC, USA	Oct 03 (Completed)	Present to SSP Program Requirements Control Board
SSP, KSC, USA	Nov 03 (Completed)	Design Review
SSP, KSC, USA	Dec 03 (Completed)	Wire Design Engineering
NASA Headquarters IA Team	Jul 04 (Completed)	Independent Assessment Final Report
SSP, KSC, USA	Mar 04 (Completed)	Wire Installation Engineering
Orbiter Project	Apr 04 (Completed)	Provide redundant firing path in the Orbiter for HDP separation
Shuttle Integration	Aug 04 (Completed)	Evaluate cross-strapping for simultaneous NSI detonation
SSP	Mar 05 (Completed)	Respond to IA team findings
SSP	Mar 05 (Completed)	Address potential common-cause failures across the other flight and launch separation interfaces
SSP	Mar 05 (Completed)	Approve new Operations and Maintenance Requirements and Specifications Document requirements for specific ground cable inspections as a condition for mating
Shuttle Processing Project	Mar 05 (Completed)	Modify, install, and certify the ground cabling to protect against damage and degradation and to implement a redundant ground electrical path to match Orbiter commands

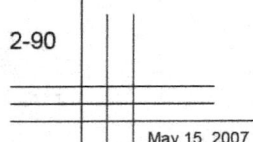

Columbia Accident Investigation Board
Observation 10.10-1

NASA should reinstate a safety factor of 1.4 for the Attachment Rings—which invalidates the use of ring serial numbers 16 and 15 in their present state—and replace all deficient material in the Attachment Rings.

Note: NASA has closed this observation through the formal Program Requirements Control Board (PRCB) process. The following summary details NASA's response to the observation and any additional work NASA performed beyond the *Columbia* Accident Investigation Board (CAIB) observation.

This response also addresses Recommendation D.a-11, SRB ETA Ring.

BACKGROUND

The External Tank Attach (ETA) rings are located on the Solid Rocket Boosters (SRBs) on the forward end of the aft motor segment (figure O10.10-1). The rings provide the aft attach points for the SRBs to the External Tank. Approximately two minutes after liftoff, the SRBs separate from the Shuttle vehicle.

In late 2002, Marshall Space Flight Center (MSFC) engineers were performing tensile tests on ETA ring web material prior to the launch of STS-107 and discovered the ETA ring material strengths were lower than the design requirement. The ring material was from a previously flown and subsequently scrapped ETA ring representative of current flight inventory material. A one-time waiver was granted for the STS-107 launch based on an evaluation of the structural strength factor of safety requirement for the ring of 1.4 and adequate fracture mechanics safe-life at launch. The most probable cause for the low strength material was an off-nominal heat treatment process. Following SRB retrieval, the STS-107 rings were inspected as part of the normal postflight inspections, and no issues were identified with flight performance. Subsequent testing revealed lower than expected fracture properties; as a result, the scope of the initial investigation of low material strength was expanded to include a fracture assessment of the ETA ring hardware.

NASA IMPLEMENTATION

NASA used a nonlinear analysis method to determine whether the rings met Program strength requirements for a factor of safety of 1.4 or greater (figure O10.10-1-2). The nonlinear analysis method is a well-established technique employed throughout the aerospace industry that addresses the entire material stress-strain response and more accurately represents the material's ultimate strength capability by allowing load redistribution. The hardware materials characterization used in this analysis includes ring web thickness measurements and hardness testing (figure O10.10-1-3) of the splice plates and ring webs. Hardware inspections were performed on low-strength material rings that supported the first two Return to Flight (RTF) missions (STS-114 and STS-121); there were no reportable problems, and all areas of the rings met factor of safety requirements.

In addition to strength analysis, a fracture mechanics analysis on the ETA ring hardware was performed to determine the minimum mission life for the rings and to define the necessary inspection interval. Serial number 15 and 16 ETA rings exhibited undesirable material variability and were set aside as the initial candidates for upgrade/replacement. The first replacement rings were delivered in March 2005 and subsequently flown on STS-115. Fracture

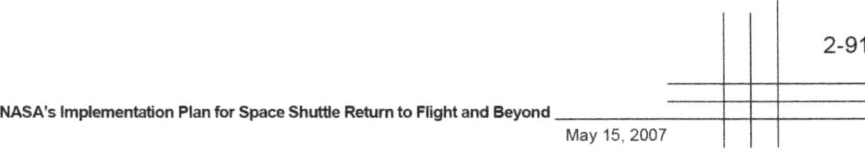

Figure O10.10-1. ETA ring location.

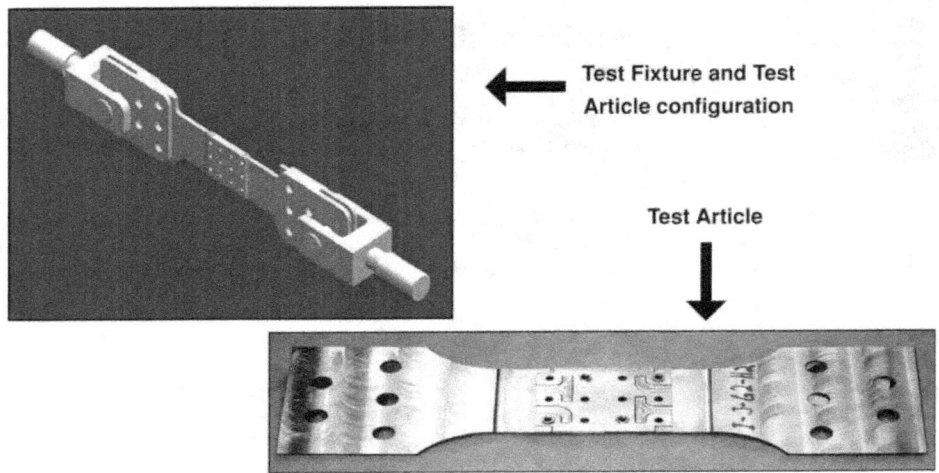

Figure O10.10-1-2. Test articles.

property testing for the splice plates resulted in unacceptable material properties. Replacement splice plates were fabricated under controlled processes and lot acceptance testing. Fracture Control Plan requirements compliance was ensured by performing extensive nondestructive inspections to rebaseline all areas of the ETA ring hardware.

NASA used testing, inspection, and analyses of flight hardware to fully characterize the material for the low-strength materials rings that supported the first two RTF missions. This provided assurance that the flight hardware met program strength and fracture mechanics requirements.

FINAL UPDATE

STS-115 marked the first flight of the newly manufactured 4340 steel ETA rings. The 4340 steel rings performed flawlessly and will support all future Space Shuttle flights.

Figure O10.10-1-3. Harness testing.

SCHEDULE

Responsibility	Due Date	Activity/Deliverable
SRB Project	Mar 04 (Completed)	New ring procurement funding approved (7 flight sets + 1 test firing)
SRB Project	Jul 04 (Completed)	CAIB observation PRCB action (S064039 MSFC-SRB Action 1-1 and 2-1) closure
SRB Project	Aug 04 (Completed)	First flight set ETA rings complete
SRB Project	Mar 05 (Completed)	Delivery of first flight set of "new" ETA rings (flew as STS-115)
SRB Project	Nov 06 (Completed)	Final delivery of the seventh flight set

Columbia Accident Investigation Board
Observation 10.11-1

Assess NASA and contractor equipment to determine if an upgrade will provide the reliability and accuracy needed to maintain the Shuttle through 2020. Plan an aggressive certification program for replaced items so that new equipment can be put into operation as soon as possible.

Note: NASA has closed this observation through the formal Program Requirements Control Board process. The following summary details NASA's response to the observation and any additional work NASA performed beyond the *Columbia* Accident Investigation Board (CAIB) observation.

BACKGROUND

The CAIB review of Shuttle test equipment at NASA and contractor facilities revealed the use of antiquated and obsolete 1970s-era technology such as analog equipment. Current state-of-the-art technology is digital rather than analog. Digital equipment is less costly, easier to maintain, and more reliable and accurate. The CAIB recommended that, with the Shuttle projected to fly through 2020, upgrading the test equipment to digital technology would avoid the high maintenance, lack of parts, and questionable accuracy of the equipment currently in use. Although the new equipment would require certification for its use, the benefit in accuracy, maintainability, and longevity would likely outweigh the drawbacks of certification costs for the Program lasting until 2020.

The Vision for Space Exploration calls for NASA to retire the Shuttle following completion of International Space Station assembly, which is planned for the end of the decade. Because NASA is going to retire the Shuttle approximately ten years earlier than was planned, NASA continually reassesses whether the benefits of new equipment will outweigh the drawback of certification costs. The Space Shuttle Program (SSP) is continuing to maintain and upgrade test equipment systems to ensure that we preserve the necessary capacity throughout the life of the Shuttle. Decisions on appropriate investments in new test equipment take into consideration the projected end of Shuttle service life.

NASA IMPLEMENTATION

The SSP Manager established a Program Strategic Sustainment Office to provide stronger focus and leadership for sustainability issues such as material, hardware, and test equipment obsolescence. The Program Strategic Sustainment Office conducted reviews of all Program Elements and supporting contractors to identify risks to Program sustainability, with an emphasis on test equipment. The Manager of the Strategic Sustainment Office had assigned an Obsolescence Manager whose primary focus was on mitigating risks related to obsolete or near-obsolete test equipment.

In 2003, the logistics board approved $32M towards equipment modernization or upgrade, such as the Space Shuttle Main Engine controller special test equipment (STE), the Orbiter inertial measurement unit, and the Star Tracker STE. Additionally, the Program Strategic Sustainment Office identified and submitted through the Integrated Space Operations Summit (ISOS) process an additional requirement for sustainability to support similar test equipment and obsolescence issues. Certification costs and schedules and the associated Program risks are required elements of the total project package reviewed by the logistics board prior to authority to proceed.

The Obsolescence Manager assessed all critical Program equipment, through regular reviews, and determined where upgrades or additional funding were needed to support the Program for the remainder of the Space Shuttle's service life. Identified upgrades were submitted through the ISOS process and the normal SSP budget process to ensure funding of specific projects.

FINAL UPDATE

Based on decisions made during the fiscal year (FY) 2005 Shuttle budget reviews, sustainability funding was returned to the Space Shuttle element projects to be prioritized among their existing operational concerns. This decision was deemed appropriate because of the pending retirement of the Space Shuttle in 2010.

SCHEDULE

Responsibility	Due Date	Activity/Deliverable
SSP	Dec 03 (Completed)	Approve FY04 test equipment upgrades
Service Life Extension Program Sustainability Panel	Feb 04 (Completed)	Define FY05 test equipment upgrades
SSP Development Office	May 04 (Completed)	Provide final Summit II investment recommendations to Space Flight Leadership Council

Columbia Accident Investigation Board
Observation 10.12-1

NASA should implement an agency-wide strategy for leadership and management training that provides a more consistent and integrated approach to career development. This strategy should identify the management and leadership skills, abilities, and experiences required for each level of advancement. NASA should continue to expand its leadership development partnerships with the Department of Defense and other external organizations.

Note: NASA has closed this observation through the formal Program Requirements Control Board process. The following summary details NASA's response to the observation and any additional work NASA has performed beyond the *Columbia* Accident Investigation Board (CAIB) observation.

BACKGROUND

NASA has always considered training and development to be a cornerstone of good management. Even prior to the *Columbia* accident, the NASA Training and Development Division offered a wide curriculum of leadership development programs to the NASA workforce. The content of internally sponsored programs was developed around the NASA leadership model, which delineates six leadership competencies at four different levels. The four levels are executive leader, senior leader, manager/supervisor, and influence leader. Each level contains distinct core competencies along with a suggested curriculum. NASA also developed leadership skills in the workforce by taking advantage of training and development opportunities at the Office of Personnel Management, Federal Executive Institute, Brookings Institute, Department of Defense, and the Center for Creative Leadership, among many other resources. In addition, the Agency sponsors leadership development opportunities through academic fellowships in executive leadership and management, as well as through the NASA-wide Leadership Development Program. Also, some NASA centers offer locally sponsored leadership development programs for their first-level and/or mid-level managers and supervisors; these programs are unique to the need of each center.

Upon review of this CAIB observation, NASA agreed that the Agency could further improve the training and development programs offered to NASA employees.

NASA IMPLEMENTATION

This CAIB Observation was the inspiration behind the One NASA Strategy for Leadership and Career Development. The Associate Administrator for Institutions and Management distributed the final version of the strategy to Officials in Charge and Center Directors in October 2004. NASA's goal for the One NASA Strategy is for the Agency to develop a more integrated process that would identify the management and leadership skills, abilities, and experiences necessary for advancement through various leadership roles. The strategy, informed by data gathered from a process of meetings and benchmarking, presents an overall competency-based framework and approach for leadership development at NASA, outlining leadership roles and core and elective experiences and training.

The underpinnings of the strategy are (1) the NASA Values – safety, the NASA family, excellence, and integrity; and (2) the NASA Leadership Model with its six performance dimensions that define the competencies, knowledge, skills, and abilities necessary for demonstrating excellence in various leadership roles.

The strategy includes a framework that is intended to provide a consistent and integrated approach to leadership and management career development. Each leadership role within the framework contains components that are designed to enable employees to achieve and demonstrate the NASA values along with the identified competencies for that role. Common elements in each role include:

- Core experiences and broadening opportunities including mobility – intellectual as well as geographical.

- Core and optional courses relevant to both achieving mastery in the role as well as preparing for the next step.

- Required role-specific courses on safety and diversity.

- Assessments – analysis of feedback from subordinates, supervisors, customers, peers, and stakeholders.
- Continuing education.
- Individual Development Plans.
- Coaching and mentoring.

FINAL UPDATE

The One NASA Strategy for Leadership and Career Development has been integrated into Agency training and development frameworks and policies and will give NASA employees a framework within which they can plan their NASA careers.

SCHEDULE

Responsibility	Due Date	Activity/Deliverable
Headquarters (HQ) Office of Human Capital Management	Oct 03 (Completed)	Begin Benchmarking Activities
HQ/Office of Human Capital Management	Oct 03 (Completed)	Begin the staff work to form the Agency team
HQ/Office of Human Capital Management	Jan 04 (Completed)	Benchmarking data to date compiled
HQ/Office of Human Capital Management	Jul 04 (Completed)	Draft strategy reviewed/validated by Enterprises/Senior leadership
HQ/Office of Human Capital Management	Sep 04 (Completed)	Strategy developed and presented to the NASA Associate Deputy Administrator for Institutions and Asset Management
HQ/Office of Human Capital Management	Oct 04 (Completed)	Strategy distributed to Officials in Charge, Center Directors

CAIB Report, Volume II, Appendix D.a, "Supplement to the Report"

Volume II, Appendix D.a, also known as the "Deal Appendix," augments the CAIB Report and concerns raised by Brigadier General Duane Deal and others. The 14 recommendations contained in the Deal Appendix expand on the CAIB Report, Volume I discussions on quality assurance processes, Orbiter corrosion detection methods, Solid Rocket Booster External Tank Attach Ring factor-of-safety concerns, crew survivability, security concerns relating to the Michoud Assembly Facility, and shipment of Reusable Solid Rocket Motor segments. NASA addressed each of the concerns offered in the appendix. Many of the concerns were addressed in previous sections of the Space Shuttle RTF Implementation Plan, and in those cases the responses in this section refer to details provided in previous sections of the Implementation Plan.

Columbia Accident Investigation Board
Volume II, Appendix D.a, Quality Assurance Section, Recommendation D.a-1 Review Quality Planning Requirements Document Process

Perform an independently led, bottom-up review of the Kennedy Space Center Quality Planning Requirements Document to address the entire quality assurance program and its administration. This review should include development of a responsive system to add or delete government mandatory inspections. Suggested Government Mandatory Inspection Point (GMIP) additions should be treated by higher review levels as justifying why they should not be added, versus making the lower levels justify why they should be added. Any GMIPs suggested for removal need concurrence of those in the chain of approval, including responsible engineers.

BACKGROUND

The *Columbia* Accident Investigation Board noted the need for a responsive system for adding or deleting Government Mandatory Inspection Points (GMIPs) and the need for a periodic review of the Quality Planning Requirements Document (QPRD). The Space Shuttle Program, Shuttle Processing Element located at the Kennedy Space Center is responsible for overseeing the QPRD process and implementation of associated GMIPs.

NASA IMPLEMENTATION, STATUS, FORWARD WORK, AND SCHEDULE

This recommendation is addressed in Section 2.1, Space Shuttle Program Action 1, and Section 2.2, Observation 10.4-1, of this Implementation Plan.

Columbia Accident Investigation Board
Volume II, Appendix D.a, Quality Assurance Section, Recommendation D.a-2 Responsive System to Update Government Mandatory Inspection Points

Kennedy Space Center must develop and institutionalize a responsive bottom-up system to add to or subtract from Government Inspections in the future, starting with an annual Quality Planning Requirements Document review to ensure the program reflects the evolving nature of the Shuttle system and mission flow changes. At a minimum, this process should document and consider equally inputs from engineering, technicians, inspectors, analysts, contractors, and Problem Reporting and Corrective Action to adapt the following year's program.

BACKGROUND

The *Columbia* Accident Investigation Board noted the need for a responsive system for updating Government Mandatory Inspection Points (GMIPs), including the need for a periodic review of the Quality Planning Requirements Document (QPRD). The Space Shuttle Program's Shuttle Processing Element, located at the Kennedy Space Center, is responsible for overseeing the QPRD process and implementation of associated GMIPs.

NASA IMPLEMENTATION, STATUS, FORWARD WORK, AND SCHEDULE

This recommendation is addressed in Section 2.2, Observation 10.4-1, of this Implementation Plan.

Columbia Accident Investigation Board
Volume II, Appendix D.a, Quality Assurance Section, Recommendation D.a-3 Statistically Driven Sampling of Contractor Operations

NASA Safety and Mission Assurance should establish a process inspection program to provide a valid evaluation of contractor daily operations, while in process, using statistically driven sampling. Inspections should include all aspects of production, including training records, worker certification, etc., as well as Foreign Object Damage prevention. NASA should also add all process inspection findings to its tracking programs.

BACKGROUND

The *Columbia* Accident Investigation Board (CAIB) noted the need for a statistically valid sampling program to evaluate contractor operations. Kennedy Space Center currently samples contractor operations within the Space Shuttle Main Engine Processing Facility; however, the sample size is not statistically significant and does not represent all processing activities.

NASA IMPLEMENTATION, STATUS, FORWARD WORK, AND SCHEDULE

This recommendation is addressed in Section 2.2, CAIB Observation 10.5-3, of this Implementation Plan.

Columbia Accident Investigation Board
Volume II, Appendix D.a, Quality Assurance Section, Recommendation D.a-4 Forecasting and Filling Personnel Vacancies

The KSC quality program must emphasize forecasting and filling personnel vacancies with qualified candidates to help reduce overtime and allow inspectors to accomplish their position description requirements (i.e., more than the inspectors performing government inspections only, to include expanding into completing surveillance inspections).

Note: The Kennedy Space Center (KSC) Quality Program improvements described here have been implemented by the KSC Director and concurred upon by Space Shuttle Program management. Therefore, this is the final revision to the Return to Flight Implementation Plan regarding Recommendation D.a-4. NASA will continue to monitor and improve our Quality Assurance programs.

BACKGROUND

The *Columbia* Accident Investigation Board expressed concern regarding staffing levels of Quality Assurance Specialists (QASs) at KSC and Michoud Assembly Facility. Specifically, they stated that staffing processes must be sufficient to select qualified candidates in a timely manner. Previously, KSC hired three QASs through a step program; none of them had previous experience in quality assurance. The step program was a human resources sponsored effort to provide training and mobility opportunities to administrative staff. Of the three, only one remains a QAS. In addition to hiring qualified candidates, staffing levels should be sufficient to ensure the QAS function involves more than just inspection. Additional functions performed should include hardware surveillance, procedure evaluations, and assisting in audits.

NASA IMPLEMENTATION

NASA currently uses two methods for selecting and developing qualified QASs. First, NASA can hire a QAS at the GS-7, GS-9, GS-11, or GS-12 level if the candidate meets a predetermined list of requirements and level of experience. QAS candidates at all levels require additional training. Candidates selected at lower grades require further classroom and on-the-job training before being certified as a QAS. The second method that NASA uses is a cooperative education program that brings in college students as part of their education process. This program is designed to develop QAS or quality control technicians for NASA and the contractor. The program is an extensive two-year program, including classroom and on-the-job training. If at the end of the cooperative education program the student does not demonstrate the required proficiency, NASA will not hire the individual.

Hiring practices have also improved. NASA can hire temporary or term employees. While permanent hiring is preferred, this practice provides flexibility for short-term staffing issues. Examples include replacements for QAS military reservists who deploy to active duty and instances when permanent hiring authority is not immediately available.

KSC has addressed the hiring issue. Identified training issues are addressed in Section 2.2, Observation O10.4-3.

FINAL UPDATE

The Hardware Surveillance Program has been deployed and fully implemented in all of the Shuttle processing areas. This program defines the areas in which hardware surveillance is performed, the checklist of items assessed, the number of hardware inspections required, and the data to be collected. KDP-P-1825, SHUTTLE QUALITY ASSURANCE HARDWARE SURVEILLANCE, was approved in September 2004.

SCHEDULE

Responsibility	Due Date	Activity/Deliverable
KSC	Completed	Develop and implement processes for timely hiring of qualified candidates
KSC	Completed	Develop and implement hardware surveillance program in the Orbiter Processing Facilities
KSC	Completed	Deploy hardware surveillance program to all QAS facilities
KSC	Completed	Develop reporting metric
KSC	Apr 04 (Completed)	Develop and implement procedure evaluation

Columbia Accident Investigation Board
Volume II, Appendix D.a, Quality Assurance Section, Recommendation D.a-5 Quality Assurance Specialist Job Qualifications

Job qualifications for new quality program hires must spell out criteria for applicants, and must be closely screened to ensure the selected applicants have backgrounds that ensure that NASA can conduct the most professional and thorough inspections possible.

Note: NASA has closed this recommendation through the formal Program Requirements Control Board process. The following summary details NASA's response to the recommendation and any additional work NASA performed beyond the *Columbia* Accident Investigation Board (CAIB) recommendation.

BACKGROUND

The CAIB expressed concern regarding staffing qualifications of Quality Assurance Specialists (QASs) at Kennedy Space Center (KSC). Previously, KSC hired three QASs, none of whom had previous experience in quality assurance, through a step program. Of the three, only one remains as a QAS.

NASA IMPLEMENTATION

NASA currently uses two methods for selecting and developing qualified QAS. First, if the candidate meets a predetermined list of requirements and level of experience, NASA can hire a QAS at the GS-7, GS-9, GS-11, or GS-12 level. QAS candidates at all levels require additional training. Candidates selected at lower grades require further classroom and on-the-job training before being certified as a QAS. The second method NASA uses is a cooperative education program that brings in college students as part of their education process. This program is designed to develop QAS or quality control technicians for NASA and the contractor. The program is an extensive two-year program, including classroom and on-the-job training. If at the end of the cooperative education program the student does not demonstrate the required proficiency, NASA will not hire the individual.

NASA has benchmarked Department of Defense (DoD) and Defense Contract Management Agency (DCMA) training requirements and determined where NASA can use their training; one example is their training on the Fundamentals of Quality Assurance. A team consisting of engineers and QAS in both the Space Shuttle and International Space Station Programs was formed to develop and document a more robust training program. The team evaluated a course on Quality Assurance skills and a course on Visual Inspection Testing. The team members presented their recommendations on how to improve the overall training program. The KSC Safety and Mission Assurance (S&MA) Directorate, using the recommendations provided, documented the training requirements for all S&MA positions in a formal training records template. Additional information on the training plan is found in Section 2.2, Observation O10.4-3.

FINAL UPDATE

The Safety and Mission Assurance Directorate Development Plan was approved and published in October 2004. Training records are kept on file and available electronically. NASA QAS have completed the NASA/USA Visual Inspection/Testing and are scheduled to complete the Fundamentals of Quality Assurance training by the end of July 2007.

SCHEDULE

Responsibility	Due Date	Activity/Deliverable
KSC	Completed	Develop and implement processes for hiring and developing qualified QAS
KSC	Completed	Benchmark DoD and DCMA training programs (from O10.4-3)
KSC	Apr 04 (Completed)	Develop and document improved training requirements (from O10.4-3)

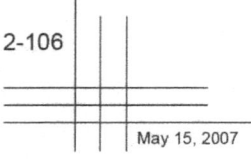

Columbia Accident Investigation Board
Volume II, Appendix D.a, Quality Assurance Section, Recommendation D.a-6 Review Mandatory Inspection Document Process

Marshall Space Flight Center should perform an independently-let bottom-up review of the Michoud Quality Planning Requirements Document to address the quality program and its administration. This review should include development of a responsive system to ad or delete government mandatory inspections. Suggested Government Mandatory Inspection Point (GMIP) additions should be treated by higher review levels as justifying why they should not be added, versus making the lower levels justify why they should be added. Any GMIPs suggested for removal should need concurrence of those in the chain of approval, including responsible engineers.

BACKGROUND

The *Columbia* Accident Investigation Board noted the need for a responsive system for adding or deleting Government Mandatory Inspection Points (GMIPs), including those at the Michoud Assembly Facility (MAF), and the need for a periodic review of the Quality Planning Requirements Document (QPRD). The Shuttle Propulsion Element at the Marshall Space Flight Center is responsible for overseeing the Mandatory Inspection Document process and implementation of associated GMIPs.

NASA IMPLEMENTATION, STATUS, FORWARD WORK, AND SCHEDULE

This recommendation is addressed in Section 2.1, Space Shuttle Program Action 1, and Section 2.2, Observation 10.4-1, of this Implementation Plan. Efforts to implement this recommendation have been in work since the issuance of the *Columbia* Accident Investigation Board Report, Volume I. NASA commissioned an assessment team, independent of the Space Shuttle Program, to review the effectiveness of the QPRD and its companion document at the MAF, referred to as the Mandatory Inspection Document, and the associated GMIPs. NASA continues efforts to improve this process through its defined implementation plan and will demonstrate its progress with this and future updates to the Return to Flight Implementation Plan.

Columbia Accident Investigation Board
Volume II, Appendix D.a, Quality Assurance Section, Recommendation D.a-7 Responsive System to Update Government Mandatory Inspection Points at the Michoud Assembly Facility

Michoud should develop and institutionalize a responsive bottom-up system to add to or subtract from Government Inspections in the future, starting with an annual Quality Planning Requirements Document review to ensure the program reflects the evolving nature of the Shuttle system and mission flow changes. Defense Contract Management Agency manpower at Michoud should be refined as an outcome of the QPRD review.

BACKGROUND

The *Columbia* Accident Investigation Board noted the need for a responsive system for updating Government Mandatory Inspection Points (GMIPs), including the need for a periodic review of the Quality Planning Requirements Document (QPRD). The Space Shuttle Program, Shuttle Processing Element, located at the Kennedy Space Center is responsible for overseeing the QPRD process and implementation of associated GMIPs.

NASA IMPLEMENTATION, STATUS, FORWARD WORK, AND SCHEDULE

This recommendation is addressed in Section 2.1, Space Shuttle Program Action 1, and Section 2.2, Observation 10.4-1, of this Implementation Plan. Efforts to implement this recommendation have been in work since the issuance of the *Columbia* Accident Investigation Board Report, Volume I. NASA commissioned an assessment team, independent of the Space Shuttle Program, to review the effectiveness of the QPRD, its companion at the Michoud Assembly Facility, referred to as the Mandatory Inspection Document, and the associated GMIPs. NASA continues efforts to improve this process through its defined implementation plan and will demonstrate progress with this and future updates to the Return to Flight Implementation Plan.

Columbia Accident Investigation Board
Volume II, Appendix D.a, Quality Assurance Section, Recommendation D.a-8 Use of ISO 9000/9001

Kennedy Space Center should examine which areas of ISO 9000/9001 truly apply to a 20-year-old research and development system like the Space Shuttle.

BACKGROUND

The *Columbia* Accident Investigation Board report highlighted Kennedy Space Center's reliance on the International Organization for Standardization (ISO) 9000/9001 certification. The report stated, "While ISO 9000/9001 expresses strong principles, they are more applicable to manufacturing and repetitive-procedure industries, such as running a major airline, than to a research-and-development, flight test environment like that of the Space Shuttle. Indeed, many perceive International Standardization as emphasizing process over product." Currently, ISO 9000/9001 certification is a contract requirement for United Space Alliance.

NASA IMPLEMENTATION, STATUS, FORWARD WORK, AND SCHEDULE

This recommendation is addressed in Section 2.2, Observation 10.4-4, of this Implementation Plan.

Columbia Accident Investigation Board
Volume II, Appendix D.a, Quality Assurance Section, Recommendation D.a-9 Orbiter Corrosion

Develop non-destructive evaluation inspections to detect and, as necessary, correct hidden corrosion.

BACKGROUND

The Space Shuttle Program has initiated an action to assess the *Columbia* Accident Investigation Board observations related to corrosion damage in the Orbiters. This action has been assigned to the Orbiter Project Office.

NASA IMPLEMENTATION, STATUS, FORWARD WORK, AND SCHEDULE

This recommendation is addressed in Section 2.2, Observations 10.7-1 through 10.7-4, of this Implementation Plan. Evaluation of this recommendation has been in work since the release of the *Columbia* Accident Investigation Board Report, Volume I. NASA demonstrates progress in the Return to Flight Implementation Plan.

Columbia Accident Investigation Board
Volume II, Appendix D.a, Quality Assurance Section, Recommendation D.a-10 Hold-Down Post Cable Anomaly

NASA should evaluate a redesign of the Hold-Down Post Cable, such as adding a cross-strapping cable or utilizing a laser initiator, and consider advanced testing to prevent intermittent failure.

This recommendation is addressed in Section 2.2, Observation 10.9-1, of this Implementation Plan.

Columbia Accident Investigation Board
Volume II, Appendix D.a, Quality Assurance Section, Recommendation D.a-11 Solid Rocket Booster External Tank Attach Ring

NASA must reinstate a safety factor of 1.4 for the Attach Rings—which invalidates the use of ring serial numbers 15 and 16 in their present state—and replace all deficient material in the Attach Rings.

This recommendation is addressed in Section 2.2, Observation 10.10-1, of this Implementation Plan.

Columbia Accident Investigation Board
Volume II, Appendix D.a, Quality Assurance Section, Recommendation D.a-12 Crew Survivability

To enhance the likelihood of crew survivability, NASA must evaluate the feasibility of improvements to protect the crew cabin on existing Orbiters.

Note: NASA has closed this recommendation through the formal Program Requirements Control Board (PRCB) process. The following summary details NASA's response to the recommendation and any additional work NASA performed beyond the *Columbia* Accident Investigation Board (CAIB) recommendation.

BACKGROUND

The CAIB found that, in both the *Challenger* and the *Columbia* accidents, the crew cabin initially survived the disintegration of the Orbiter intact.

NASA IMPLEMENTATION

Implementation of this recommendation has been in work since the release of the *Columbia* Accident Investigation Board Report, Volume I. The Space Shuttle Service Life Extension Program II Crew Survivability Sub-panel recognized the need for the Program to continue funding the vehicle forensic analysis and follow-on thermal and structural hardening analysis. This work plays a part not only as resolution to a CAIB Recommendation but also as a component of furthering the technical understanding of the space/atmosphere-aero interface and conveys knowledge capture for future programs.

On July 21, 2004, the Space Shuttle Upgrades PRCB approved the formation of the Space Craft Survival Integrated Investigation Team (SCSIIT). This multidisciplinary team, comprised of JSC Flight Crew Operations, JSC Mission Operations Directorate, JSC Engineering, Safety and Mission Assurance, the Space Shuttle Program, and Space and Life Sciences Directorate, was tasked to perform a comprehensive analysis of the two Shuttle accidents for crew survival implications. The team's focus is to combine data (including debris, video, and Orbiter experiment data) from both accidents with crew module models and analyses. After completion of the investigation and analysis, the SCSIIT will issue a formal report documenting lessons learned for enhancing crew survival in the Space Shuttle and for future human space flight vehicles, such as the Crew Exploration Vehicle.

Space Shuttle-critical flight safety issues are reported to the PRCB for disposition. Future crewed-vehicle spacecraft will use the products of the multidisciplinary team to aid in developing the crew safety and survival requirements.

FINAL UPDATE

Fiscal year 2005 (FY05) and FY06 funding was committed for this team's activities. When the SCSIIT's analysis is complete, the results will be incorporated into a final report with recommendations. Limited additional funding for FY07 was provided to complete preparation of the final report, conduct an independent review, and publish the results. The SCSIIT anticipates the final report will be issued in August 2007.

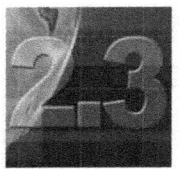

Columbia Accident Investigation Board
Volume II, Appendix D.a, Quality Assurance Section, Recommendation D.a-13 RSRM Segment Shipping Security

NASA and ATK Thiokol perform a thorough security assessment of the RSRM segment security, from manufacturing to delivery to Kennedy Space Center, identifying vulnerabilities and identifying remedies for such vulnerabilities.

Note: NASA considers this recommendation closed, and the following summary details NASA's response.

BACKGROUND

During security program assessments at the ATK Thiokol Reusable Solid Rocket Motor (RSRM) Production Facility, the *Columbia* Accident Investigation Board raised concerns about several elements of the overall security program. Most notable of these concerns was protection of completed segments prior to rail shipment to the Kennedy Space Center (KSC).

NASA IMPLEMENTATION

NASA has conducted a full security program vulnerability assessment of the ATK Thiokol RSRM Production Facility, with the goal of identifying and mitigating security vulnerabilities.

NASA security officials, together with ATK Thiokol Security Program officials, performed an assessment of the RSRM security program from RSRM manufacturing to delivery, inspection, and storage at KSC. The assessment included a review of the ATK Thiokol manufacturing plant to the railhead; participation in the rail shipment activities of RSRM segment(s) to or from KSC; regional and local threats; and rotation, processing, and storage facility security at KSC.

STATUS

NASA conducted assessments of several key elements of the ATK Thiokol RSRM operation: December 8–12, 2003, ATK Thiokol RSRM Facilities; January 26–27, 2004, KSC RSRM Facilities; and January 30–February 9, 2004, RSRM Railway Transport Route and Operations.

An RSRM Security Assessment briefing was provided by the assessment team lead to both Marshall Space Flight Center Security and RSRM Project in March 2004. The written report was submitted at a later date. The team's assessment concluded that "threat" and "vulnerability" were low and no critical findings were noted.

A number of recommendations to enhance RSRM security were provided for RSRM Project consideration. These recommendations were grouped into three categories: Corinne Site (where RSRM segments are loaded onto rail cars), rail transport, and general operations. The Project assessed the impact and viability of noted recommendations. Those recommendations the Project agreed would effectively enhance RSRM security were implemented prior to the shipment of flight hardware to KSC (December 2004).

FINAL UPDATE

Based on NASA's assessment of the RSRM security program, the RSRM Project has implemented those recommendations that would effectively enhance RSRM segment security.

SCHEDULE

Complete.

Columbia Accident Investigation Board
Volume II, Appendix D.a, Quality Assurance Section, Recommendation D.a-14 Michoud Assembly Facility Security

NASA and Lockheed-Martin complete an assessment of the Michoud Assembly Facility security, focusing on items to eliminate vulnerabilities in its current stance.

NOTE: NASA has closed this recommendation through the formal Program Requirements Control Board process. The following summary details NASA's response to the recommendation and any additional work NASA performed beyond the *Columbia* Accident Investigation Board (CAIB) recommendation.

BACKGROUND

During security program assessments at the Michoud Assembly Facility (MAF), the CAIB expressed concerns about several elements of the overall security program. Most notable of these concerns is the adequacy of particular security equipment and staffing.

NASA IMPLEMENTATION

NASA conducted a full security program vulnerability assessment of the MAF and External Tank (ET) production activity, with the goal of identifying and mitigating security vulnerabilities.

They assessed the MAF and the ET production security programs from ET manufacturing to delivery, inspection, and storage at Kennedy Space Center (KSC). The assessment included a review of the MAF to the shipping port; shipping activities of the ET to and from KSC; regional and local threats; and Vehicle Assembly Building security at KSC. Based on the assessment, NASA conducted a vulnerability mitigation activity.

STATUS

The NASA assessment was conducted from January 26 through January 30, 2004. A comprehensive Report of Findings and a separate Executive Summary, both administratively controlled documents, were prepared by the assessment team and presented to the NASA Office of Security Management and Safeguards and to the Marshall Space Flight Center (MSFC) Security Director.

In June 2004, MSFC Protective Services assigned a Civil Service Security Specialist to the MAF to review and assess the Lockheed Martin-Michoud Operations approach and assure the proposed enhancements are compatible with NASA security standards.

In July 2004, Lockheed Martin submitted a detailed and prioritized security enhancement plan. The priorities were determined based on discussions with MSFC Protective Services, MAF NASA Management, and Lockheed Martin Management.

Lockheed Martin initiated implementation of the improvements that were considered within the scope of the current contract, and staffing needs were addressed. NASA has budgeted the appropriate funding. Other improvements were implemented by authorization of proposals that preceded the security plan. These include an integrated Security Control system that includes closed circuit television, access control, alarm monitoring, and identification management. Additionally, the total modernization of the Security Dispatch Center was completed.

All elements of the security plan that were not within the scope of contract at the time of the vulnerability assessment were reviewed by NASA, and budgetary approval was granted in March 2005. A Lockheed Martin proposal to correct the deficiencies was negotiated.

FINAL UPDATE

The integrated security control system project, including the electronic access system (LENEL) with 88 cameras, and the Security Dispatch Center modifications have been completed. Security staffing levels have been increased to a full complement of security officers, and the complement of three emergency medical technicians has been maintained.

The Lockheed Martin proposal to address all findings in the vulnerability assessment was on contract by July 2005. All contract actions requested by NASA to improve security have been addressed.

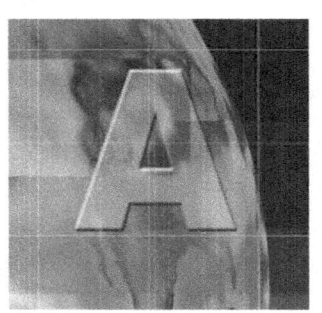

Appendix A: NASA's Return to Flight Process

The following information describes the process NASA used to Return to Flight. It is included here now, without updates, for historical reference.

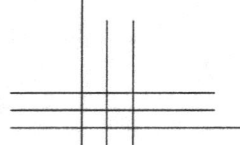 NASA's Implementation Plan for Space Shuttle Return to Flight and Beyond

Note: The following information describes the process NASA used to Return to Flight. It is included now, without updates, for historical reference.

BACKGROUND

The planning for Return to Flight (RTF) began even before the Agency received the first two *Columbia* Accident Investigation Board (CAIB) preliminary recommendations on April 16, 2003. Informally, activities started in mid-February as the Space Shuttle projects and elements began a systematic fault-tree analysis to determine possible RTF constraints. In a more formal sense, the RTF process had its beginnings in a March 2003 Office of Space Flight (OSF) memorandum.

Mr. William F. Readdy, the Associate Administrator for Space Flight, initiated the Space Shuttle Return to Flight planning process in a letter to Maj. Gen. Michael C. Kostelnik, the Deputy Associate Administrator for International Space Station and Space Shuttle Programs, on March 12, 2003. The letter gave Maj. Gen. Kostelnik the direction and authority "to begin focusing on those activities necessary to expeditiously return the Space Shuttle to flight."

Maj. Gen. Kostelnik established a Return to Flight Planning Team (RTFPT) under the leadership of astronaut Col. James Halsell. The RTF organization is depicted in figure A-1.

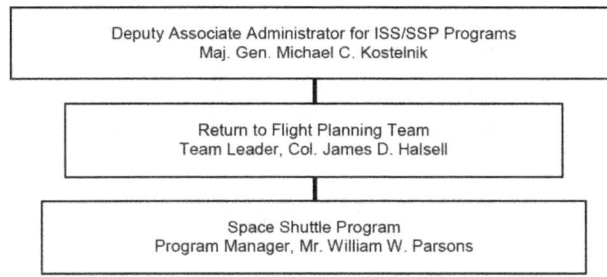

Figure A-1. Original RTFPT organization.

Space Shuttle Program (SSP) Role in Return to Flight

The SSP provided the analyses required to determine the NASA Return to Flight constraints (RTFCs). SSP project and element fault-tree analyses combined with technical working group documentation and analyses provided the database needed to create a list of potential RTFCs.

For example, the SSP's Orbiter Project organized first as the Orbiter Vehicle Engineering Working Group (OVEWG) to develop fault-tree analyses, and later as the Orbiter Return-to-Flight Working Group to recommend implementation options for RTFCs. The OVEWG structure and its subgroups are listed in figure A-2.

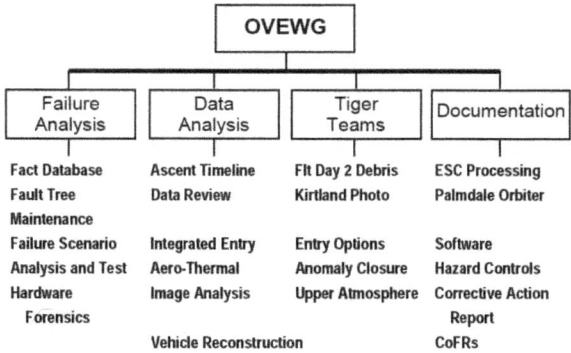

Figure A-2. OVEWG organization.

Once analyses were complete, the working groups briefed the CAIB on their findings and solicited the Space Shuttle Program Requirements Control Board's (SSPRCB's) approval of identified corrective actions.

Each SSP project and element formed similar organizations to accomplish thorough fault-tree analysis and closure.

Return to Flight Planning Team

The RTFPT was formed to address those actions needed to comply with formal CAIB recommendations and NASA initiatives ("Raising the Bar"), and to determine the fastest path for a safe RTF. The approximately 30-member team was assembled with representatives from NASA Headquarters and the OSF Field Centers, crossing the Space Shuttle Operations, Flight Crew Operations, and Safety and Mission Assurance disciplines.

Starting in early April 2003, the RTFPT held weekly teleconferences to discuss core team processes and product delivery schedules. Weekly status reports, describing the progress of RTF constraints, were generated for Maj. Gen. Kostelnik and Dr. Michael

Greenfield, one of the Space Flight Leadership Council (SFLC) co-chairs. These reports were also posted on a secure Web site for the RTFPT membership and other senior NASA officials to review. The RTFPT often previewed RTF briefing packages being prepared for SSPRCBs. The leader of the RTFPT, Col. Halsell, became a voting member of the SSPRCB for all RTF issues. The RTFPT also arranged for all recommended SSPRCB RTF issues to be scheduled for SFLC review and approval. These RTFPT tasks were primarily assessment, status, and scheduling activities. The team's most significant contribution has been preparing and maintaining this Implementation Plan, which is a living document chronicling NASA's RTF.

As the Implementation Plan has matured and obtained SFLC approval, NASA has transitioned from planning for RTF to implementing the plan. As intended, the lead role has transitioned from the RTFPT to the Space Shuttle Program, which is now responsible to the SFLC for executing the plan to successful completion. Accordingly, Maj. Gen. Kostelnik decommissioned the RTFPT on June 7, 2004, and transferred all remaining administrative and coordination duties to the Management Integration and Planning Office (MG) of the Space Shuttle Program, under the direction of former astronaut Col. (Ret.) John Casper. The MG office has established a Return to Flight Branch that is responsible for the coordination of RTF constraint closures with the RTF Task Group.

These changes reflect the real progress toward RTF that has been made in the last few months, and NASA's commitment to optimizing our processes and organization as we execute the RTF Plan.

Space Flight Leadership Council

The SFLC was co-chaired by the Associate Administrator for Space Flight (Mr. William F. Readdy) and the Associate Deputy Administrator for Technical Programs (Dr. Michael Greenfield) until August 2004. As NASA moved to an organization of Mission and Support Directorates, the co-chairs became the Associate Administrator for Space Operations (Mr. William Readdy's post-transformation title) and the Deputy Chief Engineer for Independent Technical Authority (Adm. Walt Cantrell). The purpose of the SFLC (figure A-3) remains unchanged and they continue to receive and disposition the joint RTFPT/SSPRCB recommendations on RTF issues. The SFLC is charged with approving RTF items and directing the implementation of specific corrective actions. The SFLC can also direct independent analysis on technical issues related to RTF issues or schedule (e.g., the category of wiring inspection on Orbiter Vehicle (OV)-103/*Discovery*.

The membership of the SFLC includes the OSF Center Directors (Johnson Space Center, Kennedy Space Center, Marshall Space Flight Center, and Stennis Space Center) and the Associate Administrator for Safety and Mission Assurance. SFLC meetings are scheduled as needed.

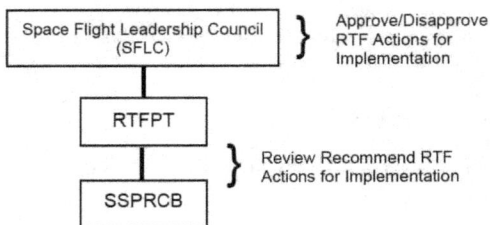

Figure A-3. Space Flight Leadership Council organization for Return to Flight issue review.

Members of the Return-to-Flight Task Group (RTFTG) are invited to attend the SFLC meetings.

Return to Flight Task Group

Also known as the Stafford Covey Task Group, the RTFTG was established by the NASA Administrator to perform an independent assessment of NASA's actions to implement the CAIB recommendations. The RTFTG was chartered from the existing Stafford International Space Station Operations Readiness Task Force (Stafford Task Force), a Task Force under the auspices of the NASA Advisory Council. The RTFTG is comprised of standing members of the Stafford Task Force, other members selected by the co-chair, and a nonvoting ex-officio member: the Associate Administrator for Safety and Mission Assurance. The RTFTG is organized into three panels: technical, operations, and management. The team held its first meeting, primarily for administrative and orientation purposes, in early August 2003, and has been meeting periodically since. The RTFTG has issued two Interim Reports—one in January 2004, and one in May 2004.

Operational Readiness Review

The SFLC will continue to convene meetings to resolve NASA's internal handling of RTFPT/SSPRCB recommendations and Return to Flight issues. The first operational readiness review meeting, a Flight Certification Review, was held at the Marshall Space Flight Center on December 11–12, 2003. As the Space Shuttle Program prepares for Return to Flight, they will conduct element, project, and finally Program Design Certification Reviews (DCRs) in preparation for the STS-114 Flight Readiness Review. To date, completed project/element DCRs are the Space Shuttle Main Engine (September 2004) and the Reusable Solid Rocket Motor project (October 2004).

RTF Schedule

See figure A-4.

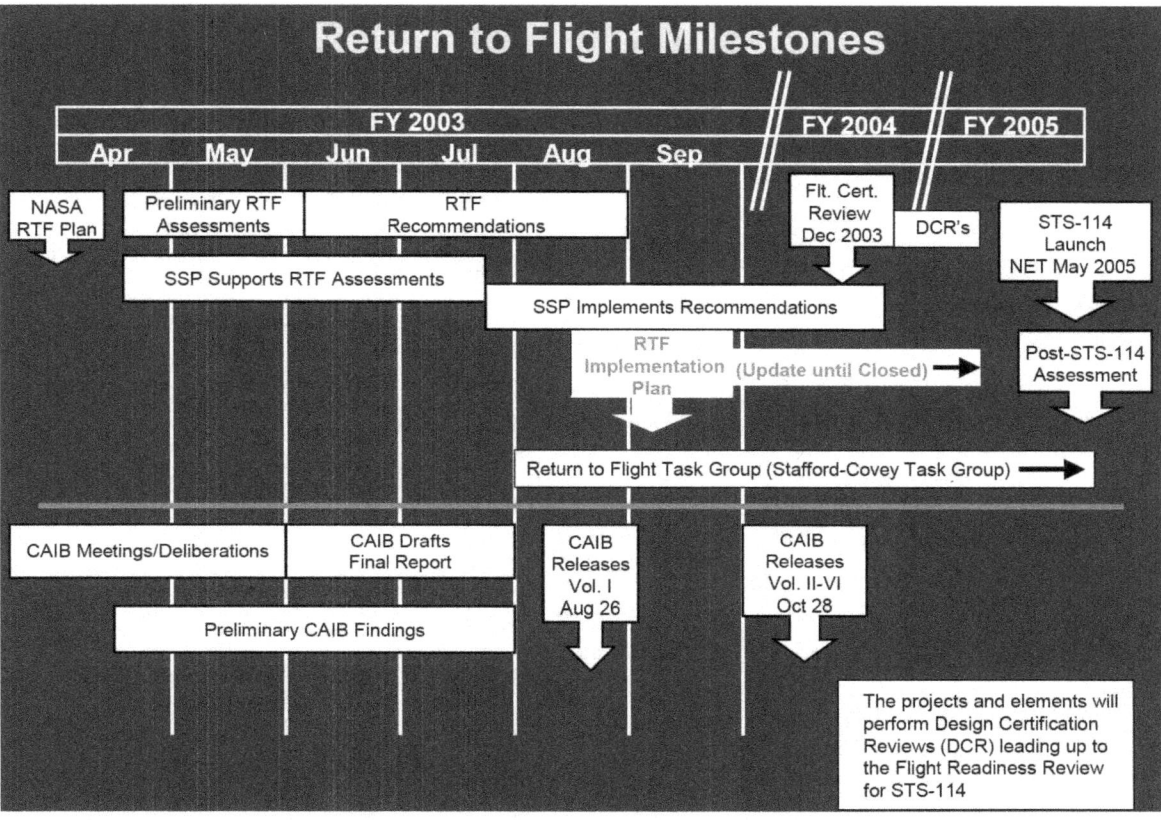

Figure A-4. RTF and RTFTG schedules overlaid with the schedule for release of the CAIB final report.

Historical Reference Only

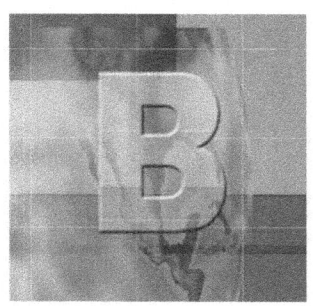

Appendix B:
Return to Flight Task Group

The following information describes the process NASA used to Return to Flight. It is included here now, without updates, for historical reference.

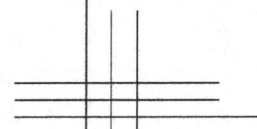 NASA's Implementation Plan for Space Shuttle Return to Flight and Beyond

INTRODUCTION

The Return to Flight Task Group, co-chaired by Thomas P. Stafford and Richard O. Covey, was formed to address the Shuttle Program's Return to Flight effort. The Task Group is chartered to perform an independent assessment of NASA's actions to implement the *Columbia* Accident Investigation Board (CAIB), as they relate to the safety and operational readiness of STS-114.

The Stafford/Covey Task Group will report on the progress of NASA's response to the CAIB report and may also make other observations on safety or operational readiness as it believes appropriate.

The Task Group will formally and publicly report their results to NASA on a continuing basis, and their recommendations will be folded into NASA's formal planning for Return to Flight. The paragraphs below describe the charter and membership for the Task Group.

RETURN TO FLIGHT TASK GROUP CHARTER ESTABLISHMENT AND AUTHORITY

The NASA Administrator, having determined that it is in the public interest in connection with performance of the Agency duties under the law, and with the concurrence of the General Services Administration, establishes the NASA Return to Flight Task Group ("Task Group"), pursuant to the Federal Advisory Committee Act (FACA), 5 U.S.C. App. §§1 et seq.

PURPOSE AND DUTIES

1. The Task Group will perform an independent assessment of NASA's actions to implement the CAIB recommendations as they relate to the safety and operational readiness of STS-114. As necessary to their activities, the Task Group will consult with former members of the CAIB.

2. While the Task Group will not attempt to assess the adequacy of the CAIB recommendations, it will report on the progress of NASA's response to meet their intent.

3. The Task Group may make other observations on safety or operational readiness as it believes appropriate.

4. The Task Group will draw on the expertise of its members and other sources to provide its assessment to the Administrator. The Task Group will hold meetings and make site visits as necessary to accomplish its fact finding. The Task Group will be provided information on activities of both the Agency and its contractors as needed to perform its advisory functions.

5. The Task Group will function solely as an advisory body and will comply fully with the provisions of the Federal Advisory Committee Act.

ORGANIZATION

The Task Group is authorized to establish panels in areas related to its work. The panels will report their findings and recommendations to the Task Group.

MEMBERSHIP

1. In order to reflect a balance of views, the Task Group will consist of non-NASA employees and one NASA nonvoting, ex-officio member, the Deputy Associate Administrator for Safety and Mission Assurance. In addition, there may be associate members selected for Task Group panels. The Task Group may also request appointment of consultants to support specific tasks. Members of the Task Group and panels will be chosen from among industry, academia, and Government personnel with recognized knowledge and expertise in fields relevant to safety and space flight.

2. The Task Group members and Cochairs will be appointed by the Administrator. At the request of the Task Group, associate members and consultants will be appointed by the Associate Deputy Administrator (Technical Programs).

ADMINISTRATIVE PROVISIONS

1. The Task Group will formally report its results to NASA on a continuing basis at appropriate intervals, and will provide a final written report.

2. The Task Group will meet as often as required to complete its duties and will conduct at least two public meetings. Meetings will be open to the public, except when the General Counsel and the Agency Committee Management Officer determine that the meeting or a portion of it will be closed pursuant to the Government in the Sunshine Act or that the meeting is not covered by the Federal Advisory Committee Act. Panel meetings will be held as required.

3. The Executive Secretary will be appointed by the Administrator and will serve as the Designated Federal Officer.

4. The Office of Space Flight will provide technical and staff support through the Task Force on International Space Station Operational Readiness. The Office of Space Flight will provide operating funds for the Task Group and panels. The estimated operating costs total approximately $2M, including 17.5 work-years for staff support.

5. Members of the Task Group are entitled to be compensated for their services at the rate equivalent to a GS 15, step 10. Members of the Task Group will also be allowed per diem and travel expenses as authorized by 5 U.S.C. § 5701 et seq.

DURATION

The Task Group will terminate two years from the date of this charter, unless terminated earlier or renewed by the NASA Administrator.

RECENT STATUS

The Task Group delivered its Executive Summary on June 28, 2005, and released its Final Report on August 17, 2005. NASA continues to evaluate its findings and recommendations, as well as its Minority Reports, and to assess the Task Group's impact on the Space Shuttle Program. NASA is also working with the Aerospace Safety Advisory Panel to define its responsibilities in tracking the completion of the CAIB's 14 non-Return to Flight recommendations and NASA's Raising the Bar Initiatives.

STAFFORD-COVEY TASK GROUP MEMBERS

Col. James C. Adamson, U.S. Army (Ret.):
CEO, Monarch Precision, LLC, consulting firm

Col. Adamson, a former astronaut, has an extensive background in aerodynamics and business management. He received his Bachelor of Science degree in Engineering from the U.S. Military Academy at West Point and his Master's degree in Aerospace Engineering from Princeton University. He returned to West Point as an Assistant Professor of Aerodynamics until he was selected to attend the Navy Test Pilot School at Patuxent River, Md. in 1979. In 1981 he became Aerodynamics Officer for the Space Shuttle Operational Flight Test Program at the Johnson Space Center's Mission Control Center. Col. Adamson became an astronaut in 1984 and flew two missions, the first aboard *Columbia* (STS-28) and the second aboard *Atlantis* (STS-43).

After retiring from NASA in 1992, he created his own consulting firm, Monarch Precision, and was then recruited by Lockheed as President/Chief Executive Officer (CEO) of Lockheed Engineering and Sciences Company. In 1995 he helped create United Space Alliance and became their first Chief Operating Officer, where he remained until 1999. In late 1999, Col. Adamson was again recruited to serve as President/CEO of Allied Signal Technical Services Corporation, which later became Honeywell Technology Solutions, Inc. Retiring from Honeywell in 2001, Col. Adamson resumed part-time consulting with his own company, Monarch Precision, LLC. In addition to corporate board positions, he has served as a member of the NASA Advisory Council Task Force on Shuttle-Mir Rendezvous and Docking Missions and is currently a member of the NASA Advisory Council Task Force on International Space Station Operational Readiness.

Maj. Gen. Bill Anders, U.S. Air Force Reserve (Ret.):

Maj. Gen. Anders graduated in 1955 as an electrical engineer from the United States Naval Academy and earned his pilot's wings in 1956. He received a graduate degree in nuclear engineering from the U.S. Air Force (USAF) Institute of Technology while concurrently graduating with honors in aeronautical engineering from Ohio State University. In 1963 he was selected for the astronaut corps. He was the Lunar Module Pilot of Apollo 8 and backup Command Module Pilot for Apollo 11. Among other successful public and private endeavors, Maj. Gen. Anders has served as a Presidential appointee to the Aeronautics & Space Council, the Atomic Energy Commission, and the Nuclear Regulatory Commission (where he was the first chairman), and as U.S. Ambassador to Norway.

Subsequent to his public service, he joined the General Dynamics Corporation, as Chairman and CEO (1990–1993), and was awarded the National Security Industrial Association's "CEO of the Year" award.

During his distinguished career, Maj. Gen. Anders was the co-holder of several world flight records and has received numerous awards including the USAF, NASA, and Atomic Energy Commission's Distinguished Service Medals. He is a member of the National Academy of Engineering, the Society of Experimental Test Pilots, and the Experimental Aircraft Association. He is the founder and President of the Heritage Flight Museum.

Dr. Walter Broadnax:

Dr. Broadnax is President of Clark Atlanta University in Atlanta, Ga. Prior to accepting the Presidency at Clark Atlanta University, Broadnax was Dean of the School of Public Affairs at American University in Washington. Previously, he was Professor of Public Policy and Management in the School of Public Affairs at the University

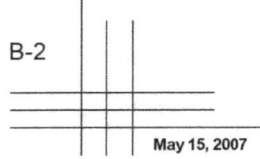

of Maryland, College Park, Md., where he also directed The Bureau of Governmental Research. Before joining the University of Maryland faculty, Dr. Broadnax served as Deputy Secretary and Chief Operating Officer of the U.S. Department of Health and Human Services; President, Center for Governmental Research, Inc., in Rochester, N.Y.; President, New York State Civil Service Commission; Lecturer and Director, Innovations in State and Local Government Programs in the Kennedy School of Government at Harvard University; Senior Staff Member, The Brookings Institution; Principal Deputy Assistant Secretary for Planning and Evaluation, U.S. Department of Health, Education and Welfare; Director, Children, Youth and Adult Services, State of Kansas; and Professor, The Federal Executive Institute, Charlottesville, Va.

He is one of America's leading scholar-practitioners in the field of public policy and management. He has published widely in the field and served in leadership positions in various professional associations: American Political Science Association, American Public Personnel Association, Association of Public Policy and Management, National Association of Schools of Public Affairs and Administration, National Association of State Personnel Executives, and the American Society for Public Administration.

Broadnax received his Ph.D. from the Maxwell School at Syracuse University, his B.A. from Washburn University, and his M.P.A from the University of Kansas. He is a Fellow of the National Academy of Public Administration and a former trustee of the Academy's Board. In March, he was installed as President of the American Society for Public Administration for 2003–2004. He is a member of the Syracuse University Board of Trustees, Harvard University's Taubman Center Advisory Board, and United States Comptroller General Advisory Board. He has also served on several corporate and nonprofit boards of directors including the CNA Corporation, Keycorp Bank, Medecision Inc., Rochester General Hospital, Rochester United Way, and the Ford Foundation/Harvard University Innovations in State and Local Government Program, the Maxwell School Advisory Board, and the National Blue Ribbon Commission on Youth Safety and Juvenile Justice Reform in the District of Columbia.

Dr. Kathryn Clark:

Dr. Clark is the President of Docere, a consulting company that specializes in science and education. She consults for the Jean-Michel Cousteau Society, the Argos Foundation, the National Marine Sanctuaries, and the Sea World Hubbs Institute to enhance the study of oceans and marine wildlife and use the data for education and awareness of the environment of the seas.

She recently completed a job for the Michigan Virtual High School to aid in the development of the Math, Science, and Technology Academy. She worked on the vision and mission of the Academy as well as the development of partners as they increase the scope and reach of the program to a national and international scale. She recently resigned from her job as NASA's Chief Scientist for the Human Exploration and Development of Space Enterprise (HEDS), a position she accepted in August 2000 after completing a 2-year term as NASA's Chief Scientist for the International Space Station Program. While on leave from the University of Michigan Medical School, she worked in the Chief Scientist position with scientists from all other areas of NASA to communicate research needs and look for possible collaboration among the science programs at NASA. She also assisted with education and outreach activities related to any human space flight endeavors, including the International Space Station, the Shuttle, any expendable launch vehicles intended to further human endeavors in space, and future missions to the Moon and Mars. Her particular interest is in "Human Factors;" all the elements necessary for the health, safety, and efficiency of crews involved in long-duration space flight. These include training, interfacing with machines and robotics, biological countermeasures for the undesirable physical changes associated with space flight, and the psychological issues that may occur in response to the closed, dangerous environments while traveling in space or living on other planets.

She received both her Master's and Doctoral degrees from the University of Michigan and then joined the faculty in the Department of Cell and Developmental Biology in 1993. She also served as the Deputy Director of the NASA Commercial Space Center, The Center for Microgravity Automation Technology (CMAT) from 1996 to 1998. CMAT provides imaging technology for use on the International Space Station. The primary commercial focus of that Center is on using high-fidelity imaging technology for science and education.

Dr. Clark's scientific interests are focused on neuromuscular development and adaptation to altered environments. Her experiments are performed at the tissue level and include immunocytochemistry and in situ hybridization of skeletal muscle and spinal cord grown both in vivo and in vitro. Her experience with NASA began with a neuromuscular development study (NIH.R1) that flew on STS-66 in November 1994. These experiments were repeated and augmented (NIH.R2) on STS-70 in July 1995. She was also involved in the Neurolab project flown on STS-90 in May 1998 and the ladybug experiment that flew on STS-93 with Commander Eileen Collins.

Dr. Clark is the Chair of the Academic Affairs Committee of Board of Control of Michigan Tech University, the Chair of the Board of Visitors of Western Reserve Academy, and serves on the boards of The Space Day Foundation and Orion's Quest, both education oriented not-for-profit organizations.

She is a former member of the Board of Directors of Women in Aerospace, is an airplane pilot and member of the 99's (the International Society of Women Pilots), and is an avid cyclist, swimmer, and cross-country skier. She owns a jazz club in Ann Arbor, Michigan. She is married to Dr. Robert Ike, a rheumatologist at the University of Michigan Medical School.

Mr. Benjamin A. Cosgrove:
Consultant

Mr. Cosgrove has a long and distinguished career as an engineer and manager associated with most of Boeing jet aircraft programs. His extensive background in aerospace stress and structures includes having served as a stress engineer or structural unit chief on the B-47, B-52, KC-135, 707, 727, 737, and 747 jetliners. He was Chief Engineer of the 767.

Mr. Cosgrove was honored by Aviation Week and Space Technology for his role in converting the Boeing 767 transport design from a three-man to a two-man cockpit configuration and received the Ed Wells Technical Management Award for addressing aging aircraft issues. He received the National Aeronautics Association's prestigious Wright Brothers Memorial Trophy in 1991 for his lifetime contributions to commercial aviation safety and for technical achievement. He is a member of the National Academy of Engineering and a fellow of both the AIAA and England's Royal Aeronautical Society. After retiring from his position as Senior Vice President of the Boeing Commercial Airplane Group in 1993 after 44 years of service, he became a consultant. He holds a Bachelor of Science degree in Aeronautical Engineering and received an honorary Doctorate of Engineering degree from the University of Notre Dame in 1993. Mr. Cosgrove is a member of the NASA Advisory Committee's Task Force on International Space Station Operational Readiness.

Col. Richard O. Covey, U.S. Air Force (Ret.):
Cochair, Return to Flight Task Group
Vice President, Support Operations, Boeing Homeland Security and Services

Col. Covey, a veteran of four Space Shuttle flights, has over 35 years of aerospace experience in both the private and public sectors. He piloted STS-26, the first flight after the *Challenger* accident, and was commander of STS-61, the acclaimed *Endeavour*/Hubble Space Telescope first service and repair mission.

Covey is a highly decorated combat pilot and Outstanding Graduate of the Air Force Test Pilot School, holds a Bachelor of Science degree in Engineering Sciences from the U.S. Air Force Academy, and has a Master of Science degree in Aeronautics and Astronautics from Purdue University.

He served as the U.S. Air Force Joint Test Force Director for F-15 electronic warfare systems developmental and production verification testing. During his distinguished 16-year career at NASA, he held key management positions in the Astronaut Office and Flight Crew Operations Directorate at Johnson Space Center (JSC). Covey left NASA and retired from the Air Force in 1994.

In his position at Boeing, his organization provides system engineering, facility/system maintenance and operations, and spacecraft operations and launch support to commercial, Department of Defense, and other U.S. Government space and communication programs throughout the world. Prior to his current position, Covey was Vice President of Boeing's Houston Operations.

He has been the recipient of numerous awards such as two Department of Defense Distinguished Service Medals, the Department of Defense Superior Service Medal, the Legion of Merit, five Air Force Distinguished Flying Crosses, 16 Air Medals, the Air Force Meritorious Service Medal, the Air Force Commendation Medal, the National Intelligence Medal of Achievement, the NASA Distinguished Service Medal, the NASA Outstanding Leadership Medal, the NASA Exceptional Service Medal, and the Goddard and Collier Trophies for his role on STS-61.

Dan L. Crippen, Ph.D.:
Former Director of the Congressional Budget Office

Dr. Crippen has a strong reputation for objective and insightful analysis. He recently served as the fifth Director of the Congressional Budget Office. His public service positions also include Chief Counsel and Economic Policy Adviser to the Senate Majority Leader (1981–1985); Deputy Assistant to the President for Domestic Policy (1987–1988); and Domestic Policy Advisor and Assistant to the President for Domestic Policy (1988–1989), where he advised the President on all issues relating to domestic policy, including the preparation and presentation of the federal budget. He has provided service to several national commissions, including membership on the National Commission on Financial

Institution Reform, Recovery, and Enforcement. He presently serves on the Aerospace Safety Advisory Panel.

Dr. Crippen has substantial experience in the private sector as well. Before joining the Congressional Budget Office, he was a principal with Washington Counsel, a law and consulting firm. He has also served as Executive Director of the Merrill Lynch International Advisory Council and as a founding partner and Senior Vice President of The Duberstein Group.

He received a Bachelor of Arts degree from the University of South Dakota in 1974, a Master of Arts from Ohio State University in 1976, and a Doctor of Philosophy degree in Public Finance from Ohio State in 1981.

Mr. Joseph W. Cuzzupoli:
Vice President and K-1 Program Manager, Kistler Aerospace Corporation

Mr. Cuzzupoli brings more than 40 years of aerospace engineering and managerial experience to the Task Group. He began his career with General Dynamics as Launch Director (1959–1962), and then became Manager of Manufacturing/Engineering and Director of Test Operations for Rockwell International (1962–1966). Cuzzupoli directed all functions in the building and testing of Apollo 6, Apollo 8, Apollo 9, and Apollo 12 flights as Rockwell's Assistant Program Manager for the Apollo Program; he later was Vice President of Operations. In 1978, he became the Vice President and Program Manager for the Space Shuttle Orbiter Project and was responsible for 5000 employees in the development of the Shuttle.

He left Rockwell in 1980 and consulted on various aerospace projects for NASA centers until 1991, when he joined American Pacific Corporation as Senior Vice President. In his current position at Kistler Aerospace (Vice President and Program Manager, 1996–present) he has primary responsibility for design and production of the K-1 reusable launch vehicle.

He holds a Bachelor of Science degree in Mechanical Engineering from the Maine Maritime Academy, a Bachelor of Science degree in Electrical Engineering from the University of Connecticut, and a Certificate of Management/Business Administration from the University of Southern California.

He was a member of the NASA Advisory Council's Task Force on Shuttle-Mir Rendezvous and Docking Missions and is a current member of the NASA Advisory Council's Task Force on International Space Station Operational Readiness.

Charles C. Daniel, Ph.D.:
Engineering Consultant

Dr. Daniel has over 35 years experience as an engineer and manager in the fields of space flight vehicle design, analysis, integration, and testing; and he has been involved in aerospace programs from Saturn V to the International Space Station. In 1968, he began his career at Marshall Space Flight Center (MSFC) where he supported Saturn Instrument Unit operations for Apollo 11, 12, and 13. In 1971, he performed avionics integration work for the Skylab Program and spent the next decade developing avionics for the Solid Rocket Boosters (SRBs). He was SRB flight operations lead in that activity.

Dr. Daniel worked as part of the original Space Station Skunk Works for definition of the initial U.S. space station concept and developed the master engineering schedule for the station.

Following the *Challenger* accident, he led the evaluation of all hazards analyses associated with Shuttle and coordinated acceptance analyses associated with the modifications to the Solid Rocket Motors (SRMs) and SRBs. During Space Station Freedom development, he was the avionics lead and served as MSFC lead for Level II assembly and configuration development. He was part of the initial group to define the concept for Russian participation in the Space Station Restructure activity and later returned to MSFC as Chief Engineer for Space Station.

Dr. Daniel holds a Doctorate degree in Engineering and has completed postgraduate work at the University of California, Berkeley, and MIT. He was a member of the NASA Advisory Council Task Force on Shuttle-Mir Rendezvous and Docking Operations and is a member of the NASA Advisory Council Task Force, ISS Operational Readiness.

Amy K. Donahue, Ph.D.:
Assistant Professor of Public Administration at the University of Connecticut Institute of Public Affairs

Dr. Amy K. Donahue is Assistant Professor of Public Policy at the University of Connecticut, where she teaches in the Master of Public Administration and Master of Survey Research programs. Her research focuses on the productivity of emergency services organizations and on the nature of citizen demand for public safety services. She is author of published work about the design, management, and finance of fire departments and other public agencies. For the past two years, Dr. Donahue has served as a technical advisor to the Department of Homeland Security's Science and Technology Directorate, helping to develop research and development programs to meet the needs of emergency responders. Dr. Donahue also served as Senior Advisor to the Administrator at NASA from 2002–2004. In this capacity, she worked within NASA to discern opportunities to contribute to homeland security efforts government-wide, including evaluating existing projects and identifying new opportunities for interagency collaboration targeted at homeland security. Dr. Donahue has 20 years of field experience and training in an array of emergency services-related fields, including managing a 911 communications center and working as a firefighter and emergency medical technician in Fairbanks, Alaska, and upstate New York. In addition, she has served on active duty as an officer in the U.S. Army's Medical Service Corps. In 2003, Dr. Donahue spent three months in the field in Texas managing the Space Shuttle *Columbia* recovery operation. Dr. Donahue holds a Ph.D. in Public Administration, an M.P.A. from the Maxwell School of Citizenship and Public Affairs at Syracuse University, and a B.A. in Geological and Geophysical Sciences from Princeton University.

Gen. Ron Fogleman, U.S. Air Force (Ret.):
President and Chief Operating Officer of Durango Aerospace Incorporated

Gen. Fogleman has vast experience in air and space operations, expertise in long-range programming and strategic planning, and extensive training in fighter and mobility aircraft. He served in the Air Force for 34 years, culminating in his appointment as Chief of Staff, until his retirement in 1997. Fogleman has served as a military advisor to the Secretary of Defense, the National Security Council, and the President of the United States.

Among other advisory boards, he is a member of the National Defense Policy Board, the NASA Advisory Council, the Jet Propulsion Laboratory Advisory Board, the Council on Foreign Relations, and the congressionally directed Commission to Assess United States National Security Space Management and Organization. He recently chaired a National Research Council Committee on Aeronautics Research and Technology for Vision 2050: An Integrated Transportation System.

Gen. Fogleman received a Master's Degree in Military History from the U.S. Air Force Academy, a Master's Degree in Political Science from Duke University, and graduated from the Army War College. He has been awarded several military decorations including: Defense Distinguished Service Medal with two oak leaf clusters; the Air Force Distinguished Service Medal with oak leaf cluster; both the Army and Navy Distinguished Service Medals, Silver Star; Purple Heart; Meritorious Service Medal, and two Distinguished Flying Crosses.

Ms. Christine H. Fox:
Vice President and Director, Operations Evaluation Group, Center for Naval Analyses

Christine H. Fox is President of the Center for Naval Analyses, a federally funded research and development center based in Alexandria, Va. Ms. Fox was the Vice President and Director, Operations Evaluation Group responsible for approximately 45 field representatives and 45 Washington-based analysts whose analytical focus is on helping operational commanders execute their missions.

Ms. Fox has spent her career as an analyst; assisting complex organizations like the U.S. Navy assess challenges and define practical solutions. She joined the Center for Naval Analysis in 1981 where she has served in a variety of analyst, leadership, and management positions.

Her assignments at the Center include serving as Team Leader, Operational Policy Team; Director, Anti-air Warfare Department; Program Director, Fleet Tactics and Capabilities; Team Leader of Third Fleet Tactical Analysis Team; Field Representative to Tactical Training Group – Pacific; Project Director, Electronic Warfare Project; Field Representative to Fighter Airborne Early Warning Wing-U.S. Pacific Fleet; and Analyst, Air Warfare Division, Operations Evaluation Group.

Before joining the Center, Ms. Fox served as a member of the Computer Group at the Institute for Defense Analysis in Alexandria, where she participated in planning and analyses of evaluations of tactical air survivability during close air support and effectiveness of electronic warfare during close air support.

Ms. Fox received a Bachelor of Science degree in mathematics and a Master of Science degree in applied mathematics from George Mason University.

Col. Gary S. Geyer, U.S. Air Force (Ret.):
Consultant

Col. Geyer has 38 years of experience in space engineering and program management, primarily in senior positions in the government and industry that emphasize management and system engineering. He has been responsible for all aspects of systems' success, including schedule, cost, and technical performance.

He served for 26 years with the National Reconnaissance Office (NRO) and was the NRO System Program Office Director for two major programs, which encompassed the design, manufacture, test, launch, and operation of several of our nation's most important reconnaissance satellites. Col. Geyer received the NRO Pioneer Award 2000 for his contributions as one of 46 pioneers of the NRO responsible for our nation's information superiority that significantly contributed to the end of the Cold War.

Following his career at the NRO, Col. Geyer was Vice President for a major classified program at Lockheed Martin and responsible for all aspects of program and mission success. His other assignments have included Chief Engineer for another nationally vital classified program and Deputy for Analysis for the Titan IV Program. Col. Geyer is teaching a Space Design course and a System Engineering/Program Management course at New Mexico State University in Las Cruces, N.M. He has a Bachelor of Science degree in Electrical Engineering from Ohio State University, and a Master's in Electrical Engineering and Aeronautical Engineering from the University of Southern California.

Col. Susan J. Helms, U.S. Air Force
Chief, Space Control Division, Requirements Directorate, Air Force Space Command

Colonel Susan J. Helms is Vice Commander of the 45th Space Wing at Patrick Air Force Base, Fla. She oversees military space launch operations from Cape Canaveral Air Force Station, Fla. (CCAFS), and Eastern Range support for commercial, NASA and military space launches from CCAFS and Kennedy Space Center, Fla., as well as ballistic missile tests at sea.

Colonel Helms is a veteran of five Space Shuttle flights as well as serving aboard the International Space Station as a member of the Expedition 2 crew for a total of 163 days. She received a Bachelor of Science degree in aeronautical engineering from the U.S. Air Force Academy in 1980 and a Master of Science degree in aeronautics/astronautics from Stanford University in 1985.

Col. Helms graduated from the U.S. Air Force Academy in 1980. She received her commission and was assigned to Eglin Air Force Base, Florida, as an F-16 weapons separation engineer with the Air Force Armament Laboratory. In 1982, she became the lead engineer for F-15 weapons separation. In 1984, she was selected to attend graduate school. She received her degree from Stanford University in 1985 and was assigned as an assistant professor of aeronautics at the U.S. Air Force Academy. In 1987, she attended the Air Force Test Pilot School at Edwards Air Force Base, California. After completing one year of training as a flight test engineer, Col. Helms was assigned as a USAF Exchange Officer to the Aerospace Engineering Test Establishment, Canadian Forces Base, Cold Lake, Alberta, Canada, where she worked as a flight test engineer and project officer on the CF-18 aircraft. She was managing the development of a CF-18 Flight Control System Simulation for the Canadian Forces when selected for the astronaut program.

Colonel Helms was selected by NASA in January 1990 and became an astronaut in July 1991. She flew on STS-54 (1993), STS-64 (1994), STS-78 (1996), and STS-101 (2000), and served aboard the International Space Station as a member of the Expedition 2 crew (2001). Colonel Helms has logged 5,064 hours in space, including an extravehicular activity of 8 hours and 56 minutes—a world record.

After a 12-year NASA career that included 211 days in space, Colonel Helms returned to the U.S. Air Force in July 2002 as the Division Chief of the Space Superiority Division of the Requirements Directorate of Air Force Space Command in Colorado Springs, Colorado.

Mr. Richard Kohrs
Chief Engineer, Kistler Aerospace Corporation

Richard Kohrs has over 40 years of experience in aerospace systems engineering, stress analysis, and integration. He has held senior management positions in major NASA programs from Apollo to the Space Station.

As a member of the Apollo Spacecraft Program's Systems Engineering and Integration Office, he developed the Spacecraft Operations Data Book system that documented systems and subsystem performance and was the control database for developing flight rules, crew procedures, and overall performance of the Apollo spacecraft.

After Apollo, he became Manager of System Integration for the Space Shuttle Program; Deputy Manager, Space Shuttle Program; and then Deputy Director of the Space Shuttle Program at JSC. As Deputy Director, he was responsible for the daily engineering, processing, and operations activities of the Shuttle Program, and he developed an extensive background in Shuttle systems integration. In 1989, he became the Director of Space Station Freedom, with overall responsibility for its development and operation.

After years of public service, he left NASA to become the Director of the ANSER Center for International Aerospace Cooperation (1994–1997). Mr. Kohrs joined Kistler Aerospace in 1997 as Chief Engineer. His primary responsibilities include vehicle integration, design specifications, design data books, interface control, vehicle weight, performance, and engineering review board matters. He received a Bachelor of Science degree from Washington University, St. Louis, in 1956.

Susan Morrisey Livingstone:

Ms. Livingstone has served her nation for more than 30 years in both government and civic roles. From July 2001 to February 2003, she served as Under Secretary of the Navy, the second highest civilian leadership position in the Department of the Navy. As "COO" to the Secretary of the Navy, she had a broad executive management portfolio (e.g., programming, planning, budgeting, business processes, organizational alignment), but also focused on Naval space, information technology and intelligence/compartmented programs; integration of Navy-Marine Corps capabilities; audit, IG and criminal investigative programs; and civilian personnel programs.

Livingstone is a policy and management consultant. Currently, she is a member of the National Security Studies Board of Advisors (Maxwell School, Syracuse University), a board member of the Procurement Round Table (for the second time), and an appointee to NASA's Return to Flight Task Group for safe return of Shuttle flight operations.

Prior to serving as Under Secretary of the Navy, Livingstone was CEO of the Association of the United States Army and deputy chairman of its Council of Trustees. She was also a vice president and board member of the Procurement Round Table, and acted as a consultant and panel chairman to the Defense Science Board (on "logistics transformation").

From 1993 to 1998, Ms. Livingstone served the American Red Cross HQ as Vice President of Health and Safety Services, Acting Senior Vice President for Chapter Services and as a consultant for Armed Forces Emergency Services.

As Assistant Secretary of the Army for Installations, Logistics and Environment from 1989 to 1993, she was responsible for a wide range of programs including military construction, installation management, Army logistics programs, base realignment and closures, energy and environmental issues, domestic disaster relief, and restoration of public infrastructure to the people of Kuwait following operation Desert Storm. She also was decision and acquisition management authority for the DoD chemical warfare materiel destruction program.

From 1981 to 1989, Ms. Livingstone served at the Veterans Administration in a number of positions including Associate Deputy Administrator for Logistics and Associate Deputy Administrator for Management. She was then the VA's Senior Acquisition Official and also directed and managed the nation's largest medical construction program. Prior to her Executive Branch service, she worked for more than nine years in the Legislative branch on the personal staffs of both a Senator and two Congressmen.

Livingstone graduated from the College of William and Mary in 1968 with an A.B. degree and completed an M.A. in political science at the University of Montana in 1972. She also spent two years in postgraduate studies at Tufts University and the Fletcher School of Law and Diplomacy.

Livingstone has received numerous awards for her community and national service, including the highest civilian awards from the National Reconnaissance Office, the VA, and the Departments of the Army and Navy. She is also a recipient of the Secretary of Defense Award for Outstanding Public Service.

Mr. James D. Lloyd:
Deputy Associate Administrator for Safety and Mission Assurance, NASA
Ex-Officio Member

Mr. Lloyd has extensive experience in safety engineering and risk management, and has supported a number of Blue Ribbon panels relating to mishaps and safety problems throughout his career. He began his career after an intern training period as a system safety engineer with the U.S. Army Aviation Systems Command in St. Louis.

He transferred to its parent headquarters, the Army Materiel Command (AMC) in 1973 and, after serving several safety engineering roles, was appointed as the Chief of the Program Evaluation Division in the Command's Safety Office, where he assured the adequacy of safety programs for AMC organizations.

In 1979, he continued his career as a civilian engineer with the AMC Field Safety Activity in Charlestown, IN, where he directed worldwide safety engineering, evaluation, and training support. In 1987, a year after the Shuttle *Challenger* disaster, Lloyd transferred from the U. S. Army to NASA to help the Agency rebuild its safety mission assurance program. He was instrumental in fulfilling several of the recommendations issued by the Rogers' Commission, which investigated the *Challenger* mishap. After the Shuttle returned to flight with the mission of STS-26, Lloyd moved to the Space Station Freedom Program Office in Reston, Va., where he served in various roles culminating in being appointed as the Program's Product Assurance Manager.

In 1993, he became Director, Safety and Risk Management Division in the Office of Safety and Mission Assurance, serving as NASA's "Safety Director" and was appointed to his present position in early 2003. He serves also as an ex-officio member of the NASA Advisory Council Task Force on ISS Operational Readiness. Lloyd holds a Bachelor of Science degree in Mechanical Engineering, with honors, from Union College, Schenectady, N.Y., and a Master of Engineering degree in Industrial Engineering from Texas A&M University, College Station.

Lt. Gen. Forrest S. McCartney, U.S. Air Force (Ret.): *Vice Chairman of the Aerospace Safety Advisory Panel*

During Lt. Gen. McCartney's distinguished Air Force career he held the position of Program Director for several major satellite programs, was Commander of the Ballistic Missile Organization (responsible for Minuteman and Peacekeeper development), Commander of Air Force Space Division, and Vice Commander, Air Force Space Command.

His military decorations and awards include the Distinguished Service Medal, Legion of Merit with one oak leaf cluster, Meritorious Service Medal, and Air Force Commendation Medal with three oak leaf clusters. He was recipient of the General Thomas D. White Space Trophy in 1984 and the 1987 Military Astronautical Trophy.

Following the *Challenger* accident, in late 1986 Lt. Gen. McCartney was assigned by the Air Force to NASA and served as the Director of Kennedy Space Center until 1992. He received numerous awards, including NASA's Distinguished Service Medal and Presidential Rank Award, the National Space Club Goddard Memorial Trophy, and AIAA Von Braun Award for Excellence in Space Program Management.

After 40 years of military and civil service, he became a consultant to industry, specializing in the evaluation of hardware failure/flight readiness. In 1994, he joined Lockheed Martin as the Astronautics Vice President for Launch Operations. He retired from Lockheed Martin in 2001 and was formerly the Vice Chairman of the NASA Aerospace Safety Advisory Panel.

Lt. Gen. McCartney has a Bachelor's degree in Electrical Engineering from Auburn University, a Master's degree in Nuclear Engineering from the Air Force Institute of Technology, and an honorary doctorate from the Florida Institute of Technology.

Rosemary O'Leary, J.D., Ph.D.:

Dr. Rosemary O'Leary is professor of public administration and political science, and coordinator of the Ph.D. program in public administration at the Maxwell School of Citizenship and Public Affairs at Syracuse University. An elected member of the U.S. National Academy of Public Administration, she was recently a senior Fulbright scholar in Malaysia. Previously Dr. O'Leary was Professor of Public and Environmental Affairs at Indiana University and cofounder and co-director of the Indiana Conflict Resolution Institute. She has served as the director of policy and planning for a state environmental agency and has worked as an environmental attorney.

She has consulted for the U.S. Department of the Interior, the U.S. Environmental Protection Agency, the Indiana Department of Environmental Management, the International City/County Management Association, the National Science Foundation, and the National Academy of Sciences.

Dr. O'Leary is the author/editor of five books and more than 75 articles on environmental management, environmental policy public management, dispute resolution, bureaucratic politics, and law and public policy. She has won seven national research awards, including Best Book in Public and Nonprofit Management for 2000 (given by the Academy of Management), Best Book in Environmental Management and Policy for 1999 (given by the American Society for Public Administration), and the Mosher Award, which she won twice, for best article by an academician published in Public Administration Review.

Dr. O'Leary was recently awarded the Syracuse University Chancellor's Citation for Exceptional Academic Achievement, the highest research award at the university. She has won eight teaching awards as well, including the national Excellence in Teaching Award given by the National Association of Schools of Public Affairs and Administration, and she was the recipient of the Distinguished Service Award given by the American Society for Public Administration's Section on Environment and Natural Resources Administration. O'Leary has served as national chair of the Public Administration Section of the American Political Science Association, and as the national chair of the Section on Environment and Natural Resources Administration of the American Society for Public Administration. She is currently a member of the NASA Aerospace Safety Advisory Panel

Dr. Decatur B. Rogers, P.E.:
Dean Tennessee State University College of Engineering, Technology and Computer Science

Since 1988, Dr. Rogers has served as the Dean, College of Engineering, Technology and Computer Science, and Professor of Mechanical Engineering at Tennessee State University in Nashville. Rogers served in professorship and dean positions at Florida State University, Tallahassee; Prairie View A&M University, Prairie View, Texas, and Federal City College, Washington, D.C.

Dr. Rogers holds a Ph.D. in Mechanical Engineering from Vanderbilt University; Master's degrees in Engineering Management and Mechanical Engineering from Vanderbilt University; and a Bachelor's in Mechanical Engineering from Tennessee State University.

Mr. Sy Rubenstein:
Aerospace Consultant

Mr. Rubenstein was a major contributor to the design, development, and operation of the Space Shuttle and has been involved in commercial and Government projects for more than 35 years. As an employee of Rockwell International, the prime contractor for the Shuttle, he was the Director of System Engineering, Chief Engineer, Program Manager, and Division President during 20 years of space programs.

He has received the NASA Public Service Medal, the NASA Medal for Exceptional Engineering, and the AIAA Space Systems Award for his contributions to human spacecraft development. Mr. Rubenstein, a leader, innovator, and problem solver, is a fellow of the AIAA and the AAS.

Mr. Robert Sieck:
Aerospace Consultant

Mr. Sieck, the former Director of Shuttle Processing at the Kennedy Space Center (KSC), has an extensive background in Shuttle systems, testing, launch, landing, and processing. He joined NASA in 1964 as a Gemini Spacecraft Systems engineer and then served as an Apollo Spacecraft test team project engineer. He later became the Shuttle Orbiter test team project engineer, and in 1976 was named the Engineering Manager for the Shuttle Approach and Landing Tests at Dryden Flight Research Facility in California. He was the Chief Shuttle Project Engineer for STS-1 through STS-7, and became the first KSC Shuttle Flow Director in 1983. He was appointed Director, Launch and Landing Operations, in 1984, where he served as Shuttle Launch Director for 11 missions.

He served as Deputy Director of Shuttle Operations from 1992 until January 1995 and was responsible for assisting with the management and technical direction of the Shuttle Program at KSC. He also retained his position as Shuttle Launch Director, a responsibility he had held from February 1984 through August 1985, and then from December 1986 to January 1995. He was Launch Director for STS-26R and all subsequent Shuttle missions through STS-63. Mr. Sieck served as Launch Director for 52 Space Shuttle launches.

He earned his Bachelor of Science degree in Electrical Engineering at the University of Virginia in 1960 and obtained additional postgraduate credits in mathematics, physics, meteorology, and management at both Texas A&M and the Florida Institute of Technology. He has received numerous NASA and industry commendations, including the NASA Exceptional Service Medal and the NASA Distinguished Service Medal. Sieck is a former consultant with the Aerospace Safety Advisory Panel.

Lt. Gen. Thomas Stafford, U.S. Air Force (Ret.):
Cochair, Return to Flight Task Group

President, Stafford, Burke and Hecker Inc., technical consulting

Lt. Gen. Stafford, an honors graduate of the U.S. Naval Academy, joined the space program in 1962 and flew four missions during the Gemini and Apollo programs. He piloted Gemini 6 and Gemini 9, and traveled to the Moon as Commander of Apollo 10. He was assigned as head of the astronaut group in June 1969, responsible for the selection of flight crews for projects Apollo and Skylab.

In 1971, Lt. Gen. Stafford was assigned as Deputy Director of Flight Crew Operations at the NASA Manned Spaceflight Center. His last mission, the Apollo-Soyuz Test Project in 1975, achieved the first rendezvous between American and Soviet spacecrafts.

He left NASA in 1975 to head the Air Force Test Flight Center at Edwards Air Force Base and, in 1978, assumed duties as Deputy Chief of Staff, Research Development and Acquisition, U.S. Air Force Headquarters in Washington. He retired from government service in 1979 and became an aerospace consultant.

Lt. Gen. Stafford has served as Defense Advisor to former President Ronald Reagan; and headed The Synthesis Group, which was tasked with plotting the U.S. return to the Moon and eventual journey to Mars.

Throughout his careers in the USAF and NASA space program, he has received many awards and medals including the Congressional Space Medal of Honor in 1993. He served on the National Research Council's Aeronautics and Space Engineering Board, the Committee on NASA Scientific and Technological Program Reviews, and the Space Policy Advisory Council.

He was Chairman of the NASA Advisory Council Task Force on Shuttle-Mir Rendezvous and Docking Missions.

He is currently the Chairman of the NASA Advisory Council Task Force on International Space Station Operational Readiness.

Mr. Tom Tate:

Mr. Tate was vice president of legislative affairs for the Aerospace Industries Association (AIA), a trade association representing the nation's manufacturers of commercial, military, and business aircraft, helicopters, aircraft engines, missiles, spacecraft, and related components and equipment. Joining AIA in 1988, Tate directed the activities of the association's Office of Legislative Affairs, which monitors policy issues affecting the industry and prepares testimony that communicates the industry's viewpoint to Congress.

Before joining AIA, Tate served on the staff of the House of Representative's Committee on Science and Technology for 14 years. He joined the staff in 1973 as a technical consultant and counsel to the House Subcommittee on Space Science and Applications. He was then appointed deputy staff director of the House Subcommittee on Energy Research and Development in 1976. In 1978, Tate returned to the space subcommittee as chief counsel; and in 1981, he became special assistant to the chairman of the committee until joining AIA.

Mr. Tate worked for the Space Division of Rockwell International in Downey, Calif., from 1962 to 1973 in various engineering and marketing capacities and was director of space operations when he departed the company in 1973. He worked on numerous programs, including the Gemini Paraglider, Apollo, Apollo/Soyuz, and Shuttle Programs.

He worked for RCA's Missile and Surface Radar Division in Moorestown, N.J. from 1958 to 1962 in the project office of the Ballistic Missile Early Warning System (BMEWS) that was being built for the USAF. From 1957 to 1958, Tate served in the Army as an artillery and guided missile officer at Fort Bliss, Texas.

He received a Bachelor's degree in marketing from the University of Scranton in 1956 and a law degree from Western State University College of Law in Fullerton, Calif., in 1970. In his final year of law school, his fellow students awarded him the Gold Book Award as the most outstanding student. In 1991, he received the Frank J. O'Hara award for distinguished alumni in science and technology from the University of Scranton.

Mr. Tate is a member of numerous aerospace and defense associations including the AIAA, the National Space Club, and the National Space Institute, where he serves as an advisor. He also served as a permanent civilian member of the NASA Senior Executive Service Salary and Performance Review Board.

Dr. Kathryn C. Thornton:
Faculty, University of Virginia

Dr. Kathryn Thornton is a Professor at the University of Virginia in the School of Engineering and Applied Science in the Division of Science, Technology and Society, and in the Department of Mechanical and Aerospace Engineering. She is also the Associate Dean for Graduate Programs. Thus, her time is divided between teaching and managing the Graduate Studies Office. Selected as an astronaut in May 1984, Dr. Thornton is a veteran of four Space Shuttle flights between 1989 and 1995, including the first Hubble Space Telescope service mission. She has logged over 975 hours in space, including more than 21 hours of extravehicular activity.

Prior to becoming an astronaut, Dr. Thornton was employed as a physicist at the U.S. Army Foreign Science and Technology Center in Charlottesville, Va. She holds a Bachelor of Science degree in physics from Auburn University and a Master of Science degree and Doctorate of Philosophy degree in physics from the University of Virginia.

Mr. William Wegner:
Consultant

Mr. Wegner graduated from the U.S. Naval Academy in 1948. He subsequently received Master's degrees in Naval Architecture and Marine Engineering from Webb Institute in New York. In 1956 he was selected by Adm. Hyman Rickover to join the Navy's nuclear program and was sent to the Massachusetts Institute of Technology, where he received his Master's degree in Nuclear Engineering. After serving in a number of field positions, including that of Nuclear Power Superintendent at the Puget Sound Naval Shipyard, he returned to Washington. He served as deputy director to Adm. Rickover in the Naval Nuclear Program for 16 years and was awarded the DoD Distinguished Service Award and the Atomic Energy Commission's distinguished service award.

In 1979, he retired from Government service and formed Basic Energy Technology Associates with three fellow naval retirees. During its 10 successful years of operation, it provided technical services to over 25 nuclear utilities and other nuclear-related activities. Wegner has served on a number of panels including the National Academy of Sciences that studied the safety of Department of Energy nuclear reactors. From 1989 to 1992, he provided technical assistance to the Secretary of Energy on nuclear-related matters. He has provided technical services to over 50 nuclear facilities. Mr. Wegner served as a Director of the Board of Directors of Detroit Edison from 1990 until retiring in 1999.

Mr. Vincent D. Watkins:
Executive Secretary, Return to Flight Task Group

Mr. Vincent Watkins is Executive Secretary to the Return to Flight Task Group (RTFTG), a federal advisory committee appointed to perform an independent assessment of NASA's Return to Flight actions to implement the recommendations of the *Columbia* Accident Investigation Board.

Prior to joining the RTFTG in May 2004, he was Assistant Chief of the Flight Equipment Division in the Safety and Mission Assurance Directorate at the Johnson Space Center (JSC) in Houston, Texas. His responsibilities included managing Safety and Mission Assurance engineering activities pertaining to the definition, design, development, and operation of JSC government furnished equipment (GFE) and extravehicular activity equipment and tools. These activities included flight readiness verification, risk assessments, hazard analysis, nonconformance tracking, and product delivery.

His 25-year career at NASA included a six-month tour at NASA Headquarters from April to December 2003. There he served as Executive Officer to the Chief of Staff, providing management oversight and technical expertise to the Office of the NASA Administrator. During this assignment, Mr. Watkins was instrumental in the development and implementation of several key Headquarters initiatives including the *Columbia* Families First Team and the *Columbia* Accident Rapid Reaction Team.

Mr. Watkins joined NASA in 1980 as a Control System Engineer on the Shuttle Training Aircraft in the Flight Crew Operations Directorate at JSC. From 1997 to 2003, he served as Chief of the GFE Assurance Branch in the Flight Equipment Division. He completed a NASA Fellowship with The Anderson School of Management at UCLA on Creativity and Innovation in the Organization in November 2003. He was selected as an inaugural member of the two-year JSC Leadership and Development Program in April 2002.

Mr. Watkins is a graduate of Albany State University with a Bachelor of Science degree in mathematics and a minor in physics and computer science. He received the Mark D. Heath Aircraft Engineering Award in 1987, the NASA Exceptional Service Medal in 1996, and numerous NASA Group Achievement Awards throughout his career at NASA.

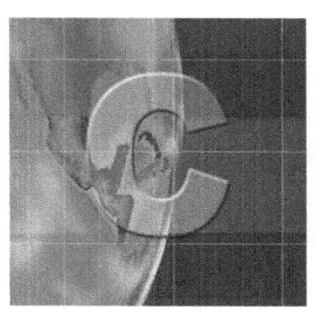

Appendix C:
Return to Flight Summary Overview

The following information describes the process NASA used to Return to Flight. It is included here now, without updates, for historical reference.

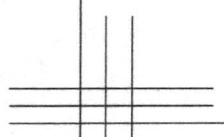 NASA's Implementation Plan for Space Shuttle Return to Flight and Beyond

Note: The Return to Flight Summary was originally written in August 2003 (for the first edition of NASA's Implementation Plan for Space Shuttle Return to Flight and Beyond) to reflect NASA's initial approach for responding to the Columbia Accident Investigation Board (CAIB) Recommendations and Observations as well as the Space Shuttle Program's Raising the Bar Actions. It has not been updated since its initial publication; therefore, it may contain outdated information. It is included as Appendix C for historical reference only.

The CAIB Report has provided NASA with the roadmap for moving forward with our Return to Flight efforts. The CAIB, through its diligent work, has determined the causes of the accident and provided a set of comprehensive recommendations to improve the safety of the Space Shuttle Program. NASA accepts the findings of the CAIB, we will comply with the Board's recommendations, and we embrace the report and all that is included in it. This implementation plan outlines the path that NASA will take to respond to the CAIB recommendations and safely Return to Flight, while taking into account the Vision for Space Exploration.

At the same time that the CAIB was conducting its assessment, NASA began pursuing an intensive, Agency-wide effort to further improve our human space flight programs. We are taking a fresh look at all aspects of the Space Shuttle Program, from technical requirements to management processes, and have developed a set of internally generated actions that complement the CAIB recommendations.

NASA will also have the benefit of the wisdom and guidance of an independent, advisory Return to Flight Task Group, led by two veteran astronauts, Apollo commander Thomas Stafford and Space Shuttle commander Richard Covey. Members of this Task Group were chosen from among leading industry, academia, and government experts. Their expertise includes knowledge of fields relevant to safety and space flight, as well as experience as leaders and managers of complex systems. The diverse membership of the Task Group will carefully evaluate and publicly report on the progress of our response to implement the CAIB's recommendations.

The space program belongs to the nation as a whole; we are committed to sharing openly our work to reform our culture and processes. As a result, this first installment of the implementation plan is a snapshot of our early efforts and will continue to evolve as our understanding of the action needed to address each issue matures. This implementation plan integrates both the CAIB recommendations and our self-initiated actions. This document will be periodically updated to reflect changes to the plan and progress toward implementation of the CAIB recommendations, and our Return to Flight plan.

In addition to providing recommendations, the CAIB has also issued observations. Follow-on appendices may provide additional comments and observations from the Board. In our effort to raise the bar, NASA will thoroughly evaluate and conclusively determine appropriate actions in response to all these observations and any other suggestions we receive from a wide variety of sources, including from within the Agency, Congress, and other external stakeholders.

Through this implementation plan, we are not only fixing the causes of the *Columbia* accident, we are beginning a new chapter in NASA's history. We are recommitting to excellence in all aspects of our work, strengthening our culture and improving our technical capabilities. In doing so, we will ensure that the legacy of *Columbia* guides us as we strive to make human space flight as safe as we can.

Key CAIB findings

The CAIB focused its findings on three key areas:

- Systemic cultural and organizational issues, including decision making, risk management, and communication;
- Requirements for returning safely to flight; and
- Technical excellence.

This summary addresses NASA's key actions in response to these three areas.

Changing the NASA culture

The CAIB found that NASA's history and culture contributed as much to the *Columbia* accident as any technical failure. NASA will pursue an in-depth assessment to identify and define areas where we can improve our culture and take aggressive corrective action. In order to do this, we will

- Create a culture that values effective communication and empowers and encourages employee ownership over work processes.
- Assess the existing safety organization and culture to correct practices detrimental to safety.
- Increase our focus on the human element of change management and organizational development.
- Remove barriers to effective communication and the expression of dissenting views.

- Identify and reinforce elements of the NASA culture that support safety and mission success.
- Ensure that existing procedures are complete, accurate, fully understood, and followed.
- Create a robust system that institutionalizes checks and balances to ensure the maintenance of our technical and safety standards.
- Work within the Agency to ensure that all facets of cultural and organizational change are continually communicated within the NASA team.

To strengthen engineering and safety support, NASA

- Is reassessing its entire safety and mission assurance leadership and structure, with particular focus on checks and balances, line authority, required resources, and funding sources for human space flight safety organizations.
- Is restructuring its engineering organization, with particular focus on independent oversight of technical work, enhanced technical standards, and independent technical authority for approval of flight anomalies.
- Has established a new NASA Engineering and Safety Center to provide augmented, independent technical expertise for engineering, safety, and mission assurance. The function of this new Center and its relationship with NASA's programs will evolve over time as we progress with our implementation of the CAIB recommendations.
- Is returning to a model that provides NASA subsystem engineers with the ability to strengthen government oversight of Space Shuttle contractors.
- Will ensure that Space Shuttle flight schedules are consistent with available resources and acceptable safety risk.

To improve communication and decision making, NASA will

- Ensure that we focus first on safety and then on all other mission objectives.
- Actively encourage people to express dissenting views, even if they do not have the supporting data on hand, and create alternative organizational avenues for the expression of those views.
- Revise the Mission Management Team structure and processes to enhance its ability to assess risk and to improve communication across all levels and organizations.

To strengthen the Space Shuttle Program management organization, NASA has

- Increased the responsibility and authority of the Space Shuttle Systems Integration office in order to ensure effective coordination among the diverse Space Shuttle elements. Staffing for the Office will also be expanded.
- Established a Deputy Space Shuttle Program Manager to provide technical and operational support to the Manager.
- Created a Flight Operations and Integration Office to integrate all customer, payload, and cargo flight requirements.

To continue to manage the Space Shuttle as a developmental vehicle, NASA will

- Be cognizant of the risks of using it in an operational mission, and manage accordingly, by strengthening our focus on anticipating, understanding, and mitigating risk.
- Perform more testing on Space Shuttle hardware rather than relying only on computer-based analysis and extrapolated experience to reduce risk. For example, NASA is conducting extensive foam impact tests on the Space Shuttle wing.
- Address aging issues through the Space Shuttle Service Life Extension Program, including midlife re-certification.

To enhance our benchmarking with other high-risk organizations, NASA is

- Completing a NASA/Navy benchmarking exchange focusing on safety and mission assurance policies, processes, accountability, and control measures to identify practices that can be applied to NASA programs.
- Collaborating with additional high-risk industries such as nuclear power plants, chemical production facilities, military flight test organizations, and oil-drilling operations to identify and incorporate best practices.

To expand technical and cultural training for Mission Managers, NASA will

- Exercise the Mission Management Team with realistic in-flight crisis simulations. These simulations will bring together the flight crew, flight control team, engineering staff, and Mission Management Team, and other appropriate personnel to improve communication and to teach better problem recognition and reaction skills.

- Engage independent internal and external consultants to assess and make recommendations that will address the management, culture, and communications issues raised in the CAIB Report.

- Provide additional operational and decision-making training for mid- and senior-level program managers. Examples of such training include, Crew Resource Management training, a U.S. Navy course on the *Challenger* launch decision, a NASA decision-making class, and seminars by outside safety, management, communications, and culture consultants.

Returning safely to flight

The physical cause of the *Columbia* accident was insulation foam debris from the External Tank left bipod ramp striking the underside of the leading edge of the left wing, creating a breach that allowed superheated gases to enter and destroy the wing structure during entry. To address this problem, NASA will identify and eliminate critical ascent debris and will implement other significant risk mitigation efforts to enhance safety.

Critical ascent debris

To eliminate critical ascent debris, NASA

- Is redesigning the External Tank bipod assembly to eliminate the large foam ramp and replace it with electric heaters to prevent ice formation.

- Will assess other potential sources of critical ascent debris and eliminate them. NASA is already pursuing a comprehensive testing program to understand the root cause of foam shedding and develop alternative design solutions to reduce the debris loss potential.

- Will conduct tests and analyses to ensure that the Shuttle can withstand potential strikes from noncritical ascent debris.

Additional risk mitigation

Beyond the fundamental task of eliminating critical debris, NASA is looking deeper into the Shuttle system to more fully understand and anticipate other sources of risk to safe flight. Specifically, we are evaluating known potential deficiencies in the aging Shuttle, and are improving our ability to perform on-orbit assessments of the Shuttle's condition and respond to Shuttle damage.

Assessing Space Shuttle condition

NASA uses imagery and other data to identify unexpected debris during launch and to provide general engineering information during missions. A basic premise of test flight is a comprehensive visual record of vehicle performance to detect anomalies. Because of a renewed understanding that the Space Shuttle will always be a developmental vehicle, we will enhance our ability to gather operational data about the Space Shuttle.

To improve our ability to assess vehicle condition and operation, NASA will

- Implement a suite of imagery and inspection capabilities to ensure that any damage to the Shuttle is identified as soon as practicable.

- Use this enhanced imagery to improve our ability to observe, understand, and fix deficiencies in all parts of the Space Shuttle. Imagery may include

    – ground-, aircraft-, and ship-based ascent imagery

    – new cameras on the External Tank and Solid Rocket Boosters

    – improved Orbiter and crew handheld cameras for viewing the separating External Tank

    – cameras and sensors on the International Space Station and Space Shuttle robotic arms

    – International Space Station crew inspection during Orbiter approach and docking

- Establish procedures to obtain data from other appropriate national assets.

- For the time being we will launch the Space Shuttle missions in daylight conditions to maximize imagery capability until we fully understand and can mitigate the risk that ascent debris poses to the Shuttle.

Responding to Orbiter damage

If the extent of the *Columbia* damage had been detected during launch or on orbit, NASA would have done everything possible to rescue the crew. In the future, we will fly with plans, procedures, and equipment in place that will offer a greater range of options for responding to on-orbit problems.

To provide the capability for Thermal Protection System on-orbit repairs, NASA is

- Developing materials and procedures for repairing Thermal Protection System tile and Reinforced Carbon-Carbon panels in flight. Thermal Protection System repair is feasible but technically challenging. The effort to develop these materials and procedures is receiving the full support of the Agency's resources, augmented by experts from industry, academia, and other U.S. Government agencies.

To enhance the safety of our crew, NASA

- Is evaluating a contingency concept for an emergency procedure that will allow stranded Shuttle crew to remain on the International Space Station for extended periods until they can safely return to Earth.

- Will apply the lessons learned from *Columbia* on crew survivability to future human-rated flight vehicles. We will continue to assess the implications of these lessons for possible enhancements to the Space Shuttle.

Enhancing technical excellence

The CAIB and NASA have looked beyond the immediate causes of the *Columbia* tragedy to proactively identify both related and unrelated deficiencies.

To improve the ability of the Shuttle to withstand minor damage, NASA will

- Develop a detailed database of the Shuttle's Thermal Protection System, including Reinforced Carbon-Carbon and tiles, using advanced nondestructive inspection and additional destructive testing and evaluations.

- Enhance our understanding of the Reinforced Carbon-Carbon operational life and aging process.

- Assess potential Thermal Protection System improvements for Orbiter hardening.

To improve our vehicle processing, NASA

- And our contractors are returning to appropriate standards for defining, identifying, and eliminating foreign object debris during vehicle maintenance activities to ensure a thorough and stringent debris prevention program.

- Has begun a review of existing Government Mandatory Inspection Points. The review will include an assessment of potential improvements, including development of a system for adding or deleting Government Mandatory Inspection Points as required in the future.

- Will institute additional quality assurance methods and process controls, such as requiring at least two employees at all final closeouts and at External Tank manual foam applications.

- Will improve our ability to swiftly retrieve closeout photos to verify configurations of all critical subsystems in time-critical mission scenarios.

- Will establish a schedule to incorporate engineering changes that have accumulated since the Space Shuttle's original design into the current engineering drawings. This may be best accomplished by transitioning to a computer-aided drafting system, beginning with critical subsystems.

To safely extend the Space Shuttle's useful life, NASA

- Will develop a plan to recertify the Space Shuttle, as part of the Shuttle Service Life Extension.

- Is revalidating the operational environments (e.g., loads, vibration, acoustic, and thermal environment) used in the original certification.

- Will continue pursuing an aggressive and proactive wiring inspection, modification, and refurbishment program that takes full advantage of state-of-the-art technologies.

- Is establishing a prioritized process for identifying, approving, funding, and implementing technical and infrastructure improvements.

To address the public overflight risk, NASA will

- Evaluate the risk posed by Space Shuttle overflight during entry and landing. Controls such as entry ground track and landing site changes will be considered to balance and manage the risk to persons, property, flight crew, and vehicle.

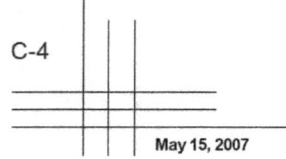

May 15, 2007

To improve our risk analysis, NASA

- Is fully complying with the CAIB recommendation to improve our ability to predict damage from debris impacts. We are validating the Crater debris impact analysis model use for a broader range of scenarios. In addition, we are developing improved physics-based models to predict damage. Further, NASA is reviewing and validating all Space Shuttle Program engineering, flight design, and operational models for accuracy and adequate scope.

- Is reviewing its Space Shuttle hazard and failure mode effects analyses to identify unacknowledged risk and overly optimistic risk control assumptions. The result of this review will be a more accurate assessment of the probability and severity of potential failures and a clearer outline of controls required to limit risk to an acceptable level.

- Will improve the tools we use to identify and describe risk trends. As a part of this effort, NASA will improve data mining to identify problems and predict risk across Space Shuttle Program elements.

To improve our Certification of Flight Readiness, NASA is

- Conducting a thorough review of the Certification of Flight Readiness process at all levels to ensure rigorous compliance with all requirements prior to launch.

- Reviewing all standing waivers to Space Shuttle Program requirements to ensure that they are necessary and acceptable. Waivers will be retained only if the controls and engineering analysis associated with the risks are revalidated. This review will be completed prior to Return to Flight.

Next steps

The CAIB directed that some of its recommendations be implemented before we Return to Flight. Other actions are ongoing, longer-term efforts to improve our overall human space flight programs. We will continue to refine our plans and, in parallel, we will identify the budget required to implement them. NASA will not be able to determine the full spectrum of recommended Return to Flight hardware and process changes, and their associated cost, until we have fully assessed the selected options and completed some of the ongoing test activities.

Conclusion

The American people have stood with NASA during this time of loss. From all across the country, volunteers from all walks of life joined our efforts to recover *Columbia*. These individuals gave their time and energy to search an area the size of Rhode Island on foot and from the air. The people of Texas and Louisiana gave us their hospitality and support. We are deeply saddened that some of our searchers also gave their lives. The legacy of the brave Forest Service helicopter crew, Jules F. Mier, Jr., and Charles Krenek, who lost their lives during the search for *Columbia* debris will join that of the *Columbia*'s crew as we try to do justice to their memory and carry on the work for the nation and the world to which they devoted their lives.

All great journeys begin with a single step. With this initial implementation plan, we are beginning a new phase in our Return to Flight effort. Embracing the CAIB Report and all that it includes, we are already beginning the cultural change necessary to not only comply with the CAIB recommendations, but to go beyond them to anticipate and meet future challenges.

With this and subsequent iterations of the implementation plan, we take our next steps toward return to safe flight. To do this, we are strengthening our commitment to foster an organization and environment that encourages innovation and informed dissent. Above all, we will ensure that when we send humans into space, we understand the risks and provide a flight system that minimizes the risk as much as we can. Our ongoing challenge will be to sustain these cultural changes over time. Only with this sustained commitment, by NASA and by the nation, can we continue to expand human presence in space—not as an end in itself, but as a means to further the goals of exploration, research, and discovery.

The *Columbia* accident was caused by collective failures; by the same token, our Return to Flight must be a collective endeavor. Every person at NASA shares in the responsibility for creating, maintaining, and implementing the actions detailed in this report. Our ability to rise to the challenge of embracing, implementing, and perpetuating the changes described in our plan will ensure that we can fulfill the NASA mission—to understand and protect our home planet, to explore the Universe and search for life, and to inspire the next generation of explorers.

Historical Reference Only

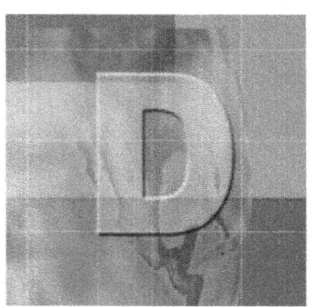

Appendix D:
The Integrated Accepted Risk Approach for Return to Flight

The following information describes the process NASA used to Return to Flight. It is included here now, without updates, for historical reference.

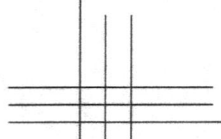 NASA's Implementation Plan for Space Shuttle Return to Flight and Beyond

Note: The following information describes NASA's approach to risk reduction and acceptance of residual risk for Return to Flight. It is included now, without updates, for historical reference.

NASA has come a long way in our journey to reduce the risks of operating the Space Shuttle system. The External Tank bipod Thermal Protection System has been redesigned to eliminate the proximate cause of the *Columbia* accident. In all areas we have applied the collective knowledge and capabilities of our Nation to comply with the *Columbia* Accident Investigation Board recommendations and to raise the bar beyond that. We have taken prudent technical action on potential threats to review and verify the material condition of all critical areas where failure could result in catastrophic loss of the crew and vehicle. We are satisfied that critical systems and elements should operate as intended—safely and reliably. While we will never eliminate all the risks from our human space flight programs, we have eliminated those we can and reduced, controlled, and/or mitigated others. The remaining identified risks will be evaluated for acceptance.

Our approach to launching, operating on orbit, and safely returning the Space Shuttle *Discovery* to flight on the planned STS-114 mission is based on a rigorous process to achieve the capabilities needed to meet our objectives. Greater capabilities may be achievable with more time and resources; however, the current primary Space Shuttle mission is to assemble and support the ongoing operation of the International Space Station. The missions and risks of the International Space Station and Space Shuttle are, for the near term, inseparable. As we look forward to the limited launch window opportunities in 2005, we must ask ourselves if the risks of Space Shuttle flight are acceptable. Although we will never eliminate all the risks from our Space Shuttle missions, we are confident that we have addressed those that constituted the proximate cause of the loss of *Columbia* and have eliminated, reduced, controlled, and/or mitigated other risks, including engineering, operational, and programmatic risks. We acknowledge that there is more that can be done over the long haul to further reduce risk, but the marginal risk return is getting smaller and smaller. With deliberate forethought, we now choose to assess the risk associated with the achievable capabilities consistent with the 2005 launch windows that are available. Before we commit to launching the STS-114 mission, we will assure that the residual risk is at an acceptable level to safely Return to Flight. If we cannot collectively decide that the risk to the Space Shuttle is acceptable for a 2005 Return to Flight, we will continue to work those technical issues until the risk is acceptable. We clearly demonstrated our commitment by our recent decision concerning the risk associated with the potential for ice formation in the forward liquid oxygen bellows. Based on the unanimous position of the Design Verification Review Board, which deemed the risk unacceptable, we now choose to install the bellows heater prior to launch. The only milestone that is irreversible is liftoff; we can choose to stop at any time before T-0.

Our risk reduction approach has its roots in the system safety engineering hierarchy for hazard abatement long employed in aerospace systems engineering. The components of the hierarchy are, in order of precedence, to: design/redesign; eliminate the hazard/risk; reduce the hazard/risk; and control the hazard/risk and/or mitigate the consequence of the remaining hazard/risk through warning devices, special procedures/capabilities, and/or training. This proven approach to risk reduction has been applied to potential hazards and risks in all critical areas of the Space Shuttle and has guided us through the technical challenges, failures, and successes present in Return to Flight endeavors. This approach, as shown in the figure at the top of the following page, provides the structured deliberation process required to verify and form the foundation for accepting any residual risk across the entire Space Shuttle Program by NASA leadership.

Space flight and operations are endeavors that could not be undertaken without accepting high levels of risk. Throughout history, humans have accepted risk to achieve the great rewards that exploration offers. Many have bravely faced the hazards and dangers of exploration and failed. NASA has had many more successes than failures, but we make every attempt to learn as much as possible from every failure before continuing. We choose to continue space exploration as an endeavor that is worthy of the risks to achieve our mission, to acquire the ultimate rewards, and to expand our knowledge of the universe. Accepting risk is not taken lightly.

Within the Space Shuttle Program, our system safety engineering hierarchy for hazard abatement requires that we understand and document how we deal with identified hazards. Hazards that have been eliminated through design by completely removing the hazard causal factors are documented as eliminated. Hazards that cannot be eliminated can be considered controlled when we can demonstrate that the frequency of occurrence or consequence has been reduced to a point that it is unlikely to occur during the life of the program. Where identified hazards cannot be eliminated or where controls of the hazard causes have limitations or uncertainties such that the hazard may occur in the life of the program, program management may, after considering all engineering data and opinions, accept the risk.

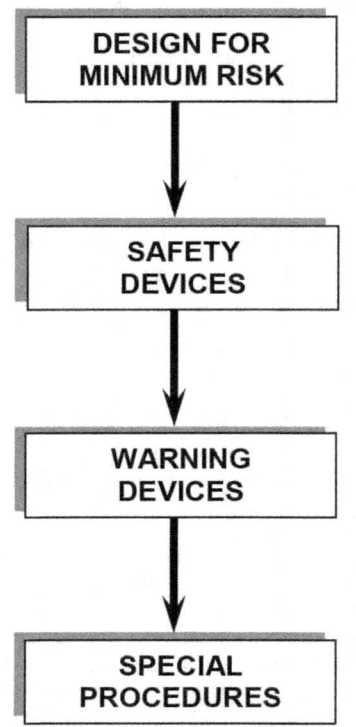

DESIGN FOR MINIMUM RISK — The major goal throughout the design phase shall be to ensure inherent safety through the selection of appropriate design features as fail operational/failure safe combinations and appropriate safety factors. Hazards shall be eliminated by design where possible. Damage control, containment, and isolation of potential hazards shall be included in design considerations.

SAFETY DEVICES — Known hazards that cannot be eliminated through design selection shall be reduced to an acceptable level through the use of appropriate safety devices as part of the system, subsystem, or equipment.

WARNING DEVICES — Where it is not possible to preclude the existence or occurrence of a known hazard, devices shall be employed for the timely detection of the condition and the generation of an adequate warning signal. Warning signals and their application shall be designed to minimize the probability of wrong signals or of improper personnel reaction to the signal.

SPECIAL PROCEDURES — Where it is not possible to reduce the magnitude of existing or potential hazard through design or the use of safety and warning devices, special procedures shall be developed to counter hazardous conditions for enhancement of ground and flight crew safety. Precautionary notations shall be standardized.

Return to Flight Requirements

Our top-level requirement for debris is the same as it was before *Columbia*: "*The SSS [Space Shuttle system], including the ground systems, shall be designed to preclude the shedding of ice and/or other debris from the Shuttle elements during prelaunch and flight operations that would jeopardize the flight crew, vehicle, mission success, or would adversely impact turnaround operations.*" The *Columbia* Accident Investigation Board determined that the primary cause of the loss of *Columbia* was the loss of the Thermal Protection System foam from the External Tank bipod that struck the Reinforced Carbon-Carbon panel on *Columbia's* left wing leading edge. Loss of foam was not an isolated incident. Over the life of the Space Shuttle Program there were seven documented cases of foam loss from the left-hand bipod and loss of foam from other areas of the External Tank on every flight. The Space Shuttle Program has compiled a comprehensive database of historical tile damage to help us understand both the sources of the damage and their effect on the integrity of the Orbiter Thermal Protection System. Since *Columbia*, we have initiated a comprehensive test and analysis program to better characterize the potential for External Tank foam loss, to understand the transport mechanisms that move liberated debris to the Orbiter, and to gain knowledge of the capabilities of the Orbiter Thermal Protection System tile and Reinforced Carbon-Carbon elements to withstand impact. From this effort, requirements for allowable debris for given sources have been established to protect the Orbiter elements from critical impact.

Design for Minimum Risk

The External Tank bipod Thermal Protection System has been redesigned to reduce the potential for loss of foam that led to the *Columbia* accident. Our far-reaching initiative to eliminate or reduce the potential for generation of critical debris has led us to the most comprehensive understanding of the overall Space Shuttle system in the history of the Program. We have identified and examined all debris sources and, where necessary, initiated redesign efforts to reduce the potential for debris formation and liberation. There are four primary areas identified on the External Tank for evaluation and redesign to reduce or eliminate the potential for critical debris generation: the bipod foam, the liquid oxygen feedline bellows ice formation, the liquid hydrogen intertank flange foam closeout, and the protuberance air load foam ramps. All have been addressed with respect to the Orbiter debris damage tolerance capabilities and will be verified for flight. In addition to the External Tank, we have assessed the Solid Rocket Booster separation motor plumes and Thermal Protection System elements, as well as potential Orbiter debris sources such as thruster plumes and butcher paper covers. In the forward portion of the Orbiter, butcher paper was previously used to cover thruster nozzles to prevent rain from entering prior to launch. This butcher paper is being replaced with a less dense material (Tyvec) that will be combined with a release capability to make the material fall off at low speeds and reduce the potential for damage to the windows. Our solid rocket bolt catcher system has been redesigned to eliminate a potential failure point, the housing weld, and has been tested and proven to meet design requirements.

Safety Devices

Although redesigning the External Tank Thermal Protection System to reduce the potential for major foam loss is our primary goal, we have crafted a wide-ranging approach for reducing the overall risk of operating the Space Shuttle system. NASA agrees that *Columbia* clearly demonstrated that the Orbiter Thermal Protection System, including Reinforced Carbon-Carbon panels, is vulnerable to impact damage from the existing debris environment. Through tests and analysis, we have a new appreciation for the potential sources and size of debris that might be present during ascent. We have a new understanding of the capability of the Orbiter Thermal Protection System to withstand debris hits in all flight regimes. Our test program forms the basis for our newly developed debris transport analysis, providing improved knowledge of the multitude of paths debris might travel to impact the Orbiter and forming the basis for a validated computerized model for future near-real-time evaluation.

Our fundamental Return to Flight rationale is based on the necessary reduction in risk of ascent debris damage to be accomplished primarily through modifications to the External Tank. The allowable debris limits for the External Tank were set, and the External Tank was delivered meeting those limits. The definition of critical debris is derived from the ability of the current Orbiter, not the hardened Orbiter, to withstand impact damage. Therefore, Orbiter hardening provides an additional level of risk mitigation above and beyond our primary control. We have initiated an Orbiter hardening program that will be implemented as feasible—an approach consistent with the *Columbia* Accident Investigation Board recommendation to initiate a program of Orbiter hardening prior to Return to Flight. The Orbiter hardening options are being implemented in three phases. Four projects were identified as Phase I, based on maturity of design and schedule for implementation, and will be implemented before Return to Flight. These include: front spar "sneak flow" protection for the most vulnerable and critical wing leading edge panels 5 through 13; main landing gear corner void elimination; forward Reaction Control System carrier panel redesign to eliminate bonded studs; and replacing side windows 1 and 6 with thicker outer thermal panes. We accept the risk associated with not having improved Orbiter hardening capability beyond what has been put in place for Return to Flight and will reduce this risk over the long haul by continuing to pursue additional hardening measures.

Warning Devices

In the unlikely event the Orbiter experiences a debris hit, we have greatly expanded capabilities to detect debris liberation during ascent, to identify locations on the External Tank where debris may have originated, and to identify impact sites on the Orbiter Thermal Protection System for evaluation. Our ability to identify debris release during the first few minutes of ascent is enhanced through the addition of high-speed cameras, aircraft-mounted cameras, and radar. A camera installed on the External Tank will provide real-time, on-vehicle views during ascent.

Video cameras on the Solid Rocket Boosters will record the condition of the External Tank intertank areas for later review after booster recovery. In addition to the External Tank umbilical film cameras that will be examined after the mission, the images gathered from a digital camera, added in the umbilical area on the Orbiter at the External Tank interface, will be downlinked soon after the Orbiter achieves orbit. The crew will also take images of the External Tank using digital cameras shortly after separation to later downlink.

In the near term, we are committed to daylight launches and External Tank separation in lighted conditions on orbit to improve our ability to identify debris releases during ascent and assess the condition of the External Tank after separation, and to determine that our debris reduction efforts were successful. Requirements for daylight launches and lighted External Tank separation will be reevaluated after the second mission, STS-121. To further augment impact detection capabilities, we are installing an impact detection sensor system on the interior of the wing leading edge to identify whether the Reinforced Carbon-

Carbon panels were struck during ascent. Once on orbit, the crew will use the new Orbiter Boom Sensor System to examine the condition of the wing leading edge and nose cap for signs of critical impact. The Orbiter Boom Sensor System is grappled by the Shuttle Remote Manipulator System, known as the arm, and will have a combination of a camera and a laser depth detection system to characterize the surface of the Reinforced Carbon-Carbon elements. When approaching the International Space Station, the Orbiter will be turned to present its underside to the Expedition crew, who will use digital cameras with telephoto lenses to capture images of the Orbiter's Thermal Protection System.

Individually, each warning device/inspection method listed above will not provide the total information needed to accurately determine the condition of the Orbiter prior to committing to entry. They are not redundant systems per se, in that each provides a different piece of the puzzle, offering overlapping information to improve our knowledge of the Orbiter's condition. We can accept failure of one or more warning devices and have the confidence that we will be able to characterize potential debris liberation and possible damage to the Thermal Protection System tile and Reinforced Carbon-Carbon components. NASA considers the limitations of this approach to be reasonable.

Special Procedures

During Space Shuttle missions, data collected from multiple ground-based, on-vehicle, and space-based sources will be immediately evaluated through an integrated imagery evaluation process. Although we made great strides in reducing the potential for debris generation, a small potential for impacts to Orbiter tile and Reinforced Carbon-Carbon elements remains. Based on our expanded understanding of debris transport mechanisms and the capability of the Orbiter Thermal Protection System, we have established criteria for further on-orbit imagery and evaluation of potential tile damage. Data collected from multiple ground-based, on-vehicle, and space-based sources will be immediately evaluated through an integrated imagery evaluation process. Should tile damage exceed our criteria, plans are in place for further evaluation and repair, if necessary. These plans include: a focused inspection using the Orbiter Boom Sensor System, a spacewalk to get close-up images and make a visual evaluation, and/or implementing our limited Thermal Protection System repair capability. In any case, the appropriate risk assessment of each course of action will be conducted and presented to the Mission Management Team for evaluation and an implementation decision. Together with the work of the inspection tiger team, the Shuttle Systems Engineering and Integration Office began development of a Thermal Protection System Readiness Determination Operations Concept document. Most critically, this document will specify the process for collecting, analyzing, and applying the diverse inspection data in a way that ensures effective and timely risk assessment and mission decision-making. This risk assessment process will provide the Mission Management Team with the most comprehensive evaluation of the Orbiter's condition prior to committing to entry in the history of the Program.

We are mindful that our newfound capabilities have both built-in conservatism and limitations in completely identifying all unknowns. In many cases, the determination of debris sources and the resulting definition of potential debris environment during ascent have led to a conservative assessment of the risk. The accuracy of ascent and on-orbit imagery is dependent on the systems working as designed, weather conditions, and lighting. Potential damage to Orbiter Thermal Protection System elements has been closely scrutinized and extensively tested with the expectation that margin is available. Our limited Thermal Protection System repair techniques remain to be demonstrated on orbit, then analyzed and tested upon return to Earth in an effort to provide evidence of capability.

Despite our best efforts, it is not possible to completely reproduce on Earth the integrated environment experienced during a Space Shuttle mission. NASA also recognizes that it is not possible to totally eliminate the possibility of a debris impact like that which doomed *Columbia*; however, the actions taken have reduced to well below any previous Space Shuttle flight the risk of a repeat of that event. Unprecedented mitigation actions for that remote eventuality are in place. In the unlikely event that all of our efforts to reduce risk and safely return the Space Shuttle to flight fail, we have made plans to keep the Space Shuttle crew on the International Space Station and mount a rescue mission. Through the flight readiness review process, we will periodically evaluate the capability of the International Space Station to accommodate the Space Shuttle crew with food, water, and breathable oxygen. This capability, known as the Contingency Shuttle Crew Support, will only be used in the direst of circumstances and will not be used to justify flying unsafely. The Contingency Shuttle Crew Support and rescue mission requirements will be evaluated after the first two Return to Flight missions.

Additional Risk Reduction Efforts

In addition to the technical and operational improvements to the Shuttle system, NASA has improved our overall approach to safety and mission success. Early on, we set up the NASA Engineering and Safety Center at Langley

Research Center to provide the Agency with a cadre of highly qualified and experienced engineers to deal with tough technical issues and to leverage the best talent our Nation has to offer. Through the implementation of our Agency Independent Technical Authority and the establishment of an independent Safety and Mission Assurance organizational structure, we have invigorated the critical checks and balances needed to provide for safe and reliable operations. Our Space Shuttle System Integration and Engineering Office has broader responsibilities and advanced tools with which to evaluate and define the critical environment in which the Space Shuttle operates in ways never before put into practice. The growth and strength of this office has been instrumental in providing greater understanding and knowledge of the interaction between elements as we prepared for return to safe and reliable Space Shuttle operations. We further defined the roles and responsibilities of the Mission Management Team and provided critical training through courses and mission simulations to assure that team members are ready for the challenges and uncertainty ahead. We have enhanced the integrity of closeout inspections by requiring a minimum of two people at each inspection, improved our digital closeout photography system and processes, and brought our foreign object debris definition processes in line with industry practices.

We are attentive to the fact that we were criticized for focusing on schedule and not heeding the warning signs that we were overtaxing available resources in the system. Our risk management system has been enhanced and strengthened by balancing technical, schedule, and resource risks to successfully achieve safe and reliable operations. Safe and reliable operations are assured by first focusing on the technical risks and taking the needed time and resources to properly resolve technical issues. Once technical risks are eliminated or reduced to an acceptable level, program managers turn to the management of schedule and resource risks to preserve safety.

Schedules are integral parts of program management and provide for the integration and optimization of resource investments across a wide range of connected systems. The Space Shuttle Program must have a visible schedule with clear milestones to effectively achieve its mission. Schedules associated with all activities generate very specific milestones that must be completed for mission success. Nonetheless, schedules of milestone-driven activities will be extended when necessary to ensure safety, as we have demonstrated numerous times during the Return to Flight process. NASA will not compromise safe and reliable operations in an effort to optimize schedules or costs.

For now, there will be a level of residual risk that will be presented to NASA leadership for acceptance prior to Return to Flight. Our risk assessment/risk management process does not end with STS-114. We are committed to continuous risk evaluation of our experiences gained through each mission and will continue to factor in ongoing enhancements over time. We understand that the primary Space Shuttle mission is to assemble and support the ongoing operation of the International Space Station. The missions and risks of the International Space Station and Space Shuttles are, for the near term, inseparable.

We have met many challenges during our journey, but we proceed with the full understanding that we have done all that is reasonably achievable; and the result of our efforts offers *Discovery's* crew, led by Commander Eileen Collins, risks that have been reduced and mitigated to the extent possible. We are committed to safely returning to flight and safely flying the Space Shuttle fleet until its retirement. To do less would inappropriately diminish the contributions of the STS-107 crew and our astronauts who will honor that crew's legacy.

Historical Reference Only

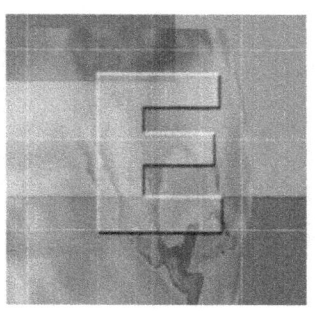

Appendix E: Return to Flight Suggestions

The following information describes the process NASA used to Return to Flight. It is included here now, without updates, for historical reference.

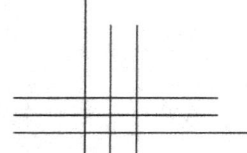
NASA's Implementation Plan for Space Shuttle Return to Flight and Beyond

Note: The information below is a synopsis of the Return to Flight Suggestions received by NASA via e-mail and NASA Web page. It is included now, without updates, for historical reference.

As part of NASA's response to the *Columbia* Accident Investigation Board (CAIB) recommendations, the Administrator asked that a process be put in place for NASA employees and the public to provide their ideas to help NASA safely Return to Flight. With the first public release of NASA's Implementation Plan for Space Shuttle Return to Flight and Beyond on September 8, 2003, NASA created an electronic mailbox to receive RTF suggestions. The e-mail address is "RTFsuggestions@nasa.gov." A link to the e-mail address for RTF suggestions is posted under the Return to Flight link on the NASA Web page "www.nasa.gov."

The first e-mail suggestion was received on September 8, 2003. Since then, NASA has received a total of 2683 messages, averaging 56 messages per week. NASA has provided a personal reply to each message. When applicable, information was provided as to where the message was forwarded for further review and consideration.

As NASA approaches our planned RTF date, it is critical that we move from development to implementation. As a part of this effort, we are now baselining all critical RTF activities. As a result, although we will continue to maintain the RTFsuggestions@nasa.gov e-mail box, beginning on September 1, 2004, NASA addressees will receive an automated response. NASA will periodically review the suggestions received for future use. We appreciate all of the interest and thoughtful suggestions received to date and look forward to receiving many more suggestions to both improve the Space Shuttle system and apply to exploration systems.

Many of the messages received are provided for review to a Project or Element Office within the Space Shuttle Program, the International Space Station Program, the Safety and Mission Assurance Office, the Training and Leadership Development Office, the newly established NASA Engineering and Safety Center, or to the NASA Team formed to address the Agencywide implications of the CAIB Report for organization and culture.

NASA organizations receiving suggestions are asked to review the message and use the suggestion as appropriate in their RTF activities. When a suggestion is forwarded, the recipient is encouraged to contact the individual who submitted the suggestion for additional information to assure that the suggestion's intent is clearly understood.

Table 1 provides a summary of the results. The table includes the following information: (1) the categories of suggestions; (2) the number of suggestions received per category; and (3) examples of RTF suggestion content from each category.

Synopsis of Return to Flight Suggestions

Category	No. of Suggestions	Example Suggestion Content
Orbiter	673	(1) Develop a redundant layer of Reinforced Carbon-Carbon panels on the Orbiter wing leading edge (WLE). (2) Cover the WLE with a titanium skin to protect it from debris during ascent.
External Tank	599	(1) Insulate the inside of the External Tank (ET) to eliminate the possibility of foam debris hitting the Orbiter. (2) Shrink wrap the ET to prevent foam from breaking loose.
General Space Shuttle Program	400	(1) Simulate Return to Launch Site scenarios. (2) Orbit a fuel tank to allow the Orbiter to refuel before entry and perform a slower entry. (3) Establish the ability to return the Shuttle without a crew on board.
Imagery/Inspection	183	(1) Use the same infrared imagery technology as the U.S. military to enable monitoring and tracking the Space Shuttle during night launches. (2) Use a remotely controlled robotic free-flyer to provide on-orbit inspection. (3) Bring back the Manned Maneuvering Unit to perform on-orbit inspection of the Orbiter.
Vision for Space Exploration	179	(1) Bring back the Saturn V launch vehicle to support going to the Moon and Mars. (2) Preposition supply/maintenance depots in orbit to reduce the need for frequently returning to Earth. (3) Construct future habitats and vehicles in space to eliminate launching large payloads from Earth.
Aerospace Technology	137	Quickly develop a short-term alternative to the Space Shuttle based on existing technology and past Apollo-type capsule designs.
Crew Rescue/Ops	127	(1) Implement a joint crew escape pod or individual escape pods within the Orbiter cockpit. (2) Have a second Shuttle ready for launch in case problems occur with the first Shuttle on orbit. (3) Have enough spacesuits available for all crewmembers to perform an emergency extravehicular activity.
Systems Integration	126	(1) Mount the Orbiter higher up on the ET to avoid debris hits during launch. (2) Incorporate temporary shielding between the Orbiter and ET that would fall away from the vehicle after lift off.
Public Affairs	85	NASA needs to dramatically increase media coverage to excite the public once again, to better convey the goals and challenges of human space flight, and to create more enthusiasm for a given mission.
NASA Culture	65	(1) Host a monthly employee forum for discussing ideas and concerns that would otherwise not be heard. (2) Senior leaders need to spend more time in the field to keep up with what is actually going on.
NASA Safety and Mission Assurance	47	(1) Learn from the Naval Nuclear Reactors Program. (2) The Government Mandatory Inspection Point review should not be limited to just the Michoud Assembly Facility and Kennedy Space Center elements of the Program.
Space Shuttle Program Safety	27	(1) Develop new Solid Rocket Boosters (SRBs) that can be thrust-controlled to provide a safer, more controllable launch. (2) Use rewards and incentives to promote the benefits of reliability and demonstrate the costs of failure.
International Space Station	20	(1) Adapt an expandable rocket booster to launch Multi-Purpose Logistics Modules to the International Space Station (ISS). (2) Add ion engines to the ISS to give it extra propulsion capability.
Leadership and Management	9	(1) Employees need to be trained while still in their current job to prepare them for increasing positions of responsibility. (2) Institute a rotational policy for senior management, similar to that of the U.S. Armed Forces.

Historical Reference Only

Category	No. of Suggestions	Example Suggestion Content
NASA Engineering and Safety Center	5	(1) Use a group brainstorming approach to aid in identifying how systems might fail. (2) NESC needs to get involved during a project's start as well as during its mission operations.
Solid Rocket Boosters	1	Ensure that the SRB hold-down bolts are properly reevaluated.
Total (As of August 9, 2004)	2683	

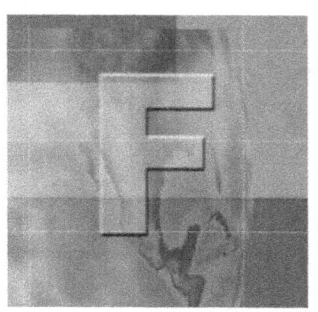

Appendix F:
CAIB Recommendations Implementation Schedule

The following information describes the process NASA used to Return to Flight. It is included here now, without updates, for historical reference.

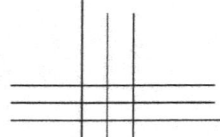 NASA's Implementation Plan for Space Shuttle Return to Flight and Beyond

CAIB Recommendations Implementation Schedule

NOTE: See legend on p. F-5 for definitions of acronyms

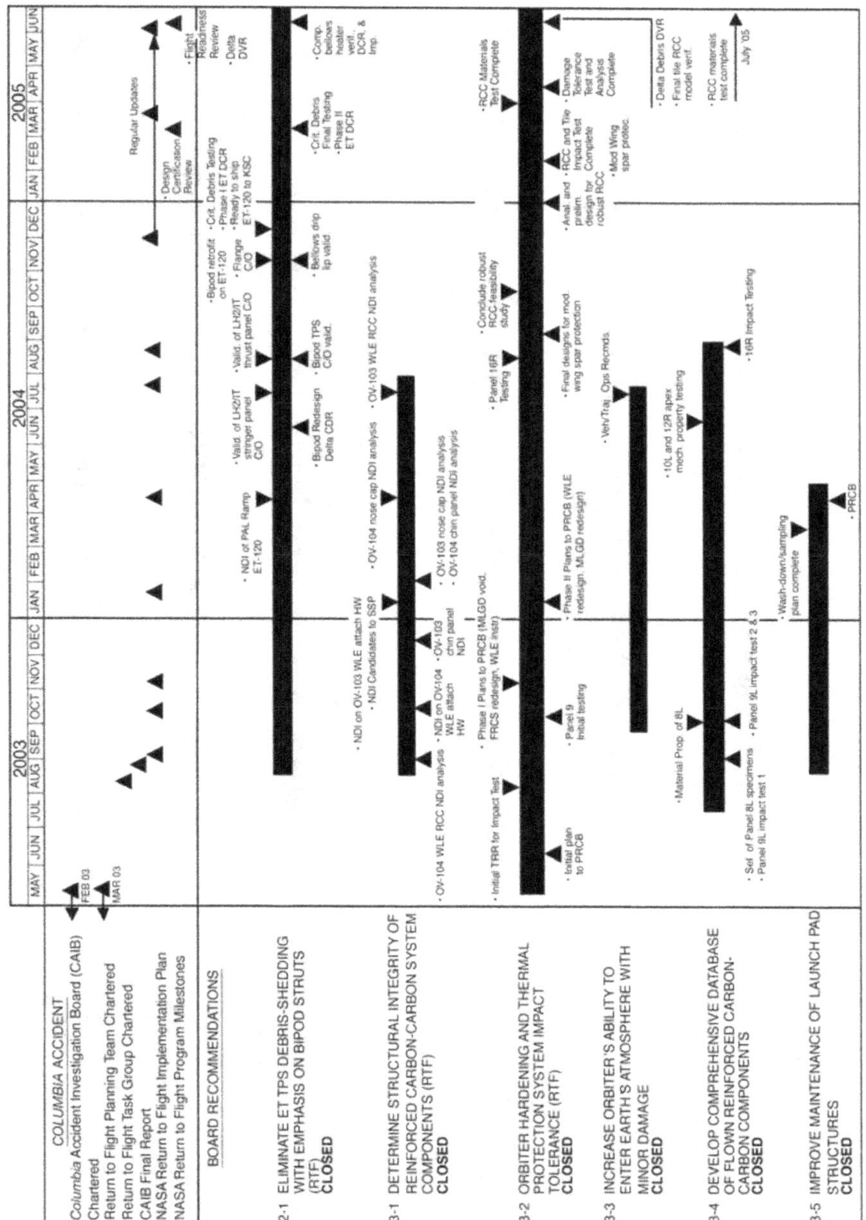

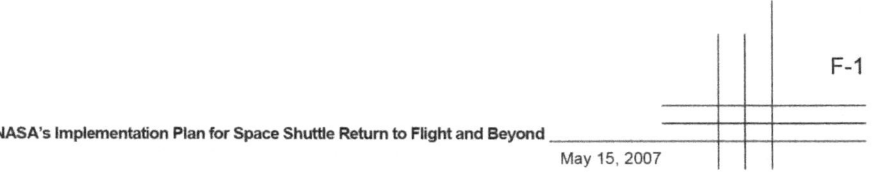

NOTE: The CAIB Recommendations Implementation Schedule was initially developed to track the Return to Flight missions discussed in Part 1 of the Return to Flight Implementation Plan. It is included now, without updates, for reference.

Historical Reference Only

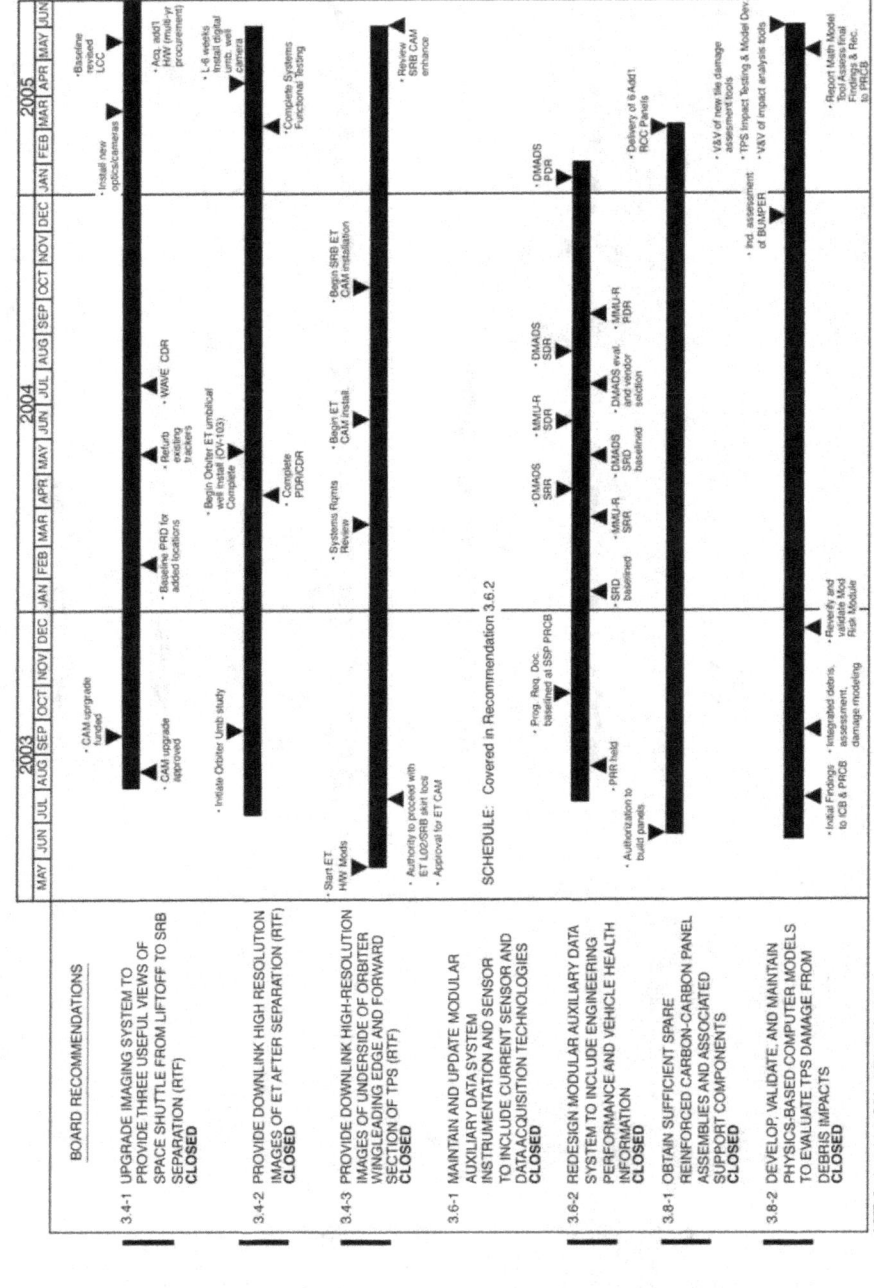

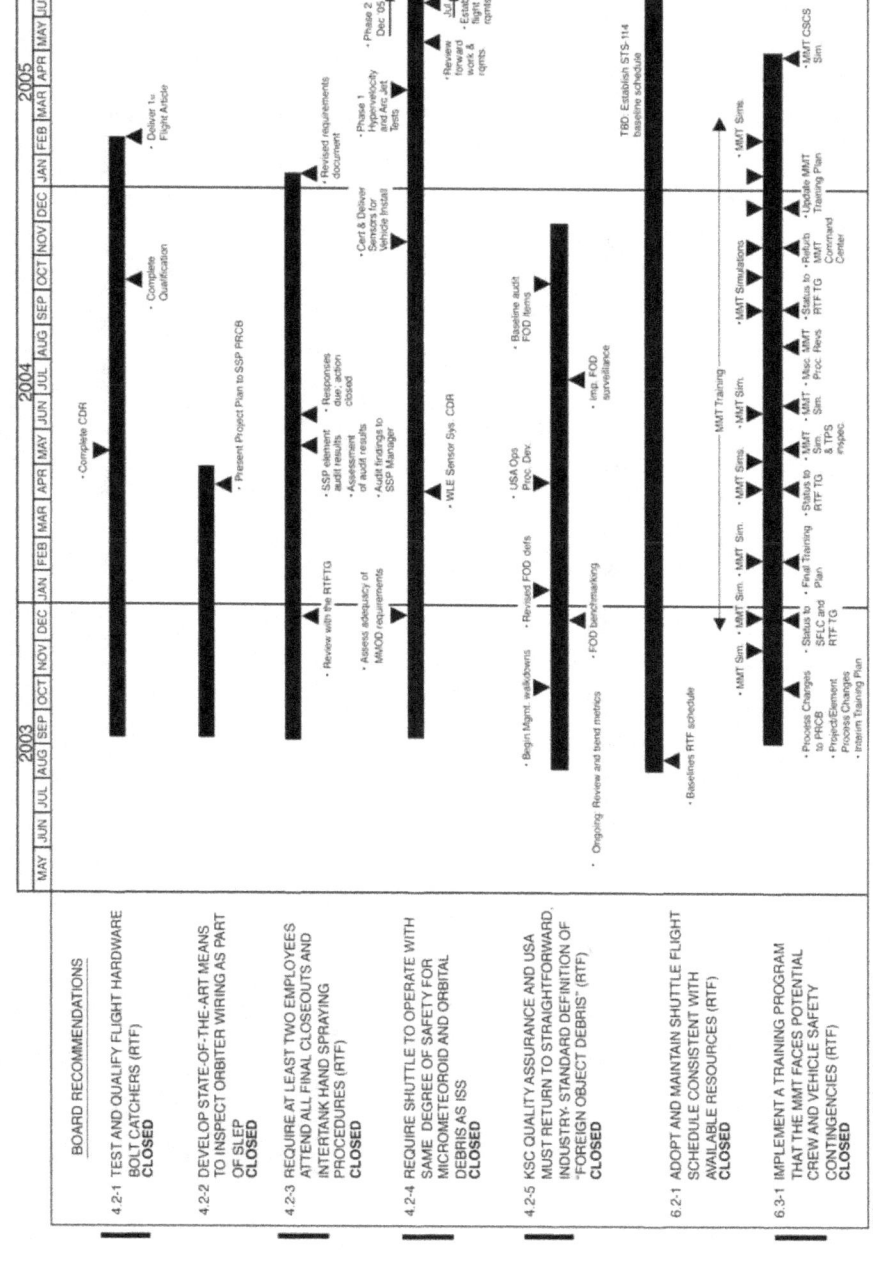

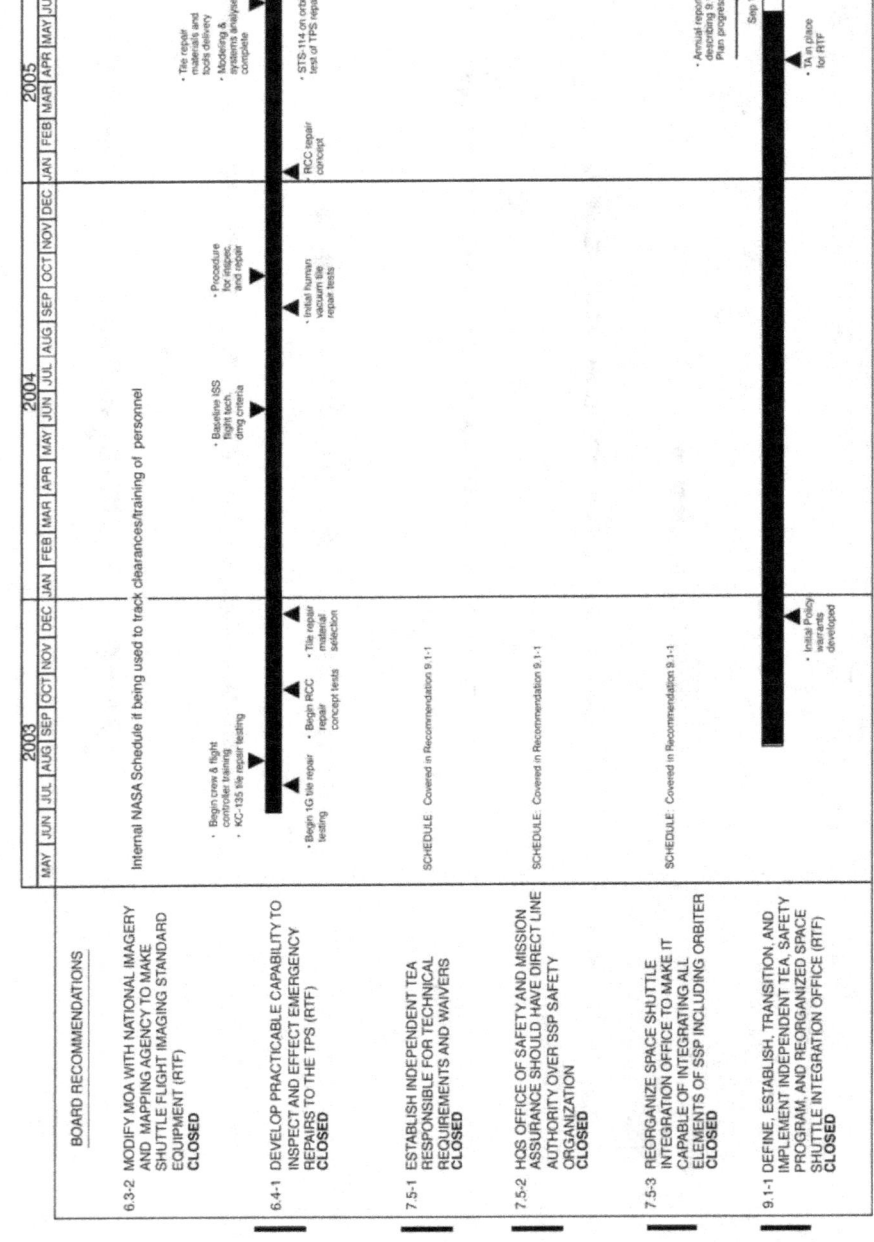

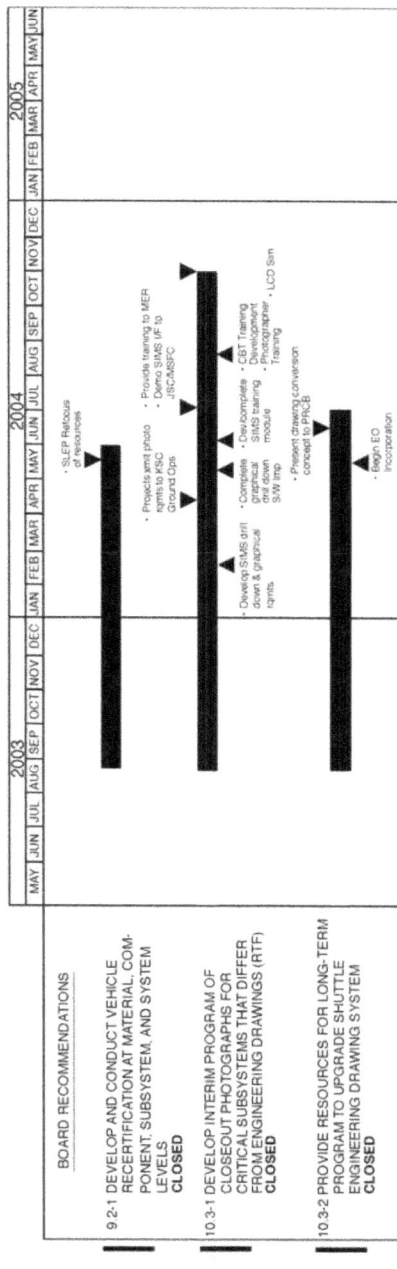